Chromatographic Methods in Metabolomics

RSC Chromatography Monographs

Series Editors:
R.M. Smith, *Loughborough University of Technology, UK*

Titles in the Series:

0: Supercritical Fluid Chromatography
1: Chromatographic Integration Methods
2: Packed Column SFC
3: Chromatographic Integration Methods, Second Edition
4: Separation of Fullerenes by Liquid Chromatography
5: Applications of Solid Phase Microextraction
6: HPLC: A Practical Guide
7: Capillary Electrochromatography
8: Hyphenated Techniques in Speciation Analysis
9: Cyclodextrins in Chromatography
10: Electrochemical Detection in the HPLC of Drugs and Poisons
11: Validation of Chromatography Data Systems: Meeting Business and Regulatory Requirements
12: Thin-layer Chromatography: A Modern Practical Approach
13: High Temperature Liquid Chromatography: A User's Guide for Method Development
14: High Performance Chelation Ion Chromatography
15: Protein and Peptide Analysis by LC-MS: Experimental Strategies
16: UHPLC in Life Sciences
17: Chromatography of Medicinal Plants
18: Chromatographic Methods in Metabolomics

How to obtain future titles on publication:
A standing order plan is available for this series. A standing order will bring delivery of each new volume immediately on publication.

For further information please contact:
Book Sales Department, Royal Society of Chemistry, Thomas Graham House, Science Park, Milton Road, Cambridge, CB4 0WF, UK
Telephone: +44 (0)1223 420066, Fax: +44 (0)1223 420247
Email: booksales@rsc.org
Visit our website at www.rsc.org/books

Chromatographic Methods in Metabolomics

Edited by

Tuulia Hyötyläinen
VTT Technical Research Centre of Finland, Finland
Email: tuulia.hyotylainen@vtt.fi

and

Susanne Wiedmer
University of Helsinki, Finland
Email: susanne.wiedmer@helsinki.fi

RSC Publishing

RSC Chromatography Monographs No. 18

ISBN: 978-1-84973-607-7
ISSN: 1757-7055

A catalogue record for this book is available from the British Library

Published by The Royal Society of Chemistry,
Thomas Graham House, Science Park, Milton Road,
Cambridge CB4 0WF, UK

Registered Charity Number 207890

For further information see our web site at www.rsc.org

Preface

Chromatographic techniques, particularly in combination with mass spectrometry, are valuable tools for metabolomics studies. Efficient techniques are required to cover the wide range of metabolites that are present in any biological sample. By global assessment of the metabolic profile, metabolomics can provide an instant snapshot of an organism's metabolic state because the metabolome contains all the biological endpoints of genomic, transcriptomic and proteomic perturbations, in addition to stimuli of environmental factors. The levels of metabolites may be strongly connected to specific conditions, such as apoptotic alterations, hypoxia, oxidative stress and inflammation, hence metabolomics can be used for finding new biomarkers for diagnosis of chronic diseases and for identifying perturbed pathways due to disease or treatment. Most frequently, metabolomics is employed in a comparative mode, where metabolic profiles of sample sets from a normal state and a perturbed state are compared to find changes related to the perturbation, such as genetic knockout, administration of a drug or change in diet or lifestyle.

One of the main factors behind the advancement of metabolomics has been the development of analytical technologies and in particular advances in mass spectrometry. However, the discovery of metabolic biomarkers remains a significant challenge because of the high number of metabolites, the wide concentration range (over eight orders of magnitude) and the chemical diversity of metabolites. It is not possible to cover the whole metabolome with a single analytical technique, but a set of multiple methods are needed. Moreover, the selection of the optimal method is not always easy. Maybe even more challenging is the mining of the relevant information among the huge amount of data produced by the novel analytical techniques. Here, efficient tools for data processing and modelling are needed.

Metabolomics is currently used within the clinical and pharmacological disciplines not only in biomarker discovery but also widely in drug discovery,

RSC Chromatography Monographs No. 18
Chromatographic Methods in Metabolomics
Edited by Tuulia Hyötyläinen and Susanne Wiedmer

Published by the Royal Society of Chemistry, www.rsc.org

drug toxicology and personalized pharmacology. Also, metabolomics is an important tool for metabolic engineering in biotechnology and food technology. In clinical disciplines, metabolic biomarkers can be used to diagnose, to select and optimize therapy (type and/or dose), to evaluate the effect of chosen therapy and to monitor disease progression. In the drug development process, on the other hand, biomarkers are being increasingly used in the early clinical development of drug candidates. Biomarkers can be helpful in the identification and validation of novel therapeutic targets, *i.e.*, in the evaluation of both the action mechanism and possible toxicity of the drug target, and also in predicting or monitoring the responsiveness of the patient to the treatment. Metabolomics also has the potential for a significant biotechnological impact in metabolic engineering: as the goal of metabolic engineering is to manipulate metabolite production, metabolomics offers tools for detailed understanding of the underlying mechanisms. Metabolic engineering can also be utilized in food technology.

Two types of approaches are typically used in metabolomics, namely targeted selective analysis and more comprehensive, non-targeted profiling methods. The third analytical approach, diagnostics applications, is then quite different. In the target analysis, only a limited number of preselected metabolites are analysed with a carefully planned analytical protocol. Although this approach allows very sensitive and robust determination of the selected metabolites, it gives relatively limited information on the global metabolome. The non-targeted analyses aim to cover as many metabolites as possible in a single analysis; however, the methods are typically only semiquantitative, hence it is not possible to optimize the method for all compounds. The diagnostic approaches are then focused on one or a few (validated) biomarkers, using a very rugged and robust technology, or even on-site measurement with a small, simple instrument. Different types of instrumentation are needed for the three types of metabolomics, *i.e.*, the targeted, non-targeted and diagnostic metabolomics.

One of the challenges in metabolomics is the lack of standardization of sampling, sample storage and sample pretreatment methods, and also quality control of the analytical methodologies. All of these steps can cause unnecessary methodological variation in the results, which then can hinder the identification of the potentially relevant, biologically significant markers. Often the differences in sampling and sample handling before any actual analysis can cause significant changes in the composition of the metabolites, especially in the levels of labile metabolites. In the whole analytical workflow, strict quality control and correction of batch-to-batch variations are also essential. Another key challenge is the identification of previously unknown metabolites. Typically it is possible to detect 600–1000 metabolites in a single profiling analysis, but usually only 10–30% of these can be identified based on spectral libraries. Future development requires the establishment of spectral databases of metabolites and associated biochemical identities, and also cross-validation of metabolites obtained by NMR or mass spectrometry.

Metabolomics has the potential to become an increasingly effective tool for clinical studies, in drug development, and in food and plant biotechnology. A broader coverage of the human metabolome can also be utilized in stratifying populations based on their metabolic phenotype or metabotype, and to use these data for developing personalized medicine, *i.e.*, to tailor a dose/type of drug or other therapy to an individual (group). This would aid the drug development by decreasing adverse drug reactions and improving disease outcomes. However, further development of more robust and sensitive analytical methodologies and advances in bioinformatics are still needed.

The aim of this book is to give the reader an overview of the chromatographic and electromigration techniques utilized in metabolomics and practical guidelines for the selection of the proper method. Microchip technologies in metabolomics are discussed with special focus on microfabrication methods utilized in microchip-based separations and microfluidic enzyme assays. In addition, critical parameters in sampling, sample preparation and data processing and analyses are discussed.

Tuulia Hyötyläinen
Susanne Wiedmer

Contents

RSC Chromatography Monographs No. 18
Chromatographic Methods in Metabolomics
Edited by Tuulia Hyötyläinen and Susanne Wiedmer

Published by the Royal Society of Chemistry, www.rsc.org.

CHAPTER 1

Selection of Analytical Methodology for Metabolomics

SUSANNE K. WIEDMER[a] AND
TUULIA HYÖTYLÄINEN*[b]

[a] Department of Chemistry, University of Helsinki, P.O. Box 55, 00014 University of Helsinki, Finland; [b] VTT Technical Research Centre of Finland, Tietotie 2, P.O. Box 1000, 02044 VTT, Espoo, Finland
*Email: tuulia.hyotylainen@vtt.fi

1.1 Introduction

Metabolomics analysis covers different strategies depending on the goal of the study (Figure 1.1). The most common strategies are based on the following types of approaches:

1. *target analysis,* which aims at the quantitative analysis of a limited number of metabolites, typically a small group of chemically similar metabolites;
2. non-targeted *global metabolic profiling*, which allows the detection of a broad range of metabolites, usually in a semiquantitative manner;
3. *metabolic fingerprinting*, which is a high-throughput, rapid methodology for the analysis of biological samples that provides fingerprints for sample classification and screening;
4. *metabolic footprinting*, which aims at the analysis of metabolites in extracellular fluids;
5. *diagnostic methods* in the clinical laboratory or with a point-of-care approach: very simple, robust methods for quick analysis of metabolic biomarkers.

RSC Chromatography Monographs No. 18
Chromatographic Methods in Metabolomics
Edited by Tuulia Hyötyläinen and Susanne Wiedmer

Published by the Royal Society of Chemistry, www.rsc.org

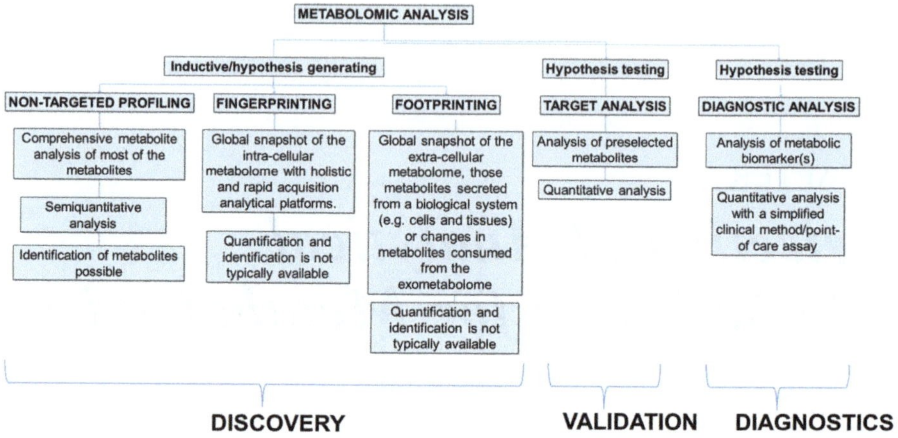

Figure 1.1 Classification of metabolic analyses.

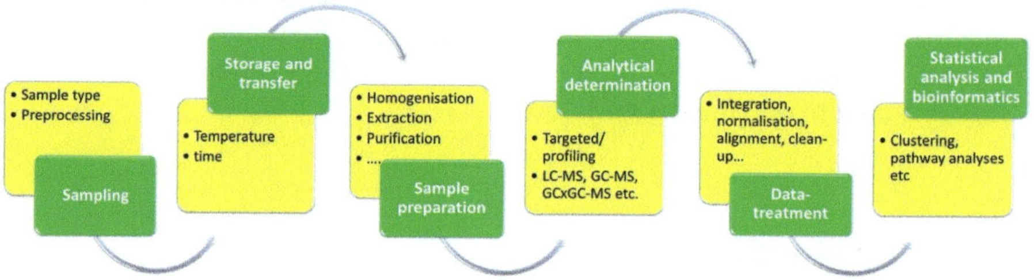

Figure 1.2 Workflow in metabolomics analysis.

The selection of a suitable analytical workflow, from sample pretreatment to analysis and data analysis, is not always straightforward. In the discovery stage, usually global profiling methods are required, and in validation, quantitative targeted methods are needed. In these two approaches, the analytical workflows are very similar. In the diagnostic stage, which may take place in a clinical laboratory or even in point-of-care facilities, the analytical workflow is very different. It also requires a different type of instrumentation, such as microfluidic techniques, which are described in Chapter 7.

The crucial steps of an analytical workflow in the discovery and validation stage are illustrated in Figure 1.2. A well-planned analytical protocol includes a proper design of experiment, optimization of both sample preparation and final analysis, data storage and manipulation, and data processing and validation. Moreover, careful quality control during the whole analytical procedure is essential, especially because many methodologies are global profiling methods for which the analytical steps cannot be optimized to the extent that optimal conditions are obtained for each group of compounds. Thus, even small variations during the analytical procedure can result in large errors in the final results. In addition, the lack of homogeneous standard protocols for the analytical workflow also introduces a barrier for comparing results among laboratories and for reproducing metabolomics experiments.[1]

An ideal analytical system would permit sufficient coverage of the metabolome, allowing simultaneous analysis of a wide concentration range in a high-throughput and rugged fashion. Moreover, it should allow the identification of unknown metabolites in a reliable manner. In practice, no single methodology is sufficient for the analysis of the global metabolic profile and each methodology has its own limitations. At present, the methodologies used in metabolomics for final separation and identification are based mainly on chromatographic methods, such as liquid chromatography (LC) and gas chromatography (GC) combined with mass spectrometric (MS) detection and shot-gun methodologies based on MS alone.[2–7] Capillary electromigration (CE) techniques have also been utilized to some extent.[8–12] In addition, nuclear magnetic resonance (NMR) spectroscopy is used fairly frequently; however, the sensitivity of NMR spectroscopy is often not sufficient ($\sim 10\,\mu mol\,L^{-1}$) for metabolomic studies and will therefore not be covered here.

The selection of a suitable technique depends on several parameters, such as the target of the analysis (target *versus* non-target profiling), the type of analytes and matrix, the amount of sample available, and the concentration of the analytes. Different approaches are needed for target-compound analysis and for global profiling. For targeted analyses, a highly selective sample pretreatment is beneficial, whereas a non-selective approach that is not biased towards any certain groups of metabolite is needed for global screening. In target analysis, on the other hand, selective sample preparation steps can be applied and optimized to improve the data quality, leading to highly sensitive and precise quantification of metabolite concentrations. In the metabolite profiling, it is difficult to find the optimal conditions for all types of analytes, so some compromises must to be made in the selection of experimental conditions.

A general guideline for selection of the method is shown in Figure 1.3. A more detailed selection of sample preparation methods is given in Chapter 2. In the selection of the method of analysis, the physicochemical characteristics of the analytes play a major role. For example, (most) lipids are non-polar and non-volatile and therefore both GC- and CE-based methods are poorly suited for their analysis. The natural choice is LC-MS and, of the LC subtypes, reversed-phase (RP) LC. The choice of the method is more difficult for small polar and semipolar compounds, which in principle can be analysed using GC, LC and CE. Each of the three separation methods have their own advantages and limitations, which should be carefully considered in the method selection.

In metabolomics, LC-MS-based methodologies have been used for the analysis of a wide variety of compound classes, including both non-polar and polar compounds such as lipids, amino acids, bile acids and eicosanoids.[6,13] LC-MS has been used both for target analysis and for non-target profiling, with high-resolution MS typically used for non-target profiling and triple-quadrupole MS systems for targeted analysis. The sensitivity in LC-MS is typically high and the identification of novel compounds is possible, although not always easy. Novel fast LC methodologies, utilizing very high pressures, elevated temperatures and novel column materials, allow high-throughput analyses. In ultra-high-performance liquid chromatography (UPLC), columns

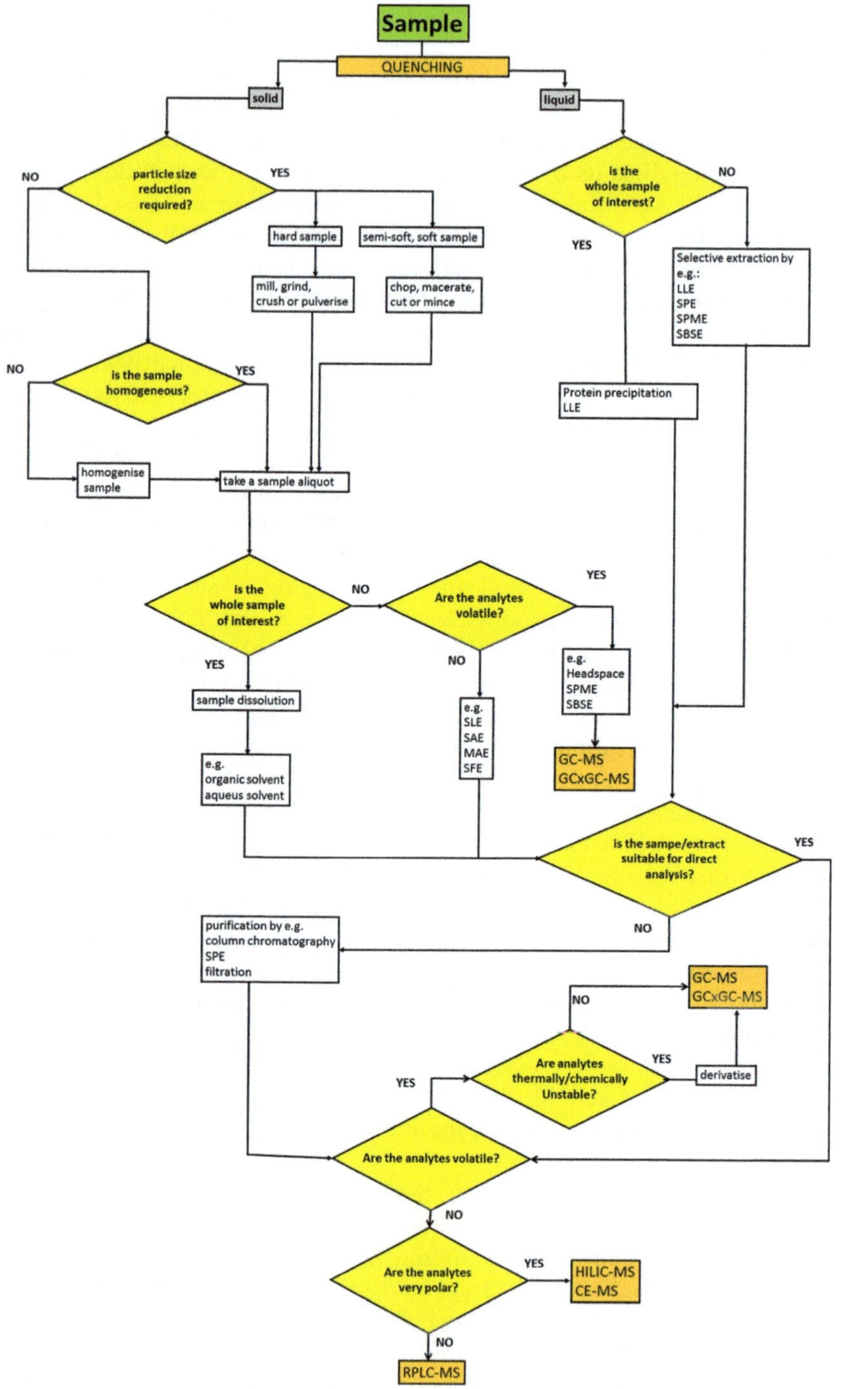

Figure 1.3 General guidelines for the selection of an analytical methodology for metabolomics.

packed with sub-2 μm particle size stationary phase are utilized and the separation efficiency is very high and the analysis time relatively short (5–15 min).

CE is an efficient analytical liquid-phase separation technique by which high-resolution separation of a wide range of analytes is possible, using only minute amounts of samples and running electrolyte solutions. The analytes that can be separated by CE range from single ions to large biomolecules. Owing to the availability of several different CE techniques, charged and uncharged and also polar and non-polar analytes can be separated. A major advantage of the technique is that only minor sample pretreatment is required, in order not to have large particles in the sample clogging the narrow-bore capillaries. The capillaries are always rinsed after each run and provided that no irreversible reactions take place, the retention of analytes is maintained. However, the real Achilles' heel of the method is the poor robustness, which in practice means poor repeatability and reproducibility of migration times.

The main drawback of CE techniques is the poor concentration sensitivity of the method, especially when conventional absorption-based detectors, such as UV-Vis, are employed. Much research has been carried out on enhancing online the concentrations of the injected analytes and various stacking methodologies have been reported. On the other hand, lower limits of detection can be obtained by using online CE-MS and in metabolomics this is more or less the only alternative. The most common ionization method by far is electrospray ionization (ESI). However, CE-ESI-MS puts some demands on the choice of the background electrolyte solution in CE, which in that case must be fairly volatile.

Both targeted and non-targeted CE approaches have been developed over the years, including miniaturized methodologies, which are covered in Chapter 6. The main technological issues when developing CE-MS methodologies have been the development of CE-MS interfaces, which would result in the highest possible coverage of metabolites. Both sheath-liquid and sheathless approaches have been reported. The most commonly used MS analyzer by far in CE-MS online studies is the time-of-flight (TOF) type. Regarding the applicability of CE-MS to metabolomics studies, the most common studies have dealt with non-targeted metabolomics approaches.

GC-based methods, on the other hand, are suitable for the analysis of sufficiently small and volatile compounds (MW <700), such as fatty acids, small polar compounds, such as carboxylic acids, sugars, amino acids and sterols. The polar metabolites require derivatization before analysis; however, the derivatization can be done in an automatic fashion in modern instruments. The main problem with derivatization is that some compounds produce multiple peaks due to incomplete derivatization. However, the GC-based methods have several advantages over the LC-MS-based approaches which also have been used for the analysis of polar metabolites. One of the main advantages is that the separation efficiency is much higher in GC than in LC methods. An additional advantage of all GC-based methods is the repeatable characteristics of retention, particularly when non-polar stationary phases are used. This makes it possible to use universal retention indices, which are very useful in the identification of unknown compounds, in combination with MS detection.

The large spectral libraries for GC (×GC)-EI-MS, combined with retention index data, is a valuable tool in the identification of unknown compounds. Even if the mass spectrum is not present in the spectral database, the fragmentation pattern can be used to obtain information about the compound class of a metabolite. In addition, the full-scan response of the EI mode for quadrupole instruments is approximately proportional to the amount of compound injected, more or less independently of the compound, hence the compounds are detected non-discriminatively.

Recently, two-dimensional GC combined with TOFMS (GC×GC-TOFMS) has been gaining popularity in metabolomics as it is one of the most powerful analytical tools for the separation of organic compounds in complex samples.[14–16] In metabolomics, GC×GC-TOFMS is superior to conventional GC-MS and to any of the LC-MS-based methods available. The sensitivity is typically 10–40 times higher and the separation efficiency is up to 20-fold better than can be achieved with GC-MS.[17] Moreover, the quality of the mass spectra obtained with GC×GC-TOFMS is much better than that obtained with GC-MS and therefore the identification is more reliable. Also, method development in GC-MS methods, including GC×GC-MS, is straightforward and much easier than in LC-MS. Further, the quantitation is more reliable, because matrix effects in GC-MS are not a critical issue. Hence GC(×GC)-MS methods are highly useful for the analysis of small metabolites.

Selection of the correct detection method is dependent on the goal of the analysis. Most often, MS methods are used in metabolomics. Several types of MS systems are used in metabolomics, as described in Chapter 3. Typically, in non-target profiling, high-resolution, accurate mass determination, together with tandem mass measurements for structural characterization, is needed for the determination of elemental composition. In LC-MS combinations, the use of hybrid quadrupole mass spectrometer TOFMS (Q-TOFMS) and Fourier trap MS (FTMS) instruments, both the cyclotron and the Orbitrap type, allow for accurate mass measurements and, subsequently, for the identification of metabolites. In GC, accurate mass TOFMS is typically used for this purpose. Lately, FT-Orbitrap-MS has gained popularity in combination with LC separation because it is a high-speed, high ion-transmission instrument, due to shorter accumulation times. Triple-quadrupole instruments, on the other hand, are utilized mainly for target compound analysis, both with LC and GC. In GC-MS, quadrupole MS is still the most common instrument, but TOFMS and other accurate mass instruments are also becoming more popular.

1.2 Comparison of Different Methodologies for Metabolomics

The key features of the analytical techniques utilized in metabolomics are shown in Table 1.1. The key features include parameters related to analytical performance (speed, sensitivity), suitability for targeted *versus* non-targeted screening, amount of sample preparation required and instrumental costs.

Table 1.1 Comparison of the key features of the chromatographic methods for metabolomics.

Feature	*UPLC-TOFMS/ Q-TOFMS/Orbitrap*	*UPLC-QqQMS*	*GC-qMS/QqQMS*	*GC-TOFMS, GC×GC-TOFMS*	*CE-TOFMS*
Approach	Profiling, semiquantitative	Targeted, quantitative	Targeted, quantitative	Profiling + targeted quantitative	Non-targeted and targeted
Sample preparation	Moderate	Tedious	Tedious	Tedious	Little
Sample amount (μl)[a]	5–100	5–100	5–100	5–100	>20
Sample throughput (per day)	40–80	20–80	20–80	20–30	40–60
Reproducibility	Moderate	Good	Good	Good	Moderate
Sensitivity	Nanomolar–picomolar	Nanomolar–picomolar	Nanomolar–picomolar	Nanomolar–picomolar	Nanomolar
Cost per sample	Medium	Medium	Low–medium	Medium	Low
Instrument cost	Moderate–high	Moderate–high	Moderate–high	Moderate–high	Low–moderate
Comments	Suitable for a wide range of metabolites, identification of unknowns requires tandem MS	Suitable for a wide range of metabolites	Only suitable for small metabolites (MW < 700), polar compounds require derivatization, huge spectral libraries available, retention indices can be used for verification of identification	Only suitable for small metabolites (MW < 700), polar compounds require derivatization, huge spectral libraries available, clean spectra due to very high separation efficiency, retention indices can be used for verification of identification	Suitable for a wide range of metabolites, coated capillaries for improving migration time repeatability
Metabolites	Lipids, polar metabolites	Acyl carnitines, amino acids, steroids, CoAs, nucleosides, eicosanoids, vitamins	Fatty acids, steroids, amino acids, small carbohydrates, tricarboxylic acid (TCA) metabolites, volatile metabolites	Fatty acids, steroids, amino acids, small carbohydrates, TCA metabolites, volatile metabolites	Amino acids, steroids, nucleotides, anionic and cationic metabolites

[a]Serum, plasma, urine.

The metabolite coverage is an important feature in the selection of a non-targeted method. A recent study compared the coverage of different analytical techniques in the analysis of human serum metabolome. This human serum metabolome study compared five different metabolic profiling methods in the characterization of the known serum metabolome using (1) NMR spectroscopy, (2) GC-MS, (3) lipid mediators by LC-ESI-MS/MS, (4) lipidomics profiling via TLC/GC-FID (flame ionization detection)-MS and (5) direct infusion MS/MS.[18] Using these five platforms, a total of 3564 distinct metabolites including several exogenous compounds could be identified from the human serum. The study showed clear differences in the ability of the platforms to identify and quantify metabolites: NMR spectroscopy was able to identify and quantify 49 compounds, GC-MS was able to identify 90 and quantify 33 compounds, targeted ESI-MS/MS identified and quantified 96 compounds, TLC/GC-FID-MS identified and quantified 3381 compounds and direct infusion MS/MS identified and quantified 139 compounds. Some metabolites could be detected with several platforms and, for example, comparison of amino acid concentrations as measured by NMR spectroscopy, GC-MS and direct fluid injection (DFI) MS/MS showed that the quantitative results are in relatively good agreement. However, some other studies have shown that there are distinct differences between the quantitative results obtained with MS-based methods and NMR spectroscopy.[19] The human metabolome serum study showed that of the known, quantifiable serum metabolome (4229 known and probable metabolites; 665 literature-derived metabolites; 3564 experimentally derived or predicted), NMR spectroscopy was able to measure ~1.2% (49/4229) of the human serum metabolome, GC-MS 2.13% (90/4229), ESI-MS/MS 2.3% (96/4229), TLC/GC-FID-MS (general lipidomics) 79.9% (3381/4229) and direct infusion MS/MS 3.3% (139/4229) of the serum metabolome. It should be noted that the study included only known metabolites. With more advanced methodologies, such as global profiling with GC×GC-TOFMS, typically 1200 metabolites can be detected from human serum samples; of these, however, only 10–30% can be identified.[15,16] Also, global profiling of lipids with UHPLC-Q-TOFMS typically allows the detection of ~600 lipids in human serum, of which ~250 can be identified.[20]

1.3 Conclusion

The selection of the optimal analytical method is not straightforward and careful consideration is necessary in the selection of all steps of the analytical workflow. In addition to analytical features of the methods, also practical aspects such as instrument price, ruggedness of the method and its suitability for high-throughput screening of large numbers of samples have an important role. In small-scale studies, when the number of samples is relatively small (<100), more tedious procedures can be used, but for large-scale studies with hundreds of samples, the whole analytical workflow must be carefully designed. In any case, no single analytical method will cover the whole metabolome, but instead, multiple parallel techniques are needed. In addition, in the discovery phase, non-targeted profiling methods give the best coverage of the metabolome, but ideally, when key metabolites, *e.g.*, biomarkers of specific disease or biological state, are identified, more accurate target methods, with a carefully optimized workflow should be used. For clinical diagnostic purposes, the methodology can then be further developed to allow simple, fast and robust measurement of the validated biomarkers. Different types of instrumentation are needed for the different types of metabolomics, *i.e.*, the targeted, non-targeted and diagnostic metabolomics.

References

1. B. Álvarez-Sánchez, F. Priego-Capote and M. D. Luque de Castro, *Trends Anal. Chem.*, 2010, **29**, 111.
2. T. Hyötyläinen, I. Bondia Pons and M. Orešič, *Mol. Nutr. Food Res.*, 2013, DOI: 10.1002/mnfr.201200759.
3. T. Hyötyläinen, *Expert Rev. Mol. Diagn.*, 2012, **12**, 527.
4. T. Hyötyläinen, *Bioanalysis*, 2010, **2**, 919.
5. S. Moco, R. J. Bino, R. C. H. De Vos and J. Vervoort, *Trends Anal. Chem.*, 2007, **26**, 855.
6. G. A. Theodoridis, H. G. Gika, E. J. Want and I. D. Wilson, *Anal. Chim. Acta*, 2012, **711**, 7.
7. K. Dettmer, P. A. Aronov and B. D. Hammock, *Mass Spectrom. Rev.*, 2007, **26**, 51.
8. R. Ramautar, G. W. Somsen and G. J. de Jong, *Electrophoresis*, 2013, **34**, 86.
9. C Barbas, E. P. Moraes and A. Villaseñor, *J. Pharm. Biomed. Anal.*, 2011, **55**, 823.
10. R. Ramautar, O. A. Mayboroda, G. W. Somsen and G. J. de Jong, *Electrophoresis*, 2011, **32**, 52.
11. R. Ramautar, G. W. Somsen and G. J. de Jong, *Electrophoresis*, 2009, **30**, 276.
12. M. R. N. Monton and T. Soga, *J. Chromatogr. A*, 2007, **1168**, 237.
13. S. Forcisi, F. Moritz, B. Kanawati, D. S. Tziotis, R. Lehmann and P. Schmitt-Kopplin, *J. Chromatogr. A*, 2013, **1292**, 51–65.
14. W. Welthagen, R. A. Shellie, J. Spranger, M. Ristow, R. Zimmermann and O. Fiehn, *Metabolomics*, 2005, **1**, 65.
15. M. Hartonen, I. Mattila, A.-L. Ruskeepää, M. Orešič and T. Hyötyläinen, *J. Chromatogr. A*, 2013, **1293**, 142–149.
16. S. Castillo, I. Mattila, J. Miettinen, M. Orešič and T. Hyötyläinen, *Anal. Chem.*, 2011, **83**, 3058.
17. T. Laitinen, S. Herrero Martín, J. Parshintsev, T. Hyötyläinen, K. Hartonen, M.-L. Riekkola, M. Kulmala and J. Luis Pérez Pavón, *J. Chromatogr. A*, 2013, **1217**, 151.
18. N. Psychogios, D. D. Hau, J. Peng, A. C. Guo, R. Mandal, S. Bouatra, I. Sinelnikov, R. Krishnamurthy, R. Eisner, B. Gautam, N. Young, J. Xia, C. Knox, E. Dong, P. Huang, Z. Hollander, T. L. Pedersen, S. R. Smith,

F. Bamforth, R. Greiner, B. McManus, J. W. Newman, T. Goodfriend and D. S. Wishart, *PLoS ONE*, 2011, **6**, e16957.
19. K. Suhre, C. Meisinger, A. Döring, E. Altmaier, P. Belcredi, C. Gieger, D. Chang, M. V. Milburn, W. E. Gall, K. M. Weinberger, H.-W. Mewes, M. Hrabé de Angelis, H.-E. Wichmann and T. Illig, *PLoS ONE*, 2010, **5**, e13953.
20. H. Nygren, P. Pöhö, T. Seppänen-Laakso, U. Lahtinen, M. Orešic and T. Hyötyläinen, *LCGC Eur.*, 2013, **March**, 142.

CHAPTER 2

Sample Collection, Storage and Preparation

TUULIA HYÖTYLÄINEN

VTT Technical Research Centre of Finland, Tietotie 2, P.O. Box 1000, 02044 VTT, Espoo, Finland
Email: tuulia.hyotylainen@vtt.fi

2.1 Introduction

The first steps in metabolomics workflow are sample collection, storage and preparation. The sample collection and storage are crucial steps in the metabolomics protocol, as failures in these steps cannot be compensated in later stages. In addition, the correct choice of sample-preparation method is extremely important in metabolomic studies because the sample preparation steps required before final analysis largely determine the quality of analytical results obtained. The sample preparation procedure has a significant impact on assay throughput, data quality and analysis cost because sampling and sample preparation steps typically account for over 60% of the total analysis time. One of the major problems in metabolomics is that the sampling and sample preparation techniques are often not well validated and therefore can produce inconsistent data. Issues that need to be addressed in metabolomics are often related to poor sample preparation methods, such as inefficient extraction methods, matrix effects and clean-up steps needed to remove the unwanted matrix compounds. In short, obtaining reliable data depends on the systematic development of techniques and methods, particularly for the sample preparation part. Despite its importance, these first steps of the workflow are often an overlooked aspect of metabolomics.

RSC Chromatography Monographs No. 18
Chromatographic Methods in Metabolomics
Edited by Tuulia Hyötyläinen and Susanne Wiedmer

Published by the Royal Society of Chemistry, www.rsc.org

Sample processing and pretreatment can take a number forms, depending on the nature of the sample.[1] Typical processes may include quenching, sample filtration, centrifugation, protein precipitation, dilution, target amplification and extraction. Successful execution of these processes is required to ensure that the analytes of interest are present in a form compatible with the analytical system. The most classical sample preparation techniques rely on extraction with solvents, including traditional techniques such as liquid–liquid extraction (LLE) and solid–liquid extraction (SLE). Various other techniques, such as solid-phase extraction (SPE) and sonication-assisted liquid extraction, are also utilized in metabolomics. In addition, methods for the analysis of volatile compounds from the vapour phase are available, from simple headspace extraction to solid-phase microextraction (SPME) and stir-bar sorptive extraction (SBSE).

Metabolomic analyses are done from a variety of sample types, such as blood (serum, plasma), urine, cerebrospinal fluid (CSF), lymph fluid, bile, faeces, saliva, cells, plants and tissues. No universal sample preparation technique suitable for all types of samples and metabolites exists. The sample preparation required is dependent on the nature of the analytes, the matrix and the final separation method. Obviously, the sample preparation must be tailored to the final analysis. The sample matrix and the type and concentration of analytes in the sample are also of primary importance. In addition, a method good for target-compound analysis is often not well suited for comprehensive chemical profiling of samples. Selectivity of the sample preparation is often a key factor for target-compound analysis whereas an exhaustive and non-selective extraction is a better choice for profiling. In the selection of a sample preparation technique, not only the effectiveness needs to be considered but also many other factors that affect the analytical scheme.[2] The major factors are ruggedness, throughput, operating costs, complexity of method development, amount of organic solvent required and level of automation. In addition, the number of samples to be analysed is also of importance – the question is whether the planned procedure will be unique or whether it will be used in carrying out routine analysis. In the latter case, techniques facilitating automation and low cost per analysis are preferred. An ideal sample preparation method for metabolomics should

- be as non-selective as possible for global metabolomics in order to ensure adequate coverage of metabolites, or be as selective as possible for targeted metabolomics;
- be simple and fast to prevent metabolite loss and/or degradation during the preparation procedure and allow high throughput;
- be reproducible and rugged; and
- incorporate a metabolism-quenching step to represent the true metabolome composition at the time of sampling.

A general scheme for sample handling is shown in Figure 2.1.

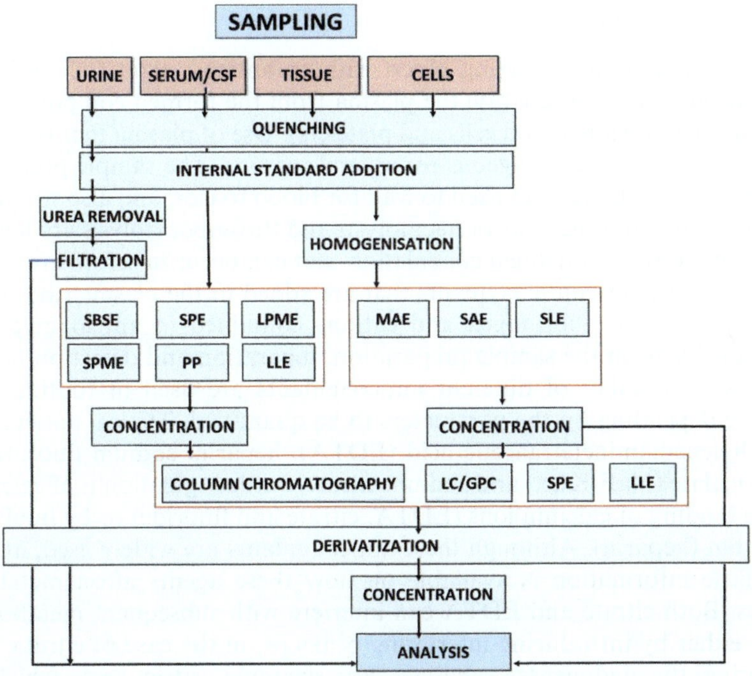

Figure 2.1 General scheme for sample handling for metabolic analyses.

2.2 Sample Type and Sample Collection

In the sampling step, it is important to take into consideration that metabolism is in constant flux, which is dependent on substrate concentration, enzyme kinetics and enzyme concentrations. Most metabolomic experiments are designed to capture a snapshot of the metabolome. Thus, in the sampling step, metabolic flux has to be stopped or inhibited in order to obtain a representative snapshot of the metabolome at the time of sampling. The time of the sampling can also be crucial.

2.2.1 Biofluids

The most common biofluids are blood plasma/serum and urine, and several protocols have been suggested for the sample preparation of these sample matrices. Blood-derived serum and plasma are common matrices utilized in human studies, both because sample collection is relatively easy and also because the blood metabolic profile reflects systemic changes in the metabolome. However, the intrinsic differences between the two biofluids are often overlooked, although serum and plasma are completely different with respect to their preparation. Many metabolites have different concentrations in plasma and serum. In addition, anticoagulants may also affect the final analysis due to matrix effects.

2.2.1.1 Plasma

Plasma is prepared by mixing blood with an anticoagulant followed by centrifugation at 4 °C to separate the plasma from the formed components of the blood (red and white blood cells and platelets). Use of plasma that still contains clotting factors is advantageous for several reasons. The sample preprocessing is rapid because there is no need to wait for blood to clot, and also the yields are usually higher and the risks of haemolysis and thrombocytolysis are lower, and losses due to postcentrifugal coagulation that can occur in serum are avoided.[3] On the other hand, anticoagulants that are added to the plasma (including the enzyme inhibitors, fibrinogen and cations contained in anticoagulants) can cause problems in the sample preparation, separation and detection of plasma samples. A number of different anticoagulants are used in routine medical practice depending on the parameters to be quantified. Typical anticoagulants are ethylenediaminetetraacetic acid (EDTA), heparin, sodium fluoride/potassium oxalate (NaF/KOx) and sodium citrate. Anticoagulation is effected either by the binding of calcium ions (EDTA, citrate and fluoride) or by inhibition of thrombin (heparin). Although these anticoagulants are widely used, at present very little information is available on how these agents affect metabolomic studies. Both citrate and EDTA can interfere with subsequent metabolic profiling, either by introducing interfering peaks or, in the case of citrate, also by obscuring the endogenous analyte. For example, citric acid, which is an important metabolite of the TCA cycle, is present in higher concentrations in EDTA plasma than in serum, most probably due to complex formation between citric acid and Ca^{2+} and Mg^{2+} ions in the serum during sample preparation, leading to precipitation in the serum.[4] For this reason, lithium heparin has been recommended for preparing plasma samples for general analysis.[5]

The anticoagulants can affect several steps of the analytical procedure. The first is the effect of a particular anticoagulant during the actual sample collection and storage. Next, the different anticoagulants can have an effect on the sample preparation such as extraction. In LLE, the anticoagulants can enhance or suppress the extraction of some of the compounds, especially if they can form complexes with the extracted compounds. Moreover, if the anticoagulant agent is present in the final extract, it can also affect the final analysis. It is obvious that the use of NaF–KOx or citrate as anticoagulant will disturb the analysis of oxalic acid (NaF/KOx as anticoagulant) or citric acid (sodium citrate as anticoagulant), both of which are important compounds in the metabolic cycle. Concerning heparin plasma, on the other hand, heparin contains sodium metabisulfite as a preservative, which can impair the quantitative analysis of two amino acids, namely cystine and homocystine, due to the formation of sulfocysteine and homosulfocysteine.[6] It should also be noted that the anticoagulants may also affect the final analysis; for example, in ESI-MS-based metabolomics, heparin plasma has been preferred because of less observable sample contamination or matrix effects compared with citrate and EDTA plasma, which showed the formation of sodium and potassium formate clusters and may additionally cause ion suppression or enhancement of metabolites coeluting with citrate and EDTA peaks.[7]

2.2.1.2 Serum

Serum is obtained by taking the blood sample and allowing it to clot naturally and then removing the clot by centrifugation. Serum is less viscous than plasma and lacks fibrinogen, prothrombin and other clotting proteins. The overall protein content in serum is lower than that in plasma, which is a benefit in the analysis of metabolites because of the reduced matrix effects. It should be noted that the temperature during the sampling may be crucial for specific metabolites, and the clotting process needed to obtain serum (30–60 min), and which is typically done at room temperature can increase the enzymatic conversion and other degradation processes, at least for some of the more labile metabolites; and it can also facilitate the possible loss of some metabolites during clot formation/precipitation. Therefore, clotting on ice and strict standardization of serum–clot contact time are recommended by the HUSERMET project.[5] However, it has also been shown in a comparison of global metabolic profiles of serum samples that were left to clot either at room temperature or at 4 °C that only a few metabolites did show drastic difference due to the clotting temperature.[8]

2.2.1.3 Urine

Urine samples are usually collected as random spot samples, timed samples, or 24 h samples, or for pharmacokinetic studies even longer sampling times may be required. Random sampling is typically performed at any time of day without taking into account diurnal variations of excretion. Timed samples are used for excretion monitoring (*e.g.*, a given food or drug). It should also be noted that the first urine of the day is generally not preferred as it is the most variable.

Urine consists of ~95% water, plus different inorganic salts (Na, Cl, K, PO_4, SO_3, I, NH_3), urea, creatinine and uric acid and mainly low molecular weight organic compounds. Urine samples have varying pH (4–6) and ionic strength, in addition to differences in metabolite concentrations. The volume of urine can vary widely based upon water consumption and other physiological factors. This causes very variable levels of urinary metabolites, hence normalizing for these effects is necessary, particularly in random sampling. The most common methods for normalization of the urine volume are based on urine volume, creatinine concentration, osmolality, and components that are common to all samples. Of these methods, measurement of the protein/creatinine ratio is one of the most commonly applied normalization procedures.

Whenever possible, urine samples should be stored preferably at −80 °C immediately after collection. Different additives are often added to urine samples, including protease inhibitors and preservatives. Protease inhibitors are used to preserve protein forms intact; however, their use in urine samples may not be advisable because it can cause changes in sample pH, dilution and denaturation of proteases in urine.[9] Other preservatives such as boric acid and sodium azide have been used to prevent bacterial contamination; however, they may form chemical complexes in the samples or cause issues in maintaining consistency and quality of sample handling.[10] A recent comparative study of the use of sodium azide and 0.20 μm filtration showed the latter to be superior for bacterial removal.[11] Also, preservatives are not mandatory if urine samples are stored at temperatures below –20 °C. In preclinical animal studies, cumulative 24 h urine is often collected at various points using metabolic cages. In this type of study, collection tubes must be placed on dry-ice to avoid both microbial growth and possible degradation of metabolites at room temperature.

Deproteinization is typically not needed for urine samples because the protein concentration in urine is relatively low (0.5–1 μg L^{-1}) compared with, *e.g.*, serum and plasma. However, if considered necessary, it can be done by addition of acetonitrile, with a 1:1–1:9 ratio of the solvent.

Centrifugation or filtration is typically applied to urine samples to remove particles and even proteins. Conventional filtration is performed on a cellulose membrane (0.45 μm pore size); however, filtration with a 0.20 μm filter has been shown to be superior to centrifugation or sodium azide addition in preventing bacterial growth during storage.[11] Ultrafiltration is also often used for removal of proteins and other macromolecules, using filters with molecular weight cut-offs of 3, 10 and 30 kDa.

2.2.1.4 Saliva

Saliva, which is secreted from three pairs of major salivary glands, consists mainly of water (99%) and several minor components such as electrolytes, proteins (*i.e.*, immunoglobulins, hydrolases or mucins) and other low molecular weight components such as urea and ammonia. A large part of the compounds present in saliva originate from the blood by passage through the spaces between cells by transcellular or paracellular routes. Therefore, most compounds found in blood are also present in saliva.

2.2.1.5 Cerebrospinal Fluid (CSF)

CSF is produced in the choroid plexus of the brain and consists of water, inorganic salts, glucose and other sugar derivatives, amino acids and ~0.8% of various proteins. CSF reflects the composition of blood plasma, although the concentrations of most proteins and metabolites in CSF are lower. CSF is obtained by lumbar puncture, which is an invasive method and therefore often only a limited amount of CSF sample is available in metabolomics. A recent study on the stability of CSF at room temperature after a low-speed centrifugation step to remove cells did not find significant changes in the measured proteome and metabolome of CSF stored at room temperature after centrifugation, except for two peptides and one metabolite, 2,3,4-trihydrobutanoic acid, of 5780 identified peptides and 93 identified metabolites.[12] Other studies have shown that blood or white blood cell contamination reduces CSF stability considerably, emphasizing the importance of the initial centrifugation step.[12]

2.2.2 Faecal Samples

Faecal samples are also utilized in metabolomic analysis. As with other biological matrices, it is important to freeze the samples as soon as possible after collection. Only a few studies have been carried out on the stability of the metabolites in focal samples under different conditions. Freezing and lyophilization have been shown to increase the branched-chain amino acid (BCAA) levels and lyophilization also decreases the levels of short-chain fatty acids (SCFAs).[13]

2.2.3 Cells

In cell samples, sampling and quenching steps are typically combined. The sampling and sample preparation for extracellular metabolites are relatively straightforward, but leakage of intracellular metabolites into the cell media during the sampling step can take place. Extracellular metabolites are usually analysed from the cell media, after a simple centrifugation step, or after simple dilution with water or suitable buffer. Most commonly, sampling and quenching are done with cold methanol; however, this has been shown to lead to drastic losses of metabolites (>60%) because of serious metabolic leakage and/or cold shock.[14–16] In particular, some types of bacterial cells are known to be highly sensitive to osmotic changes of the medium, causing leakage of intracellular metabolites by cellular membrane disruption.[17] This can be minimized by controlling the ionic strength of the quenching solution with a suitable buffer (*e.g.*, tricine, HEPES or ammonium carbonate), thus avoiding damage to the cell envelopes. Other methods proposed included quenching with cold isotonic phosphate-buffered saline (PBS) and rapid addition of liquid nitrogen.[18] However, the liquid nitrogen can cause damage to the cellular envelopes, owing to the formation of ice crystals during freeze–thaw cycles. Although a rapid heat shock has also been proposed for the deactivation of metabolism, this may cause degradation of thermolabile metabolites and increase cell permeability.

In a recent study, quenching with 60% methanol (buffered or unbuffered) was shown to result in leakage of intracellular metabolites from the cells whereas quenching with cold isotonic saline (0.9% w/v NaCl, 0.5 °C) did not damage cells and effectively halted conversion of ATP to ADP and AMP, indicative of metabolic arrest.[19]

More complex procedures are required for the analysis of intracellular metabolites. The steps include sampling, quenching and washing; followed by cell lysis and metabolite extraction from cell pellets. The washing step for the cell metabolomics is required in order to remove the extracellular matrix and the growth buffer. However, the washing step must be carried out carefully to avoid losses of metabolites because of metabolite leakage or metabolic alterations. The washing buffer should match the physiological levels to minimize the metabolite leakage. Other procedures sometimes used in cell metabolomics include trypsinization to detach the cells from the growth surface; however, this can cause alterations to metabolic profiles. Typically, quenching is followed by splitting of the sample into two fractions, *i.e.*, to supernatant and cells, by either fast filtration or cold centrifugation.

2.2.4 Tissues

In animal tissues, the whole tissue can be sampled, whereas human tissue samples are typically small biopsies. The tissue biopsies can be problematic because a random tissue biopsy may not be representative and, moreover, blood or adipose tissue sampled together with the tissue may cause additional problems. Alteration of the metabolic composition may take place during the sampling (tissue biopsies, animal sacrifice) or during tissue isolation. Microscopic investigation of the tissue biopsies can be utilized to ensure that the samples are representative. Sampling of the required tissue part should be performed as fast as possible, and the sample should be frozen immediately to avoid metabolite changes caused by enzymatic reactions. The simplest and fastest way is to freeze the sample with liquid nitrogen.

2.2.5 Plant Material

In general, preparing a plant sample for a metabolomic study involves sampling and often drying before the analytical steps. Similarly to other metabolic samples, sampling (*i.e.*, harvesting), requires special care because rapid metabolite changes (*e.g.*, enzymatic degradation and oxidation) may take place during the sampling. Therefore, separation of the required part of the original plant should be carried out very rapidly. In order to avoid metabolite changes caused by enzymatic reactions connected with the handling and wounding of the plant, the sample should be frozen immediately, *e.g.*, with liquid nitrogen. The time for harvesting is also important, because of the daily variation of the metabolic levels in plants. For example, the levels of several primary metabolites, *e.g.*, malic acid and sugars, have been shown to fluctuate following a daily cycle.[20]

2.2.6 Storage and Stability

All metabolic samples should be stored preferably at –80 °C immediately after collection. Most studies related to metabolic studies indicate that samples are relatively stable when stored at –80 °C. Also, a recent study showed that for serum and urine samples, changes in the metabolome caused by storage at 4 °C for 24 h are relatively minimal when compared with the variability observed between subjects.[21] Several stability analyses using LC/TOFMS, NMR and GC/MS methods have also shown that urine samples should be stable for up to 6 months when stored at –20 or –80 °C.[22–24] In addition, no significant difference in urinary metabolite compositions was observed when the urine was subjected to multiple freeze–thaw cycles (up to nine cycles).[22,24]

There are not many detailed reports on the long-term stability of a typical metabolomic sample at –80 °C. Current recommendations advise dividing the samples into aliquots prior to storage to minimize repeated freeze–thaw cycles.

In one study, after 1 year storage of plasma samples at –80 °C, alterations in total cholesterol and triglycerides were observed.[25] However, other study showed that the fatty acid compositions of plasma triglycerides and erythrocyte phospholipids did not change significantly after storage for 4 years at –80 °C.[26]

2.3 Sample Preparation Techniques

The optimum sample preparation method is dependent on the type of sample and the analytes of interest, the analytical method and the goal of the analysis (targeted *versus* global). In targeted methods, carefully optimized, selective methods are utilized, whereas in global, non-targeted methods an unselective sample-preparation procedure is required for the analysis of as wide a range of metabolites as possible. Non-selective sample preparation is generally accomplished by using very simple sample preparation procedures, for example, dilution, solvent precipitation, solvent extraction. In both approaches, it is advisable to minimize the time and overall number of steps in the procedure, both to avoid metabolite losses and to facilitate high sample throughput. It should also be emphasized that metabolomic studies usually compare the relative levels of metabolites in a large number of samples, so good reproducibility for a wide range of metabolites is essential to ensure that small changes in these metabolite levels can be accurately determined. For a reliable method, the sample preparation should be sufficiently quantitative and, even more importantly in metabolomics, rugged, as small changes in the initial composition of the (other compounds present in the) sample may affect the extraction efficiencies and this variation might be incorrectly misinterpreted as a significant difference. Novel approaches in this field have utilized techniques such as cryogenic homogenization of tissues, solid-phase extraction with novel formats and automated at-line sample pretreatment systems can help to minimize the challenges in the sample preparation steps.

2.3.1 Quenching

The first step of the sample preparation should be deactivation of metabolism, *i.e.*, quenching. It is particularly critical in the analysis of cells and tissues when the goal is to elucidate the metabolic profile inside (endometabolome) and outside (exometabolome) the cell.[27] Quenching is more challenging for tissues and cells than biofluids owing to the risk of leakage of metabolites from the cell or tissue through cell membranes. The stability of a specific metabolite is dependent on several parameters, such as its chemical structure and lability, its involvement in enzymatic processes and their time scale, the amount of enzymatic activity remaining in the sample during storage and/or preparation, and processes such as protein degradation, which can elevate levels of some metabolites, for example amino acids. Several procedures are utilized in quenching:

- cold solvent addition
- freezing in liquid nitrogen

- acid addition
- fast heating
- addition of enzyme inhibitors and/or antioxidants.

Freezing is an efficient means of deactivation of metabolic processes, provided that cells or tissues are not allowed to thaw partially before extraction. Therefore, before extraction, enzyme activity can be inhibited by lyophilization (freeze-drying) or by immediate addition of organic solvents. However, with freezing, some loss of metabolites may occur due to cold shock. It should also be noted that freezing of the cells and tissues may also cause permanent damage to the sample as a result of osmotic shock, causing the metabolites to come into contact with enzymes. Thawing of the sample for the analysis may enhance this deterioration, hence the tissue should be homogenized and extracted with solvents at cold temperatures. Also, lyophilization of samples before extraction may cause changes in the metabolic profile as a result of irreversible adsorption of metabolites on cell walls and membranes.[28] Freeze clamping is a faster process of freezing cells, but it is too tedious for large numbers of samples. Other quenching procedures, such as heating or acid treatment, can also cause even severer losses of metabolites due to degradation, and is therefore not well suited for non-targeted profiling. Addition of enzyme inhibitors, often together with antioxidants, is usually used together with other quenching procedures.

A quenching step is routinely utilized in cell and plant metabolomics, but it is often neglected in the analysis biological fluids, such as blood and urine, which may cause alterations to metabolic levels. Analytical data validating the omission of a quenching step for biofluids are currently lacking, so the exact consequences of including or omitting this step are not well known.[29]

It should be noted that some metabolites are very unstable, and several metabolic processes are very fast, making quenching very challenging. Quenching can also alter the metabolic profile due to degradation or loss of some metabolites of interest.

2.3.2 Drying

Drying is typically applied to plant samples, but it is rarely used for other metabolic samples. The drying can be carried out by methods such as ambient air drying, oven-drying, freeze-drying and trap-drying. Among these, freeze-drying (lyophilization or cryodesiccation) is the preferred way to dry samples for metabolic studies, as the other methods may cause substantial changes to the metabolic profiles.

2.3.3 Homogenization

Prior to sample extraction, different methods can be used to break up tissues and cells. Both chemical and mechanical/physical methods are available for disruption of the sample matrix, as shown in Figure 2.2. For many types of

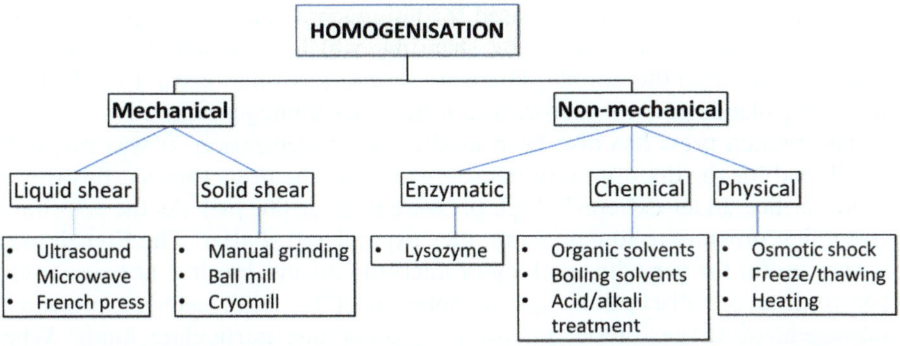

Figure 2.2 Methods for homogenization of tissues, cells and other biological samples.

samples, chemical homogenization is sufficient, but for many others, such as microorganisms, intact tissues and solid specimens, such as seeds, more efficient homogenization is required, using mechanical disruption. Mechanical and physical methods that rely on grinding, shearing, beating and shocking are then used. Mechanical homogenizers, manual homogenizers, mortar and pestles, sonicators, mixer mills and vortexers are several of common tools used for mechanical and physical disruption.

The most common mechanical methods include grinding in a liquid nitrogen-cooled mortar and pestle or homogenization by an electric homogenizer directly in the extraction solvent. Grinding frozen tissues by mortar and pestle has been the traditional gold standard. However, it has several drawbacks. In addition to being very labour intensive, it is not very well suited for very small amounts of sample, and it is challenging to prevent partial thawing of the tissues while grinding. Grinding samples frozen with liquid nitrogen using a mortar and pestle avoids the problems with partial thawing, but this method is even more tedious. Cryogenic grinding can also be performed in more efficient manner by using special instruments.

A simple homogenization method is bead beating, which relies on placing a sample and beads in a tube and rapidly shaking them. Bead beating has been used for years for the disruption of microorganisms, originally using small glass beads and dental amalgamators. The method is simple and relatively effective, but relatively time and labour consuming.

Vortex bead beating is also a widely used method for homogenization. This method relies on the addition of grinding beads to the sample tube and then repeatedly vortexing the sample. The grinding beads can be made of zirconium or stainless steel. The method is typically used for the lysis of cells and microorganisms. It should be noted that homogenizing samples by vortexing generates significant amounts of heat due to the friction created by the grinding balls, which in turn can cause alterations to the metabolic profile of the sample. Performing the vortexing in ice minimizes the problems with heat generation.

Blenders are also sometimes used for homogenization for samples available in large amounts. They work by shearing, which is created by applying a tangential force to the sample. There are several tools that disrupt by shearing, including blenders, rotor-stators and some glass homogenizers.

The French press has also been used for homogenization. It was originally developed for the disruption of microorganisms. The press forces cells through a tiny orifice under extremely high pressure (*e.g.*, 20 000 psi). As the cells move from high pressure to low pressure, they expand and shatter. The French press is very useful for periodic cracking of microorganisms, but it is not very useful for routine cell disruption. The samples must be fluid, such as previously homogenized tissue, blood, microbes or other fine particulate fluids. When disrupting microbes, cells are often harvested by centrifugation and the cell paste is processed. The advantage of the French press is that the homogenates are very uniform. However, the throughput is very low and for most samples, a pre-homogenization step is required.

Shock waves are also used for disrupting samples, with ultrasonication being the best example. Elevated pressure can also be used for homogenization, often combined with extraction and heat (*e.g.*, pressurized liquid extraction, PLE).

In a recent study, different tissue disruption techniques for the homogenization and extraction of the nematode *Caenorhabditis elegans* were compared, namely (1) manual grinding with a 10 cm ceramic mortar and pestle at liquid nitrogen temperatures; (2) homogenization with an electric homogenizer; (3) bead beating using a FastPrep 12; and beating in a reciprocating ball-mill using (4) metal spheroids, (5) glass beads or (6) metal ball bearings at cryogenic temperatures.[30] Two solvent systems (methanol–water, 80 : 20 v/v, and a monophasic methanol–chloroform–water, 1 : 2 : 0.8, v/v/v) were tested in the extraction, and the extracts were analysed by three different platforms (NMR spectroscopy, GC-MS and UPLC-MS). The results showed that rotary homogenizers did not give satisfactory results in comparison with the other homogenization techniques. For the analysis of polar metabolites, the best results were obtained with a bead beater in combination with 80% methanol or, if a bead beater is not available, a mortar and pestle.

2.3.4 Protein Precipitation and Urea Removal

Proteins typically disturb the metabolic analyses, so they have to be removed from the samples before analysis, if present. Biofluids, blood serum and plasma, and also CSF, contain substantial amounts of proteins (6–8 g dL^{-1}), whereas in urine the protein concentrations are typically low (50–100 mg dL^{-1}).[31] The most common method for protein removal is precipitation with organic solvents, such as methanol, ethanol, acetone or acetonitrile. Trichloroacetic acid and zinc sulfate have also been used. Addition of an organic solvent to a biofluid also disrupts any binding between metabolites and proteins, hence the metabolic concentrations obtained are equivalent to the sum of (protein-) bound and free metabolite concentrations. Ultrafiltration has also been used for the removal of proteins: the sample is pushed through a special filter with a specific molecular weight cut-off (3000, 10 000 and 300 000 Da) by either pressure or centrifugation. It has been recommended particularly for use with NMR spectroscopy.[32] It should be noted, however, that the method suffers from significant loss of hydrophobic species.

The performances of different protein precipitation (PP) methods have been compared extensively in terms of efficiency of protein removal, method precision, and number of metabolites observed.[33] Precipitation with acetonitrile and acetone seems to perform well in terms of protein removal.[29,33] However, precipitation with alcohol (methanol, ethanol or their mixture) results in better metabolic coverage and method precision.[33] Also, the ratio of plasma or serum to precipitant solvent is an important consideration and various ratios have been suggested. Typically, ratios from 1 : 2 to 1 : 4 are used. None of the protein removal methods remove the proteins completely, and typically 2–10% of the total protein amount is still found in the final sample.[34] The conventional manual PP methods can also be carried out automatically by membrane-based PP filter plates which allow precipitation of proteins and filtration without centrifugation and supernatant transfer steps. Most commonly, PP plates are available in 96-well format, which permits both manual and robotic automation for the PP process.

In urine samples, high concentrations of urea often disturb the final analysis; particularly for GC-based analysis, removal of urea is often needed before the analysis. The most common way to remove urea is by enzymatic degradation using urease. A concentration of 30–100 U of urease enzyme is generally used, depending on the volume of the urine sample. However, the responses for a range of other metabolites are decreased with urease pretreatment, hence the urease treatment will influence the metabolic profile in non-targeted analyses.

2.3.5 Extraction

A large number of different methods and techniques have been developed for different types of analytes and sample matrices. Extraction procedures vary in degree of selectivity, speed, and convenience, depending on the approach and conditions used. The selection of the best method and optimization of the parameters are often not easy tasks. The choice of the extraction method is largely dependent on the matrix: different methods are required for liquid and semi-solid samples, such as cells and tissues. In addition, for target compound analysis different methods are often needed than for comprehensive chemical profiling of the samples. For target compound analysis, selectivity of the sample preparation is often one of the key features in the selection whereas for profiling an exhaustive extraction method is a better choice. The targeted methods for specific groups of metabolites are summarized in Chapter 9 (Applications), and this chapter mainly describes the extraction procedures for global metabolic profiling.

In each extraction technique, the operating principle is the same: the partitioning of analytes between the sample matrix and an extracting phase. The selective extraction of analytes is based on differences in their chemical and physical properties. These typically include molecular weight, charge, solubility (hydrophobicity), polarity or differences in volatility. Some extraction methods, such as solid-phase extraction with immunoaffinity materials and imprinted polymers, utilize selectivity for specific structural groupings or mimic a biological selectivity.

In the selection and optimization of the extraction conditions, theoretical models developed for various techniques can be utilized. By optimizing the extraction procedures using experimental design and combining them with formal optimization strategies, it is possible to obtain optimum operating conditions with minimal labour, time and cost. In most cases, parameters affecting the extraction process are interdependent, hence optimization of each parameter in time is not the best option.

By increasing the extraction time and temperature, the extraction yields of the metabolites can be increased; however, the risk of artefact formation also increases. One should also take the reactivity of solvents into consideration in the case of prolonged extractions or when heating is applied in the extraction procedure. In most metabolic studies, a short extraction is preferred, often carried out at 4 °C or at room temperature. Enhanced extraction methods that are in widespread use in other applications, such as microwave-assisted extraction (MAE), supercritical fluid extraction (SFE) and PLE, are not widely used in metabolomics, mainly owing to the issue of the metabolic stability. Of these, SFE is a relatively soft extraction method and it has been utilized in metabolomics mainly for the extraction of volatile and/or lipophilic components from plant materials.[35]

Different extraction methods are required for targeted and non-targeted analyses. In targeted analyses, selective extraction is preferred whereas in non-targeted analysis, non-selective extraction is required. However, also in non-targeted analysis, some selection based on the chemical characteristics, mainly solubility, is needed in the selection of the extraction methods. For example, non-polar compounds such as lipids require non-polar solvents for their (liquid) extraction; polar metabolites are then excluded, as they have poor solubility in non-polar organic solvents. Usually, the non-polar metabolites and polar metabolites are either extracted separately from two aliquots of the samples, or sequential extraction is applied.

2.3.5.1 Non-targeted Extraction of Lipids

Lipids have two functional features that control their solubility in organic solvents: the hydrophobic hydrocarbon chains of the fatty acid or other aliphatic moieties and any polar functional groups, such as phosphate or sugar residues, which are markedly hydrophilic. Those lipids that do not have any polar groups, for example, triacylglycerols and cholesterol esters, are readily soluble in solvents such as hexane, toluene and cyclohexane and also in moderately polar solvents such as diethyl ether and chloroform, whereas they are poorly soluble in polar solvents such as methanol. The solubility of these neutral lipids in alcohols increases with increase in chain length of the hydrocarbon moiety of the alcohol, so they are more soluble in ethanol and completely soluble in butan-1-ol. Polar lipids, such as phospholipids and glycosphingolipids, are only slightly soluble in hydrocarbons, but they dissolve well in more polar solvents such as methanol, ethanol and chloroform. The pH also plays an important role in the extraction of specific lipid subclasses, due either to degradation at a specific pH or to their acid–base characteristics. For example, plasmalogens which contain a vinyl ether linkage are labile in acidic conditions, resulting in hydrolysis of the vinyl ether linkage. Thus, acidic extraction procedures result in losses of plasmalogens and production of 2-acyl lysolipids. On the other hand, acyl-CoAs are very sensitive to alkaline pH, which results in hydrolysis of the thioester linkage in the acyl-CoAs. Acidic solvents in the extraction can also cause α-hydroxyl migration of acyl chains, for example, conversion of 1,2-diacylglycerols to 1,3-diacylglycerols. To prevent autoxidation of unsaturated lipids, it is advisable to add an antioxidant (*e.g.*, butylated hydroxytoluene, BHT) to the solvent (at a level of 50–100 mg L^{-1}). On the other hand, acidic lipids, such as phosphatidylserines and phosphatidic acids, are generally more efficiently extracted under acidic conditions. In addition, careful optimization of the extraction conditions is necessary for the more polar lipids such as gangliosides and polyphosphoinositides.

An efficient extraction solvent should not only dissolve the lipids readily but also overcome the interactions between the lipids and the sample matrix. The simple lipids are generally easily extracted because they are typically part of large aggregates in storage tissues, such as oil bodies and adipose tissue. The complex lipids are more difficult to extract because they are usually constituents of membranes, where they occur in close association with membrane proteins and polysaccharides. An ideal solvent should also have low or no solubility of proteins and polar metabolites (*e.g.*, amino acids, carbohydrates). The solvents used for lipid extraction include alcohols (methanol, ethanol, propan-2-ol, *n*-butanol), acetone, acetonitrile, ethers (diethyl ether, diisopropyl ether, dioxane, tetrahydrofuran), halocarbons (chloroform, dichloromethane), hydrocarbons (hexane, benzene, cyclohexane, isooctane), or their mixtures. The best results have been obtained by using mixtures of solvents, the most common mixture being chloroform and methanol in the ratio 2:1 v/v, as suggested by Folch *et al.* in 1957,[36] and which has been modified slightly by Bligh and Dyer.[37] In the Bligh and Dyer method, the sample is homogenized with chloroform and methanol in such proportions that a miscible system is formed with water in the sample.[37] After blending the mixture, chloroform and water are added to obtain a final composition of chloroform–methanol–water (2:2:1.8). Addition of chloroform and water forms a biphasic system with a lower chloroform phase that contains extracted lipids. As a drawback of this method, it was observed that samples that contain more than 2% of lipids produced much lower estimates of lipid content than with Folch extraction.

The underestimation of lipid concentration increases with increasing lipid content of the sample.[38] If needed, traces of water in the final extract can be removed by passing the chloroform extract through a bed of anhydrous sodium sulfate. In the extraction of tissue samples, the ratio of chloroform–methanol to tissue (assumed to be mainly water) should be >17 : 1, because then the water present in the tissues can be solvated and remain in a single phase. In lipid extracts using chloroform–methanol as extractant, the coextracted matrix compound can be removed by washing with water or dilute KCl solution (0.88% w/v). Salt solution also prevents or at least minimizes the formation of an intermediate phase. When chloroform–methanol (2 : 1 v/v) is used for extraction, addition of water/dilute salt solution results in the formation of a two-phase system, where the lower phase consists of chloroform–methanol–water (86 : 14 : 1 v/v/v) and the upper phase consists of the same, but in a different ratio (3:48 : 47 v/v/v). The lower phase comprises about two-thirds of the total volume and contains mainly lipids. However, it should be noted that more polar lipids, such as some phospholipids and glycolipids and all gangliosides, may remain in the upper phase.

Although Folch extraction is widely used, it does have some features that make its use in modern robotic sample preparation challenging. Chloroform has a higher density than water–methanol mixture, hence it forms the lower phase of the two-phase partitioning system. While collecting the chloroform fraction, a glass pipette or a needle of the pipetting robot has to pass through a layer of non-extractable insoluble matrix, usually residing at the interface of the water–methanol and chloroform phases. This may lead to contamination of the organic phase with protein-containing matrix when using centrifugation to assist in the phase separation or during the removal of the lipid fraction. Therefore, use of methyl *tert*-butyl ether (MTBE)–methanol–water has been suggested and it has been shown to give recoveries the same as or slightly better than the established Bligh and Dyer method.[39] Other mixtures that have been used for lipid extraction are listed in Table 2.1. The ratio of the

Table 2.1 Summary of liquid extraction methods for lipid analysis.

Extraction solvent	*Sample types*	*Ref*
Chloroform–methanol (2:1), with extractant to sample ratio 20:1	Global lipid profiling, tissues, serum, plasma, cells	36,40
Chloroform–methanol (1:2 v/v)	Global lipid profiling, tissues, serum, plasma	37
Chloroform–propan-2-ol mixtures	Erythrocytes	41,42
Dichloromethane–methanol (2:1 v/v)	Plasma, liver	43
Propan-2-ol–hexane (3:2 v/v)	Animal tissues	44,45
Methanol–hexane (1:1 v/v)	Leaf tissue, higher concentrations of sensitive lipids such as phosphatidylinositol	46
Butan-1-ol saturated with water	Cereals, wheat-flour	47
Methyl *tert*-butyl ether (MTBE)–methanol–water (10:3:2.5 v/v/v)	Serum/plasma	48

extraction solvent to the amount of sample is also important. Usually, the ratio should be 25% w/v tissue-to-solvent to allow satisfactory recovery of the lipids. However, it is not possible to obtain quantitative recovery of all lipid subclasses using a single extraction method. The differences in the recoveries can be at least partially corrected by the use of class-specific standards. Of course, the class-specific standards do not account the different extraction behaviours within the lipid subclass due to differences in the fatty acid chain lengths and the degree of saturation, but these differences are typically relatively minor.

In a comparative study, five different extraction methods were compared for the extraction of human leukaemia cells, namely, extraction with two different ratios of chloroform–methanol, propan-2-ol–methanol and propan-2-ol–hexane and with butanol (Table 2.2).[49] The chloroform–methanol mixtures gave the best results for sphingomyelins (SM), phosphatidylcholines (PC), phosphatidylethanolamines (PA) and triglycerides (TGs), whereas both propan-2-ol–hexane mixture and butanol extracted cholesterol esters more efficiently. However, these two methods did not extract TG, and also propan-2-ol–chloroform was poorly suited for the extraction of TG.

Specific lipids can also possess either acidic or basic features, which should be taken into account in the extraction. The general procedures described above generally do not extract these acidic or basic lipids efficiently.

Other extraction methods used in the lipid extraction include SFE and PLE. With SFE, good extraction yields have been obtained for non-polar lipids, including esterified fatty acids, acylglycerols, and unsaponifiable matter. Complex polar lipids are only sparingly soluble in supercritical carbon dioxide alone and a polar modifier, such as methanol, ethanol and even water, is required to improve the extraction of polar lipids.[50] SFE has been used for the extraction of lipids especially from various food matrices, such as different nuts, edible oils and seeds.[51] The recoveries of lipids in SFE were similar to those with conventional solvent extraction methods,[52,53] and no significant differences between the extraction of fatty acids were observed. PLE has also been used in lipid extraction although only in very few applications.[54] The elevated temperatures used in PLE can cause alterations of the lipid composition.

Table 2.2 Comparison of liquid extraction methods for lipid extraction.

Extraction procedure	*SM + PC + PE*	*Cho*	*TG*
Chloroform–methanol (2:1 v/v)	++	+	++
Chloroform–methanol (1:2 v/v)	++	+	+−
Propan-2-ol–chloroform (7:111 v/v)	++	+	−
Propan-2-ol–hexane (2:3 v/v)	+	++	−
Butanol	+	++	−

SM = sphingomyelins, PC = phosphatidylcholines, PE = phosphatidylethanolamines, Cho = cholesteryl esters, TG = triglycerides.
Source: modified from Ref. 49.

2.3.5.2 Non-targeted Extraction of Polar Metabolites

The polar (and semipolar) metabolites include compounds such as amino acids, carbohydrates, acylcarnitines, nucleotides, bile acids and short-chain carboxylic acids. For the non-target profiling, simple protein precipitation (serum, plasma, urine, CSF) or liquid extraction with mixtures of water or buffer and different organic solvents (tissues, plant, food) have been utilized for the extraction of polar metabolites. For target analysis, both liquid extraction and SPE have been widely used.

In global profiling of polar compounds from liquid samples, simple protein precipitation is commonly used, hence no extraction is required. For cells, tissues and plant-based materials, liquid extraction with methanol, acetonitrile, acetone or solvent mixtures including aqueous buffer–water is commonly used. For example, in the extraction of liver tissue, methanol–water (2:1), acetonitrile–water (1:1 or 2:1) and methanol–chloroform–water have all been shown to give good extraction yields; however, methanol–water extraction gave more variable results than acetonitrile–water and methanol–chloroform–water extraction.[55] On the other hand, methanol–water has been shown to give more effective and reproducible results than acetonitrile–water for LC-MS-based metabolic profiling of serum.[56] In a recent untargeted LC-MS analysis of polar metabolites in plasma and serum, five methods were evaluated for sample preparation (Table 2.3).[57] Of the methods tested, a combination of solvent extraction with acidic acetonitrile and SPE-mediated removal of phospholipids was observed to improve the signal response of most metabolites of blood fluids, probably by reducing the background of interfering species and minimizing ion suppression phenomena.

Table 2.3 Comparison of extraction methods for non-targeted profiling of polar metabolites.

Criteria	*ACN*	*MeOH*	*MeOH–EtOH*	*Solvent-free filtration*	*Acidic ACN + SPE*
CV (%)	15	10	9	13	12
Extraction efficiency	3-fold higher	3-fold higher	3-fold higher	Lowest	4-fold higher
Detection	Moderate	Poor	Poor	Moderate	High
Amino acids	Moderate	Moderate	Moderate	Moderate	High
Purine alkaloids	Poor	Poor	Poor	Poor	High
Flavonoids/microbial					
Metabolites:	High	Moderate	Moderate	Poor	Moderate
Acylcarnitines	High	Moderate	Poor	Very poor	Poor
Fatty acids (C_8–C_{18})	Poor	Moderate	Poor	Very poor	High
Bile acids + conjugates					
Simplicity	Moderate	Moderate	Moderate	High	Moderate
Speed	Moderate	Moderate	Moderate	Low	High
Automation compatibility	Low	Low	Low	High	High

ACN = acetonitrile.
Source: modified from Ref. 57.

In a recent study, different methods were evaluated for the extraction of polar metabolites, such as nucleotides (uracil, adenine, CMP, UMP, GMP, AMP, GDP, ADP, CTP, GTP, ATP, NAD, NADP), UDP–sugars (UDP–glucose, UDP–glucuronic acid), amino acids (glycine, phenylalanine, arginine, leucine, aspartic acid, methionine, norvaline, sarcosine), organic acids (fumarate, α-ketoglutarate, citrate, malate, succinate) and glucose in cell samples.[19] A summary of the results is shown in Figure 2.3. Hot ethanol- and water-based extractions were shown to promote conversion of either ATP to ADP and AMP or UDP–sugars to UDP, hence they should not be used. Alkaline extraction of intracellular metabolites was shown to give poor results, due to the occurrence of a viscous precipitate after centrifugation, which made removal of the supernatant difficult. The best methods were based on extraction with cold 50% MeOH, MeOH freeze, MeOH/Chloro and ACN (see Figure 2.3). These four methods were superior in terms of both recovery of standards (*i.e.*, minimum degradation) and maximum cell extraction. Of these four methods, acetonitrile was considered to be most suited as an extractant for metabolomic analysis, and the optimum solvent to cell ratio was 1 mL of extraction solution per 5×10^6 cells. It was also observed that freeze–thaw cycles are not necessary to lyse mammalian cells or stop enzymatic activity.

2.3.5.3 Non-targeted Screening of Both Lipids and Polar Metabolites

For the analysis of both lipids and polar metabolites, four different approaches are typically applied:

1. single extraction with a suitable solvent mixture;
2. two separate extractions of two aliquots of the sample with solvents optimized for lipids and polar metabolites: easy with liquid samples, more difficult with tissues and cells;
3. simultaneous extraction of polar and non-polar metabolites using a combination of water, methanol and chloroform, resulting in two layers separated by a protein pellet; or
4. two sequential extractions: an aqueous extraction using polar organic solvents (*e.g.*, methanol or acetonitrile) mixed with water or perchloric acid, or followed by a lipid extraction of the remaining pellet, usually with chloroform–methanol

For untargeted analyses, methanol–water–chloroform combinations in different proportions have been shown to extract both hydrophilic and hydrophobic compounds.[58] In a multiplatform study, methanol–water–chloroform (59 : 26 : 15 v/v/v) was shown to give the best overall results.[59] However, it is impossible to optimize a single solvent mixture to obtain very high recoveries for both very hydrophilic and very hydrophobic compounds, so the coverage with this approach is rather limited and, in addition, small changes in sample

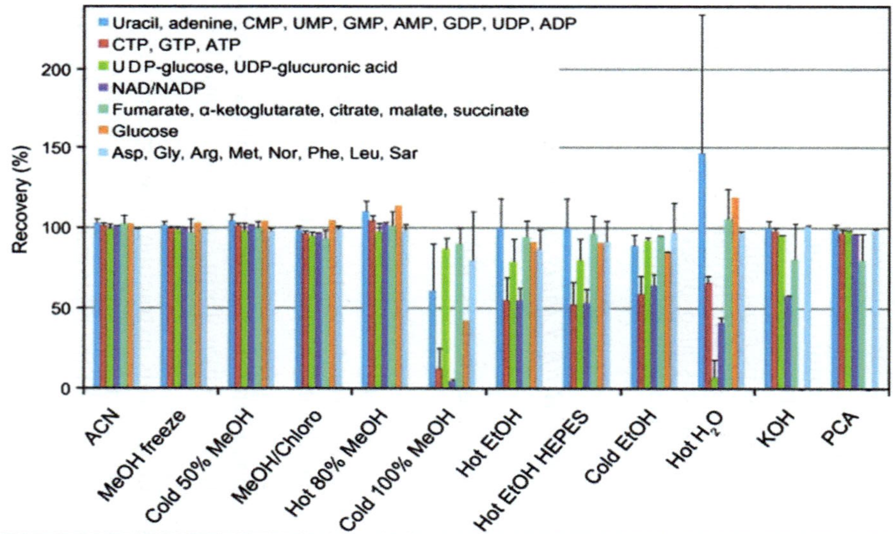

Extraction method	Procedure
ACN	Resuspend cell pellets in ice-cold 50% aqueous acetonitrile,vortex,and incubateon ice for 10 min
MeOH freeze	Resuspend cell pellets in 1 ml of 50% aqueous MeOH (≤ 40 °C),frozen in liquid N2,and thawed on ice. Repeate twice
Cold 50% MeOH	Resuspend cell pellets in 1 ml of 50% aqueous MeOH (≤ 40 °C),vortex,and incubate on ice for 10 min
MeOH/Chloro	Resuspend cell pellets in 1 ml of 50% MeOH (40°C). Add 0.5 ml of chloroform (40 °C) vortex,and incubate for 10 min on ice . Both phases were kept for analysis.
Hot 80% MeOH	Resuspend cell pellets in 1 ml of 80% MeOH (70 °C),vortex,and incubate at 70 °C for 5 min. Cool on ice for 5 min
Cold 100% MeOH	Resuspend cell pellets in 1 ml of 100% MeOH (40 °C),vortex,and incubate on ice for 10 min
Hot EtOH	Resuspend cell pellets in 1 ml of 75% aqueous EtOH (80°C),vortex,and incubated at 80 °C for 5 min. Cool on ice for 5 min
Hot EtOH HEPES	Resuspend cell pellets in 1 ml of 75% aqueous EtOH + 0.25 M HEPES (80°C),vortex,and incubate at 80 C for 5 min. Cool on ice for 5 min
Cold EtOH	Resuspend cell pellets in 1 ml of ice-cold 75% EtOH,vortex,and incubate on ice for 10 min.
Hot H2O	Resuspend cell pellets in 1 ml of water (95 °C),vortex,and incubated at 95 °C for 5 min. Samples were then cooled on ice for 5 min
KOH	Resuspend cell pellets in 1 ml of ice-cold 0.05 M KOH,vortex,and incubate on ice for 10 min. Neutralized with 150 ul of 1 M K2HPO4 and 70 ul of 0.5 M PCA
PCA	Resuspend cell pellets in 1 ml of ice-cold 0.5 M PCA,vortex,and incubate on ice for 10 min. Nneutralize with 200 ul of 1.1 M K2HPO4 in 2.5 M KOH

Figure 2.3 Extraction of polar metabolites from cell samples. Reprinted with permission from ref. 19.

composition may cause large changes in the recoveries of some critical compounds, thus making the protocol less robust. A better approach is to use two solvent mixtures, one suitable for the extraction of hydrophobic metabolites and the other for hydrophilic metabolites. If a sufficient amount of sample is available, the application of two separate extractions is the most robust way to extract the metabolites, if two different methods are used for the analysis also. In the sequential or two-phase extraction systems, part of the metabolites will be extracted into both phases, again resulting in a less robust method for these analytes, but also reducing the sensitivity of these compounds.

In a recent study, consecutive extraction and two-phase extraction were compared in the non-targeted extraction of liver samples.[60] In the first approach, methanol–water (1 : 1) and acetonitrile–water (1:1) were used for homogenization in a TissueLyser. Supernatants were collected after centrifugation and dried, followed by extraction of the remaining pellets with dichloromethane–methanol (3 : 1) in the TissueLyser. The two-phase extraction was carried out by homogenization of the tissue either first with dichloromethane–methanol (2 : 1) before adding water or *vice versa, i.e.*, the sample was first homogenized in water and the organic solvent mixture was then added to the sample. Centrifugation produced a biphasic mixture and upper (aqueous) and lower (organic) layers were collected separately. The optimum protocol for profiling both polar and non-polar metabolites was found to be an aqueous extraction with methanol–water followed by an organic extraction with dichloromethane–methanol.

For untargeted analyses, methanol–water–chloroform combinations in different proportions has been shown to extract both hydrophilic and hydrophobic compounds in plants such as *Arabidopsis thaliana*.[61] Sequential and selective extraction has also been used in non-targeted extraction of plant and food materials. The first extraction is usually effected with an aqueous solvent (typically methanol–water) followed by extraction with a non-polar solvent (typically chloroform) of the centrifuged pellet.

2.3.5.4 *Volatile Metabolites*

Volatile organic compounds (VOCs) are important components of the metabolome and include metabolites such as alcohols, alkanes, aldehydes, furans, ketones, pyrroles and terpenes. For volatile metabolites, several solvent-free extraction methods have been developed. Solid-phase microextraction has been used to extract volatile metabolites, mainly from the headspace of the samples. Other techniques involve the use of traps made of monolithic materials which are hybrid materials of silica gel, active carbon and octadodecylsilica phase. They can be used for headspace sampling, followed by either solvent elution or thermal desorption. They are a new-generation tool for sorptive extraction of volatile and semi-volatile components.

A typical headspace SPME analysis involves the extraction of the VOCs via partitioning into an SPME fibre (polymer-coated fused-silica fibre), subsequent desorption of the VOCs by heating the fibre in the injection port of a gas chromatograph, followed by GC-MS [or GC-FID (flame ionization detection)]

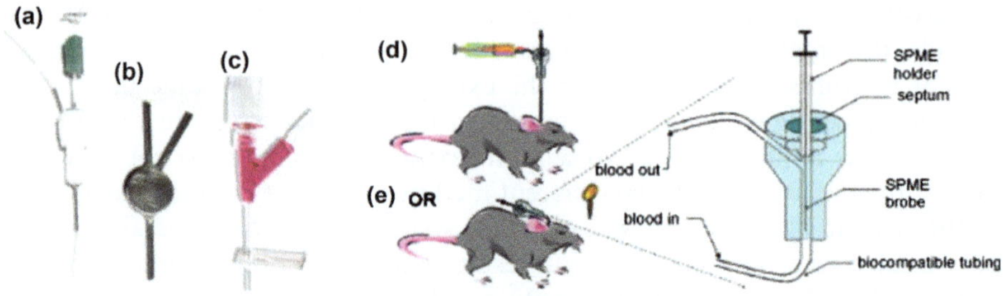

Figure 2.4 *In vivo* SPME sampling in rats. (a) Teflon Y adapter, custom made (b) stainless-steel Y adapter and custom made (c) BD Saf-T-Intima closed intravenous catheter system. (d) Schematic showing the procedures for *in vivo* SPME sampling in rats where the interface is connected to a carotid artery catheter and blood flow through the interface is established using push–pull action with a syringe. (e) Schematic showing the procedures for *in vivo* SPME sampling in rats using a Y-interface where blood is recirculated to the animal.
Reprinted with permission from Ref. 66.

analysis. However, in SPME, the extraction is not exhaustive, as the enrichment of the analytes on the fibre is dependent on two equilibrium distribution steps, sample–gaseous phase and gaseous phase–fibre coating. Moreover, this distribution is influenced by analyte–matrix interactions, making quantitative analysis challenging. SPME has been utilized in the extraction of volatile signatures of fecal samples and food and plant matrices, including wines, grapes, honey, herbs, coffee, barley coffee, strawberry, butter, olive oil, roasted hazelnut and cacao.[62] In addition, SPME has been utilized in breath gas analyses. SPME also allows integration of sampling with extraction, and *in vivo* sampling has been tested for extraction of flowing blood[63] and sampling of volatiles from humans[64] and insects.[65] An example of *in vivo* SPME is shown in Figure 2.4.[66] The choice of SPME fibre material is an important parameter, the optimum choice being dependent on the polarity of the analytes. Several SPME fibre coatings are currently commercially available, including polyacrylate (PA), polydimethylsiloxane (PDMS), Carbowax–poly(ethylene glycol) (PEG) and mixed phases of Carboxen (CAR)–PDMS, divinylbenzene (DVB)–PDMS, and CAR–DVB–PDMS. However, a single SPME fibre coating may be insufficient for a comprehensive analysis of the metabolic profile of biological samples. It is also possible to use SPME for the extraction of non-volatile compounds. In a systematic study of SPME coating materials for untargeted metabolite profiling in combination with LC-MS, over 40 fibre materials were evaluated.[67] The best performance was observed for mixed-mode, phenylboronic acid and polystyrene–divinylbenzene coatings.

2.4 Clean-up of Extracts

In the targeted analysis of biological samples, extracts often require clean-up or fractionation, or both, before the final analysis. Typical compounds that disturb the analyses include proteins, (non-target) lipids, starch/cell wall/other carbohydrates, DNA/RNA, chlorophyll and other pigments, high-abundance primary metabolites and salts.

The clean-up of the extract is important even if selective detection is used with the chromatographic technique utilized. With selective detection, matrix compounds may be undetected, but they can disturb the separation system and cause problems in the subsequent analyses. In GC, non-volatile matrix compounds can cause problems in injection, and if they enter the column they may cause contamination and even damage the column stationary phase. In split and splitless injection, a large amount of matrix compounds may change the evaporation process, leading to changes in the amount of analytes entering the column. In splitless injection, matrix compounds can cause severe matrix effects. In on-column injection, the non-volatile compounds will enter the column, causing contamination. PTV (programmed temperature vaporization) injection tolerates dirty samples relatively well, and this technique is therefore best suited to dirty samples. Also in LC, the analysis of dirty samples causes problems. Matrix compounds usually shorten the lifetime of the column. Because the separation efficiency of LC is not as good as that of GC, coelution of the target analytes with matrix compounds is a typical problem, and these matrix effects lead to problems in quantitation due to ion suppression. Hence with MS detection, careful clean-up of the samples is typically required.

The extracts of biological samples usually contain high concentrations of lipids, which may interfere with the analysis of compounds other than the lipids of interest. Particularly if GC is used in the analysis, efficient removal of lipids is crucial. Liquid–liquid partitioning with a non-polar solvent (hexane, chloroform) and SPE are the most commonly applied methods for lipid removal. Specific commercial SPE systems for the removal of phospholipids are also available; and they have been shown to be able to reduce the matrix effects substantially.[68] Phospholipid removal is often carried out in combination with protein precipitation, and the actual phospholipid removal is based on elution of the extract through a packed bed consisting of proprietary zirconia-coated silica particles. The zirconia sites exhibit Lewis acid (electron acceptor) properties that will interact strongly with Lewis bases such as the phosphate group of the phospholipids. It essential to use a suitable modifier, such as formic acid, to improve the recovery of, *e.g.*, acidic metabolites.

As pointed out earlier, lipid extracts can be washed with water or dilute KCl solution to remove coextracted matrix compounds. Non-lipid contaminants may also be removed partly or completely by evaporation of the lipid extract to dryness in vacuum or under nitrogen and then re-extraction of the lipids with a non-polar solvent, such as hexane. Acetone precipitation can also be used to separate in one step polar lipids (phospholipids and glycolipids) from all neutral or non-polar lipids (triglycerides, cholesterol, some pigments). The lipid extract is evaporated to dryness, ~20–30 volumes of acetone are added and the mixture is left on ice for ~1 h. After centrifugation, the supernatant is collected and the procedure is repeated. The acetone extracts contain more polar lipid, glycerides, sterols, sterol esters, carotenoids and lipid-soluble vitamins whereas the pellet contains phospholipids and glycolipids.

Removal of matrix components can also be carried out by liquid–liquid partition chromatography or by SPE. For lipid extracts, purification can be performed, *e.g.*, by passing the lipid extract through a Sephadex G-25 column, which retains the matrix compounds while lipids are quickly eluted through the column with methanol–water–chloroform mixture.[69] SPE has been used both for clean-up of the lipids and, more often, for fractionation of the lipids into specific lipid classes.[70–74] For example, aminopropyl-bonded phase and pure silica have been used for the fractionation of neutral lipids, glycolipids and phospholipids from lipid extracts. Silica-based cartridges used with solvent combinations of increasing polarity can be used for the fractionation of steryl esters, triglycerides, free fatty acids, 1,2-diglycerides, 1,3-diglycerides, monoglycerides, monogalactosylglycerides, digalactosylglycerides, phosphatidylcholine and lysophosphatidylcholines.[75] reversed-phase (RP) cartridges (C_2, C_8, C_{18}, cyclohexyl, phenyl) have also been used in SPE fractionation. An example of multistep lipid fractionation using three aminopropyl cartridges is shown in Figure 2.5.[76]

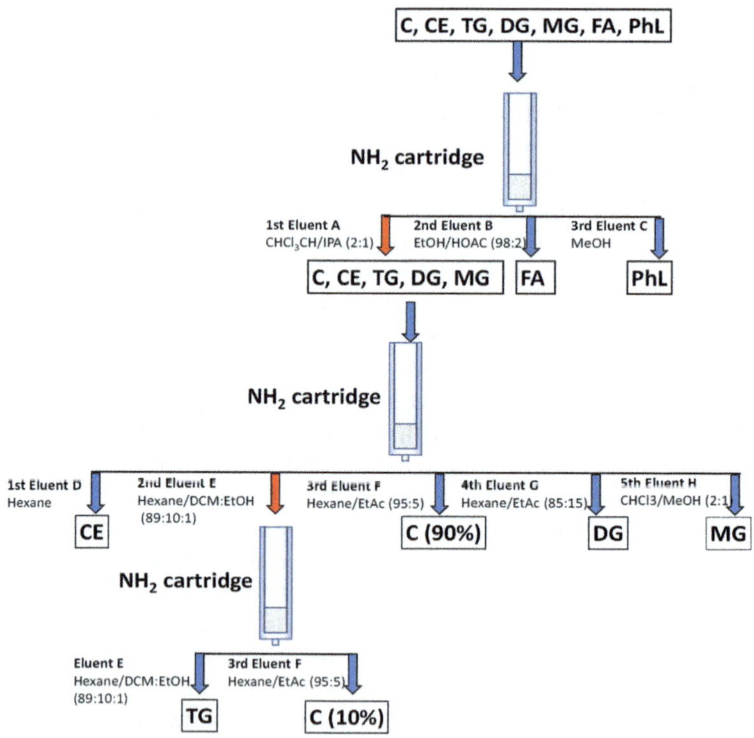

Figure 2.5 Multistep lipid fractionation. Using three NH_2 cartridges and eight different solvent mixtures, cholesterol esters (CE), triglycerides (TG), diglycerides (DG), monoglycerides (MG), cholesterol (C), fatty acids (FA) and phospholipids (PhL) could be fractionated.
Source: modified from Ref. 76.

Fractionation can also be carried out by column chromatography or liquid–liquid partitioning, or liquid chromatography can be used in the clean-up of extracts. High-performance liquid chromatography (HPLC) and gel permeation chromatography (GPC) have also been used for clean-up and fractionation of the extract. HPLC fractionation has several benefits over the conventional column chromatography. Because the particle size of the HPLC column is smaller than that in column chromatography, the separation efficiency in HPLC is clearly better. Moreover, the fractionation can be monitored with a detector and this feature allows not only the precise cutting of the fractions but also faster optimization of the fractionation conditions.

2.5 Derivatization

For GC analysis, polar compounds need to be derivatized prior their analysis. In HPLC, derivatization is seldom required, but can be used for enhancing the detection or retention.

In GC, samples that contain highly polar compounds, such as fatty acids, amino acids, carboxylic acids and carbohydrates, typically need to be derivatized before their GC analysis. The aim of the derivatization is to improve volatility and reduce the polarity of the substance, and thus improve the peak shape and reduce peak tailing of the GC analysis. Sometimes the derivatization also improves the stability of thermally labile compounds. For GC analysis, the most typical derivatization procedures utilize silylation, alkylation and acylation reactions.

In silylation, the active hydrogens from acids, alcohols, thiols, amines and other groups in the molecules are modified with an inert trimethylsilyl (TMS) group (Figure 2.6). Silylation also increases the molar mass of the compounds. The most common silylation reagents are bistrimethylsilyltrifluoroacetamide (BSTFA) or *N*-methyl-*N*-(trimethylsilyl)trifluoroacetamide (MSTFA) with 1% trimethylchlorosilane (TMCS). Both BSTFA and MSTFA are sufficiently volatile to provide little interference with early-eluting peaks, and TMCS is used as a catalyst to increase the TMS donor potential.[77] Pyridine, which is an acid scavenger, is typically used to drive the reaction forward. Other silylation reagents, such as *N*,*O*-bis(trimethylsilyl)trifluoroacetamide (BSA), have been used, but they can produce derivatization by-products that can attack the initially formed ester to yield an artefact. *tert*-Butyldimethylsilyl (TBDMS) derivatization can also be utilized, but incomplete derivatization of compounds such as polyols has been observed. In addition, as the derivative group is larger, the retention is increased, which may be a problem with compounds that contain multiple-derivatizable groups. Prior to silylation, methoximation (MOX) is typically used for specific classes of compounds (*e.g.*, keto acids, sugars) as it protects carbonyl moieties (converts them to methoximes) in order to improve their GC properties by preventing multiple reaction products (Figure 2.6).

In acylation, compounds that contain active hydrogens (*e.g.*, –OH, –SH and –NH) are converted into esters, thioesters and amides, respectively, through the

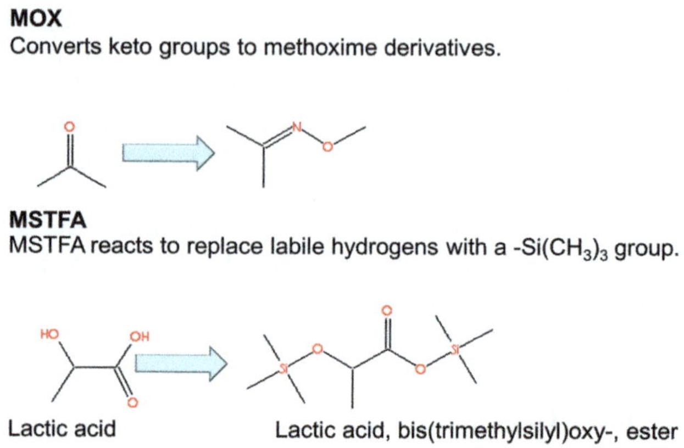

Figure 2.6 Typical two-step derivatization utilizing methoximation (MOX) and silylation.

action of a carboxylic acid or a carboxylic acid derivative. In alkylation, active hydrogen is replaced by an aliphatic or aliphatic–aromatic (*e.g.*, benzyl) group. Recently, ethyl chloroformate (ECF) derivatization has also generated interest as ECF derivatization forms more stable derivatives and is reactive in aqueous medium, is less expensive and is less time consuming than trimethylsilyl (TMS) and TBDMS derivatization.[78,79] Other methods for derivatization include reactions with alkyl chloroformates, which is a simple and rapid reaction, but it has been applied to only a limited number of biological samples.[80,81]

Usually the derivatization is performed after the sample pretreatment procedure, just before GC analysis. However, in some cases, the derivatization is carried out prior to the sample preparation in order to improve the extraction efficiency, for example, before SPME or SBSE. However, particular attention then has to be paid to the selection of the derivatization procedure because most GC derivatizations cannot be done in the presence of water. Conventional derivatization methods include heating the derivatization mixture for periods of time at elevated temperature and, more recently, also assisted by microwaves. Microwave-assisted derivatization (MAD) methods produce fewer side products, increase sensitivity and decrease the sample preparation time.[82,83]

In LC, the chemical derivatization is typically used to improve the detection sensitivity by converting a compound with a poor detector response into a highly detectable product. Sometimes derivatization is used to improve the retention of the compounds, *e.g.*, in the case of very polar metabolites such as carnitines. In LC analyses, UV chromophores and fluorophores are often introduced into sample molecules to increase their sensitivity to UV absorption and fluorescence detection. Also with MS detection, derivatization of the analytes can improve the selectivity and sensitivity of the detection. Benzoyl chloride, *m*-toluyl chloride and *p*-nitrobenzoyl chloride are reagents that can add a benzene ring to a solute molecule and turn it into a UV-absorbing compound. For example, to introduce UV chromophores into a solute containing a carbonyl group, 3,5-dinitrophenylhydrazine and *p*-nitrobenzylhydroxylamine are probably the two most common and effective reagents. To prepare fluorescent derivatives of phenols, and primary and secondary amines, dansyl chloride (5-dimethylaminonaphthalene-1-sulfonyl chloride) is often used. Another useful fluorescent derivative is 4-chloro-7-nitrobenz-2,1,3-oxadiazole (NBD chloride), which provides highly fluorescent derivatives of primary and secondary amines but aromatic amines, phenols and thiols yield only weakly or non-fluorescent derivatives.

2.6 Quality Control

Quality control (QC) is essential in metabolomics, particularly in large-scale non-targeted approaches. Whereas targeted methods can be carefully optimized for each target metabolite, compromises have to be made in non-targeted methods, both in the sample preparation and in the analysis. Moreover, typically very simple and straightforward sample preparation is performed in non-targeted methods, and this also means that the extracts are generally dirty, *i.e.*, they contain matrix compounds that may gradually contaminate the analytical system. The performance of the method has to be monitored and the possible fluctuations of the results must be corrected using a suitable set of internal standards, control samples and blanks. Randomization of both sample preparation and analysis order is important to avoid systematic bias due to possible analytical drift.

It is important to realize that different methods and instruments may require different QC procedures owing to their different intrinsic characteristics. Specific instruments may undergo a systematic drift in sensitivity over time whereas other instruments may lose specificity for only certain compounds without affecting other metabolites, and certain instrumentation may simply lack robustness for large-scale measurements. These errors are difficult to evaluate using a single analytical sequence, but in order to observe such trends identical samples have to be analysed over a sufficiently long period of time.

For maintaining the robustness of the analytical results, different methodologies may be used. A practical way of monitoring the quality of the system is to use QC charts. The analytical parameters are plotted over time on the chart, using, *e.g.*

- retention times;
- peak area/height;
- peak shape/symmetry.

Each parameter should be defined with upper and lower intervention limits: within these limits, the analytical process is in control. If the set limits are crossed (process out of control), the results should be considered unreliable and action must immediately be taken to bring the process back under control, *e.g.*, by instrument servicing/clean-up.

2.6.1 Quality Control Samples

QC samples should be as identical as possible with the actual samples, with a similar metabolic and sample matrix composition. Several types of control samples are required:

- technical samples;
- calibration samples;
- pooled QC samples;
  - pooled QC in which small aliquots of each biological sample to be studied are pooled and mixed, and
  - commercially available biofluids composed of multiple biological samples;
- batch samples;
- blanks.

Pooled QC samples have a number of advantages. They have a very similar composition to the actual biological samples. They are well suited to small, focused studies in which all of the samples are available before analysis (*e.g.*, small clinical trials or animal studies). However, pooled QC samples are not always applicable in large-scale studies in which there are many thousands of samples to be analysed and in which sample collection is not completed before sample preparation and analysis begin. It is notable that in studies involving a large number of samples that are analysed in multiple batches, a quality check of each batch requires identical QC samples pooled from samples in all batches. Also, for cells and tissues it is more difficult to pool the samples. One option is to take an aliquot of each extract and pool the extracts, and use this as a QC sample. In routine analyses, the number of QC samples should be ~5% of the total sample number (1 in 20 samples) but in more complex analyses, QC samples should be analysed more frequently (1 in 5 to 1 in 10 samples).

2.6.2 Internal Standards/Surrogates

Internal standards should be added to the samples prior the sample preparation to monitor methodological performance and correct for sources of variance introduced by sample preparation and analysis steps. Optimally, the internal standards should not be present in the sample endogenously, they should closely mimic the physicochemical properties of target compounds and they should behave like analytes throughout the method, *i.e.*, both in the sample preparation steps and in the separation and detection. For this purpose, stable isotopes are the best approach, when available. If the stable isotopes are not available, compounds that have a similar structure to the target compounds but which are not found in the actual samples should be used, as illustrated in Figure 2.7.

The concentration of internal standards should resemble that of the measured metabolites, but at concentrations that are well above that of the

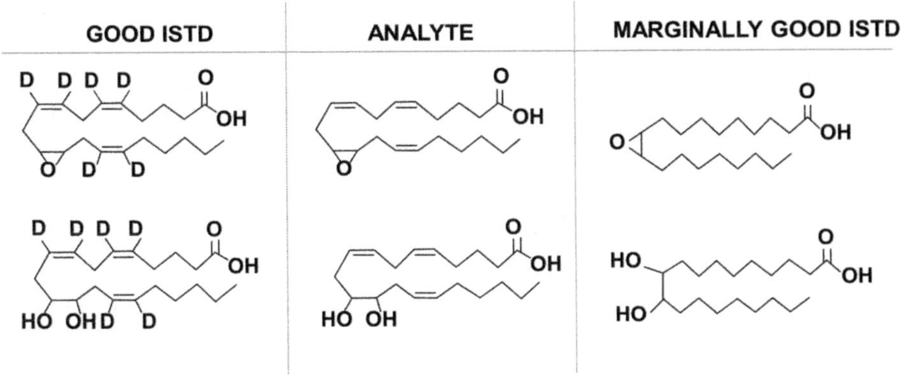

Figure 2.7 An example of selection of internal standards.

background signal (minimum ×10 of signal in blank). Often, in a multistep procedure, it is advisable to add several sets of internal standards at different stages of the work flow as this enables one to validate and control the individual experimental sub-processes such as sample extraction, fractionation, derivatization, injection and analysis.

2.7 Conclusion

The first steps of the metabolic workflow are crucial and careful optimization and QC are required in each step to guarantee successful results. Particularly in the non-targeted approaches, several compromises have to be made in the selection of sample pretreatment conditions, hence it is crucial that the optimized method is rugged and well validated. The conditions should be chosen so that small changes in sample composition do not affect the results. Careful selection of a suitable set of internal standards is crucial, both in targeted and non-targeted analyses, in addition to the use of a sufficient set of QC samples for the control of the analysis.

References

1. S. P. J. van Leeuwen and J. de Boer, *J. Chromatogr. A*, 2008, **1186**, 161.
2. J. R. Dean and G. Xiong, *Trends Anal. Chem.*, 2000, **19**, 553.
3. W. G. Guder, S. Naraynan, H. Wisser and B. Zawta, *Diagnostic Samples: from the Patient to the Laboratory*, 4th edn, Wiley-VCH, Weinheim, 2009.
4. W. Gronwald, M. S. Klein, H. Kaspar, S. R. Fagerer, N. Nürnberger, K. Dettmer, T. Bertsch and P. J. Oefner, *Anal. Chem.*, 2008, **80**, 9288.
5. W. B. Dunn, D. Broadhurst, P. Begley, E. Zelena, S. Francis-McIntyre, N. Anderson, M. Brown, J. D. Knowles, A. Halsall, J. N. Haselden, A. W. Nicholls, I. D. Wilson, D. B. Kell, R. Goodacre and Human Serum Metabolome (HUSERMET) Consortium, *Nat. Protocols*, 2011, **6**, 1060.

6. P. Parvy, J. Bardet, D. Rabier, J. P. Bonnefont and P. Kamoun, *Clin. Chem.*, 1989, **35**, 178.
7. T. Barri and L. O. Dragsted, *Anal. Chim. Acta*, 2013, **768**, 18.
8. B. Jorgenrud, S. Jäntti, I. Mattila, T. Seppänen-Laakso, M. Pflueger, A.-G. Ziegler, H. Yki-Järvinen, M. Orešič and T. Hyötyläinen, *PlosOne*, submitted.
9. N. C. van de Merbel, *Trends Anal. Chem.*, 2008, **27**, 924.
10. P. Gupta, R. Gokhale and B. S. Kirschner, *J. Pediatr. Gastroenterol. Nutr.*, 2001, **33**, 450.
11. P. Bernini, I. Bertini, C. Luchinat, P. Nincheri, S. Staderini and P. Turano, *J. Biomol. NMR*, 2011, **49**, 231.
12. T. Rosenling, M. P. Stoop, A. Smolinska, B. Muilwijk, L. Coulier, S. Shi, A. Dane, C. Christin, F. Suits, P. L. Horvatovich, S. S. Wijmenga, L. M. C. Buydens, R. Vreeken, T. Hankemeier, A. J. van Gool, T. M. Luider and R. Bischoff, *Clin. Chem.*, 2011, **57**, 1703.
13. J. Saric, Y. Wang, J. Li, M. Coen, J. Utzinger, J. R. Marchesi, J. Keiser, K. Veselkov, J. C. Lindon, J. K. Nicholson and E. Holmes, *J. Protein Res.*, 2008, **7**, 352.
14. M. H. Shin, D. Y. Lee, K. Liu, O. Fiehn and K. H. Kim, *Anal. Chem.*, 2010, **82**, 6660.
15. C. J. Bolten, P. Kiefer, F. Letisse, J. C. Portais and C. Wittmann, *Anal. Chem.*, 2007, **79**, 3843.
16. C. J. Bolten, P. Kiefer, F. Letisse, J.-C. Portais and C. Wittmann, *Anal. Chem.*, 2007, **79**, 3843.
17. S. G. Villas-Bôas and P. Bruheim, *Anal. Biochem.*, 2007, **370**, 87.
18. M. A. Lorenz, C. F. Burant and R. T. Kennedy, *Anal. Chem.*, 2011, **83**, 3406.
19. S. Dietmair, N. E. Timmins, P. P. Gray, L. K. Nielsen and J. O. Krömer, *Anal. Biochem.*, 2010, **404**, 155.
20. O. Queiroz and C. Morel, *Plant Physiol.*, 1974, **53**, 596.
21. W. B. Dunn, D. Broadhurst, D. I. Ellis, M. Brown, A. Halsall, S. O'Hagan, I. Spasic, A. Tseng and D. B. Kell, *Int. J. Epidemiol.*, 2005, **37**, i23–i30.
22. H. G. Gika, G. A. Theodoridis and I. D. Wilson, *J. Chromatogr. A*, 2008, **1189**, 314.
23. M. Lauridsen, S. H. Hansen, J. W. Jaroszewski and C. Cornett, *Anal. Chem.*, 2007, **79**, 1181.
24. K. K. Pasikanti, P. C. Ho and E. C. Chan, *Rapid Commun. Mass Spectrom.*, 2008, **22**, 2984.
25. B. B. Devanapalli, M. A. Bermingham and D. D. Mahajan, *Clin. Chim. Acta.*, 2002, **322**, 179.
26. L. Hodson, C. M. Skeaff, A. J. Wallace and G. L. B. Arribas, *Clin. Chim. Acta.*, 2002, **321**, 63.
27. B. Álvarez-Sánchez, F. Priego-Capote and M. D. Luque de Castro, *Trends Anal. Chem.*, 2010, **29**, 120.
28. O. Fiehn, *Plant Mol. Biol.*, 2002, **48**, 155.
29. D. Vuckovic, *Anal. Bioanal. Chem.*, 2012, **403**, 1523.

30. F. M. Geier, E. J. Want, A. M. Leroi and J. G. Bundy, *Anal. Chem.*, 2011, **83**, 3730.
31. M. S. Chang, Q. Ji, J. Zhang and T. A. El-Shourbagy, *Drug Dev. Res.*, 2007, **68**, 107.
32. S. Tiziani, A. Emwas, A. Lodi, C. Ludwig, C. M. Bunce, M. R. Viant and U. L. Günther, *Anal. Biochem.*, 2008, **377**, 16.
33. C. Polson, P. Sarkar, B. Incledon, V. Raguvaran and R. G. Lilly, *J. Chromatogr. B*, 2003, **785**, 263.
34. S. J. Bruce, I. Tavazzi, V. Parisod, S. Rezzi, S. Kochhar and P. A. Guy, *Anal. Chem.*, 2009, **81**, 3285.
35. S. M. Pourmortazavi and S. S. Hajimirsadeghi, *J. Chromatogr. A*, 2007, **1163**, 2.
36. J. Folch, M. Lees and G. H. Sloane Stanley, *J. Biol. Chem.*, 1957, **226**, 497.
37. E. G. Bligh and W. J. Dyer, *Can. J. Biochem. Physiol.*, 1959, **37**, 911.
38. S. J. Iverson, S. L. C. Lang and M. H. C. Cooper, *Lipids*, 2001, **36**, 1283.
39. K. Ichihara, K. Yoneda, A. Takahashi, N. Hoshino and M. Matsuda, *Lipids*, 2011, **46**, 297.
40. H. Nygren, T. Seppänen-Laakso, S. Castillo, T. Hyötyläinen and M. Orešič, *Methods Mol. Biol.*, 2011, **708**, 247.
41. R. M. Broekhuyse, *Clin. Chim. Acta*, 1974, **51**, 341.
42. G. Freyburger, A. Heape, H. Gin, M. Boisseau and C. Cassagne, *Anal. Biochem.*, 1988, **171**, 213.
43. L. A. Carlson, *Clin. Chim. Acta.*, 1985, **149**, 89.
44. A. Hara and N. S. Radin, *Anal. Biochem.*, 1978, **90**, 420.
45. N. S. Radin, *Methods Enzymol.*, 1981, **72**, 5.
46. S. Somersalo, P. Karunen and E. M. Aro, *Physiol. Plant.*, 1986, **68**, 467.
47. W. R. Morrison, S. L. Tan and K. D. Hargin, *J. Sci. Food Agric.*, 1980, **31**, 329.
48. V. Matyash, G. Liebisch, T. V. Kurzchalia, A. Shevchenko and D. Schwudke, *J. Lipid Res.*, 2008, **49** 1137.
49. M. Petkovic, A. Vocks, M. Müller, J. Schiller and J. Arnhold, *Z. Naturforsch.*, 2005, **60**, 143.
50. M. J. Cocero and L. Calvo, *J. Am. Oil Chem. Soc.*, 1996, **73**, 1573.
51. G. Brunner, *J. Food Eng.*, 2005, **67**, 21.
52. S. L. Taylor, J. W. King and G. R. List, *J. Am. Oil Chem. Soc.*, 1993, **70**, 437.
53. A. I. Hopia and V.-M. Ollilainen, *J. Liq. Chromatogr.*, 1993, **16**, 2469.
54. S. Pieber, S. Schober and M. Mittelbach, *Biomass Bioenergy*, 2012, **47**, 474.
55. C. Lin, H. Wu, R. Tjeerdema and M. Viant, *Metabolomics*, 2007, **3**, 55.
56. E. J. Want, G. O'Maille, C. A. Smith, T. R. Brandon, W. Uritboonthai, C. Qin, S. A. Trauger and G. Siuzdak, *Anal. Chem.*, 2006, **78**, 743.
57. S. Tulipani, R. Llorach, M. Urpi-Sarda and C. Andres-Lacueva, *Anal. Chem.*, 2013, **85**, 341.
58. K. Dettmer, N. Nürnberger, H. Kaspar, M. A. Gruber, M. F. Almstetter and P. J. Oefner, *Anal. Bioanal. Chem.*, 2011, **399**, 1127.

59. J. Saric, E. J. Want, U. Duthaler, M. Lewis, J. Keiser, J. P. Shockcor, G. A. Ross, J. K. Nicholson, E. Holmes and M. F. M. Tavares, *Anal. Chem.*, 2012, **84**, 6963.
60. P. Masson, A. Couto Alves, T. M. D. Ebbels, J. K. Nicholson and E. J. Want, *Anal. Chem.*, 2010, **82**, 7779.
61. J. Gullberg, P. Jönsson, A. S. Nordstrom, M. Sjöström and T. Moritz, *Anal. Biochem.*, 2004, **331**, 283.
62. E. A. Souza Silva, S. Risticevic and J. Pawliszyn, *Trends Anal. Chem.*, 2013, **43**, 24.
63. H. Lord, R. P. Grant, M. Walles, B. Incledon, B. Fahie and J. B. Pawliszyn, *Anal. Chem.*, 2003, **75**, 5103.
64. Z. Zhang, J. Cai, G. Ruyan and G. Lim, *J. Chromatogr. B*, 2005, **822**, 244.
65. D. Djozan, T. Baheri, R. Fahrsbaf and S. Azhari, *Anal. Chim. Acta.*, 2005, **554**, 197.
66. F. M. Musteata, I. de Lannoy, B. Gien and J. Pawliszyn, *J. Pharm. Biomed. Anal.*, 2008, **47**, 907.
67. D. Vuckovic and J. Pawliszyn, *Anal. Chem.*, 2011, **83**, 1944.
68. D. Neville, R. Houghton and S. Garrett, *Bioanalysis*, 2012, **4**, 795.
69. R. E. Wuthier, *J. Lipid Res.*, 1966, **7**, 558.
70. J. Bodennec, D. Pelled and A. H. Futerman, *J. Lipid Res.*, 2003, **44**, 218.
71. G.-Q. Chen, Y. Jiang and F. Chen, *Food Chem.*, 2007, **104**, 1580.
72. J. Bodennec, I. A. Koul, G. Brichon, G. Zwingelstein and J. Portoukalian, *J. Lipid Res.*, 2000, **41**, 1524.
73. F. Pernet, C. Pelletier and J. Milley, *J. Chromatogr. A*, 2006, **1137**, 127.
74. T. Perez-Palacios, J. Ruiz and T. Antequera, *Food Chem.*, 2007, **102**, 875.
75. J. A. Prieto, A. Ebri and C. Collar, *J. Am. Oil Chem. Soc.*, 1992, **69**, 387.
76. V. Ruiz-Gutierrez and M. C. Perez-Camino, *J. Chromatogr. A*, 2000, **885**, 321.
77. J. Drozd, *Chemical Derivatization in Gas Chromatography*, Elsevier, Amsterdam, 1981.
78. P. Husek, P. Matucha, A. Vrankova and P. Simek, *J. Chromatogr. B*, 2003, **789**, 311.
79. X. Tao, Y. Liu, Y. Wang, Y. Qiu, J. Lin, A. Zhao, M. Su and W. Jia, *Anal. Bioanal. Chem.*, 2008, **391**, 2881.
80. Y. Ni, M. Su, Y. Qiu, M. Chen, Y. Liu, A. Zhao and W. Jia, *FEBS Lett.*, 2007, **581**, 707.
81. Y. Qiu, M. Su, Y. Liu, M. Chen, J. Gu, J. Zhang and W. Jia, *Anal. Chim. Acta*, 2007, **583**, 277.
82. K. Kouremenos, J. Harynuk, W. L. Winniford, P. Morrison and P. J. Marriott, *J. Chromatogr. B*, 2010, **878**, 1761.
83. F. O. Silva and V. Ferraz, *Food Chem.*, 2004, **88**, 609.

CHAPTER 3

Mass Spectrometric Detection for Chromatography

PÄIVI PÖHÖ AND TUULIA HYÖTYLÄINEN*

VTT Technical Research Centre of Finland, P.O. Box 1000, 02044 VTT, Finland
*Email: tuulia.hyotylainen@vtt.fi

3.1 Introduction

Several types of mass spectrometric (MS) instruments are used as chromatographic detectors. The selection of the MS instrument depends on both the chromatographic system and the goals of the analysis. For example, in global metabolic profiling, it is usually necessary to determine the elemental composition of specific unknown metabolite and therefore MS with high-resolution accurate mass determination (HRAM), typically together with the capability of tandem mass measurements for structural characterization, is required. High resolution also permits the determination not just of the elemental composition but even the isotopic ratios of unknown compounds. However, the linear dynamic range and therefore the quantitative features of many of the HRAM MS systems are generally not as good as with low-resolution quadruple MS instruments. Therefore, for targeted, qualitative analyses, low-resolution mass spectrometers such as quadrupole systems are typically used, with triple quadrupole systems for LC-MS and single quadruple systems for GC-MS.

Table 3.1 compares the most important features of the different MS systems: the scanning speed, dynamic range, mass resolution, scan speed and mass accuracy.

RSC Chromatography Monographs No. 18
Chromatographic Methods in Metabolomics
Edited by Tuulia Hyötyläinen and Susanne Wiedmer

Published by the Royal Society of Chemistry, www.rsc.org

Table 3.1 Comparison of MS instruments in terms of accuracy, resolution and scanning speed.

Analyser	*Mass range (m/z, ×10^{-3})*	*Resolution (Δm/z)*	*Resolving power, FWHM (×10^{-3})*	*Mass accuracy (ppm)*	*Scan rate (Hz)*	*Chromatographic detector for*	*Linear range*	*Description of operation*
Single quadrupole	3	0.1–0.7	–	<100	0.5–20	GC	$10^5 - 10^6$	A low mass resolution analyser. DC and RF potentials are applied to four parallel rods, each opposite pair being electrically connected. These potentials are varied in order to provide a mass filter where ions of a chosen mass have trajectories of amplitude less than half the radius of the quadrupoles and traverse the mass analyser, whereas ions of lower and higher mass are lost by collisions with the rods. The DC and RF potentials are varied (while keeping their ratio constant) to provide a mass scan where ions of increasing mass traverse the mass analyser
QqQ	3	0.1–0.7	–	<100	0.5–4	LC, GC	$10^5 - 10^6$	Consists of three quadrupoles (Q1, Q2 and Q3). Q1 and Q3 are separated by a Q2 quadrupole (collision cell) operating at a higher pressure. Ions accelerated from Q1 to the collision cell undergo collision-induced dissociation (CID) in Q2 followed by mass analysis in Q3. A range of MS/MS experiments are possible, including product ion scanning, neutral loss scanning and single/multiple reaction monitoring (SRM/MRM)

Ion trap [quadrupole ion trap (QIT) and linear ion trap (LIT)]	4–20	0.1–0.7	–	50–200	5–10	LC	$10^4 – 10^5$	The QIT is constructed of a ring and two cap electrodes and operates with the application of DC and RF potentials to constrict ions in stable oscillatory orbits. Helium gas is used to stabilize ions. Detection occurs by destabilization of orbits and ion ejection to a detector. Allows MS^n experiments to be performed to provide structural information. Linear ion traps apply a 2D quadrupole field rather than a 3D field as applied in QIT
Orbitrap	6	–	50–100	0.5–1	0.5–2	LC	5×10^3	A coaxial inner spindle-like electrode is surrounded by two outer barrel-like electrodes with an electric field applied between electrodes. Ions orbit the central electrode in axial and radial directions, through a balance of electrostatic and centrifugal forces and the orbital frequency is detected as an image current by the outer electrodes. A Fourier transform is employed to convert it from the time to frequency domain. The instrument is a hybrid type, a linear ion trap is coupled to the Orbitrap so as to collect ions before periodic introduction into the Orbitrap mass analyser

Table 3.1 (*Continued*)

Analyser	*Mass range* (m/z, $\times 10^{-3}$)	*Resolution* ($\Delta m/z$)	*Resolving power, FWHM* ($\times 10^{-3}$)	*Mass accuracy* (*ppm*)	*Scan rate* (*Hz*)	*Chromatographic detector for*	*Linear range*	*Description of operation*
TOF	20–40	–	15–20	2–5	20	LC, GC	$10^4 - 10^5$	Measures the flight time of an ion from source to detector in a vacuum, the flight time is dependent on *m/z*, lower masses are detected first. The use of ion mirrors *i.e.* reflectrons, focuses ions of the same *m/z* but different kinetic energies which increases mass resolution
QTOF	10–40	–	15–25	3–5	20	LC, GC	$10^4 - 10^5$	A hybrid instrument of quadrupole and TOF mass analysers separated by a higher pressure collision cell used to provide CID of selected ions
IM-qTOF	10–100	–	20–40	0.5–1	30	LC	$10^4 - 10^5$	A hybrid instrument of quadrupole and TOF mass analysers combined with an ion mobility (IM) spectrometer as additional separation dimension, either prior the QTPF or between the Q and TOF analysers
FT-ICR	5–15	–	100–1500	0.1–1	0.3–1	LC	10^4	Ions orbit in a cell operating at ultra-high vacuum (10^{-10} atm) and in a high magnetic field strength (>7 T). The orbital frequency, dependent on *m/z*, is detected as an image current and is converted from time to frequency domains with a Fourier transform

When an MS system is utilized as a chromatographic detector, scanning of the mass spectrum, *i.e.*, the number of scans per second, is an important parameter, particularly when the peaks are very narrow, *e.g.*, in fast GC, in GC×GC and in CE separations where the peak widths can be in the range 20–500 ms only. Time-of-flight (TOF) MS instruments allow fast scanning (500 scans per second) over a mass range of 1000 Da with reasonable mass resolution. On the other hand, Fourier transform ion cyclotron resonance (FT-ICR) MS, which has a very high resolving power, can only sweep through one scan per second and is therefore poorly suited as a detector in GC or CE or even ultra-high-performance liquid chromatography (UPLC). Therefore, Orbitrap MS, which also has a very high resolving power, but clearly shorter accumulation times than FT-ICR-MS, has been gaining popularity as a detector for UPLC. The novel tandem MS system combining ion mobility with TOFMS allow also high scanning speed, high resolving power and even relatively good quantitative features. Further, even though FT-ICR-MS and Orbitrap systems offer superior resolution and mass accuracy, TOFMS has more stable characteristics over a broad range of sampling rates and also better sensitivity at high sampling rates.

The importance of the scanning speed is demonstrated in Figure 3.1.[1] A sufficiently high number of scans per peak is required to perform correct peak picking and mass spectral deconvolution of full-scan mass spectra. If the number of scans per peak is too low, chromatographic separation is partially lost. For example, if the peak width is ~3 s, it is possible to obtain 60 spectra per peak with TOFMS but only about six spectra per peak with quadrupole MS. The minimum requirement for quantitative determination is ~10 spectra per peak (Table 3.2). More spectra allow more efficient deconvolution and, thus, more accurate information. In addition, scanning MS systems, such as TOFMS, do not suffer from spectral skewing, *i.e.*, the effect of concentration changes occurring in the ion source during peak elution, which in turn affect spectral continuity. The non-skewed mass spectra allow optimal performance of the mass spectral deconvolution algorithms.

The choice of the MS detector depends also on the chromatographic system. In LC-MS combinations, the most common MS systems are triple quadrupole (QqQ) mass spectrometer for targeted analysis and TOFMS and hybrid quadrupole mass spectrometer time-of-flight (Q-TOF) and Orbitrap for non-targeted profiling. With CE, fast-scanning instruments such as TOFMS systems are typically applied. Targeted LC-MS analyses with QqQMS allow high duty cycle and a large dynamic range in the MS/MS mode and are therefore well suited for quantitative analyses. The QqQ system utilizes collision-induced dissociation (CID) while monitoring unique precursors to product ion transitions, affording superior sensitivity, precision and accuracy, especially for simultaneous analysis of multiple analytes (selected reaction monitoring, multiple reaction monitoring). It should be noted, however, that novel HRAM MS systems are now achieving sensitivity comparable to that of QqQ instruments while providing full-scan MS data on all analytes, which will certainly

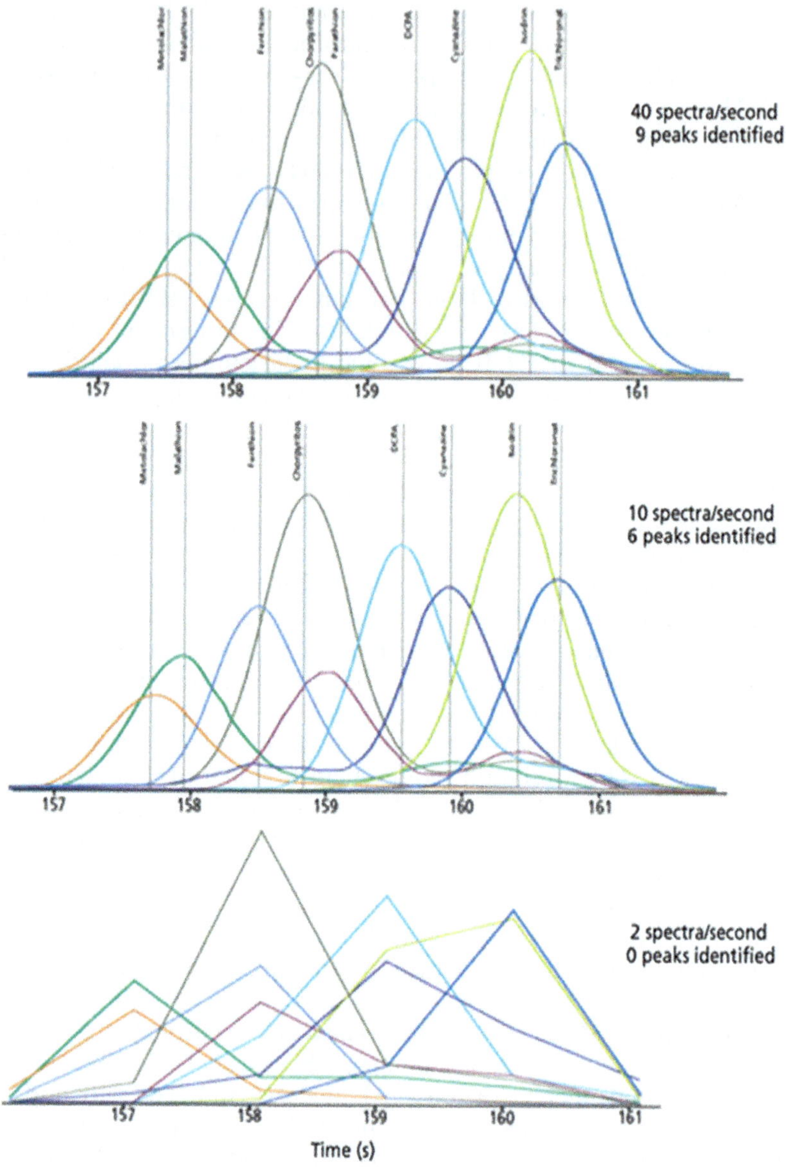

Figure 3.1 Effect of scanning speed on the deconvolution of the peaks. Reprinted with permission from ref. 1.

increase their utility for quantitative assays. An especially attractive feature of the HRAM approach is that, unlike multiple reaction monitoring experiments, data are collected over a preselected wide mass range and the ability to measure unexpected analytes is preserved. With GC, the single quadrupole is still the most common MS detector, but QqQMS; TOFMS and qTOFMS systems are also gaining popularity.

Table 3.2 Dependence of acquisition rate and peak width.

Peak width (s)	*Chromatography*	*Necessary acquisition rate for quantitative determination (Hz)*
1	HPLC, GC	10
0.5	UPLC, fast GC	20
0.2	Fast GC, CE	50
0.1–0.05	GC×GC	100–200

3.2 Mass Spectrometric Instrumentation

3.2.1 MS Interfaces for GC, LC and CE

Different interfaces are needed for combining MS with LC, GC and CE. The choice of interface and the ionization system is dependent on both the chromatographic system and the analytes of interest. The most common interfaces between a chromatographic system and an MS detector and their operating principles are listed in Table 3.3. For GC, the most widely used ionization technique is electron ionization (EI), which is a hard ionization technique. The other option with GC is chemical ionization. With LC- and CE-based techniques, soft ionization techniques such as atmospheric pressure ionization (API) and electrospray ionization (ESI) are usually employed. In addition, other soft ionization techniques such as atmospheric pressure chemical ionization (APCI) and atmospheric pressure photoionization (APPI) have also been used in metabolomics. With LC, ESI is often preferred for non-targeted profiling, since ESI as a soft ionization approach forms intact molecular ions and aids initial identification. Also, APCI and APPI typically induce little fragmentation and are considered robust and relatively tolerant to high buffer concentrations, and are suitable for the detection of non-polar and thermally stable compounds.[2] For LC-MS, a single ionization source containing combinations of ESI and APCI or ESI and APPI is available from many different manufacturers. CE-MS requires specific interfaces, due to the very low sample flow, the interfaces including both sheath-flow interfaces and sheathless interfaces, as described in more detail in Chapter 6.

3.2.1.1 Electron Ionization and Chemical Ionization

Electron ionization (also called electron impact ionization) is typically used with GC. Typically, a high electron voltage (70 eV) is used for the ionization, causing strong fragmentation of the molecules. Naturally, the ionization efficiency and production of fragment ions depend strongly also on the structure of the analytes. The fragmentation in GC-EI-MS is very rugged and not dependent on different instrumental conditions. Thus, large commercial spectral libraries are available such as NIST/EPA/NIH, Fiehnlib[3] and Golm[4], which are based on electron ionization using high internal energy (70 eV). In particular, the NIST '08 library contains more than 222 000 70 eV EI mass

Table 3.3 Summary of the range of ion sources employed in metabolomic investigations, including a brief description of their operation.

Ion source	*Description*	*Chromatographic interface*
Soft ionization		
Atmospheric pressure chemical ionization (APCI)	Operates by passing the LC/CE eluent through a heated glass tube (up to 500 °C), which produces explosive vaporization. A corona discharge needle provides ionization of the solvent molecules (which are in a large excess) followed by ion or charge transfer to metabolite molecules. Degradation of molecules can occur due to the high temperatures used, although minimal fragmentation of molecular ions is typically observed	LC-MS and CE-MS applications for ionization of polar and semipolar metabolites at atmospheric pressure
Atmospheric pressure photoionization (APPI)	Similar to APCI, but instead of a corona discharge needle, electron ejection from molecules is produced by photons emitted from UV discharge lamps	LC-MS and CE-MS applications for ionization of non-polar metabolites at atmospheric pressure and acts as a complementary tool to ESI and APCI
Electrospray ionization (ESI)	Operates by passing the LC/CE eluent through a capillary held at high voltage (2–5 kV). Ion formation occurs in the capillary followed by nebulization and desolvation, which provides transfer of ions from the liquid to the gas phase and their introduction into the MS vacuum. The source acts as an electrochemical cell. Minimal fragmentation of the molecular ion is observed; fragmentation can be produced in-source or by tandem MS	LC-MS and CE-MS for ionization of polar and semipolar metabolites at atmospheric pressure
Chemical ionization (CI)	Chemical reagent gases (methane, ammonia and others) are introduced into an EI source, in large excess over metabolite molecules, and the reagent gas molecules are ionized by electron bombardment. Ion or charge transfer induces ionization of metabolites. Produces minimal fragmentation of the molecular ion	GC–MS applications for ionization of polar and non-polar metabolites at vacuum pressures

Table 3.3 (*Continued*)

Ion source	*Description*	*Chromatographic interface*
Hard ionization		
Electron ionization (EI)	Bombardment of the metabolite molecules by high-energy electrons creates positively charged ions. Produces heavy fragmentation with highly reproducible fragmentation patterns which can be used to deduce metabolite structure. Large commercial spectral libraries are available	GC-MS applications for ionization of polar and non-polar metabolites at vacuum pressures

Table 3.4 Comparison of EI and CI.

EI	*CI*
• Molecular ion is often not observed owing to strong fragmentation • Rich fragment/rearrangement ions • 'Fingerprint' nature of fragment patterns for library searches • Standardized to 70 eV (lower energies may be used) for which extensive spectral libraries are available • Quadrupole MS and TOFMS are similar	• More abundant molecular ions • Less fragmentation • Different reagent gases can be to give different fragmentation, *e.g.*, NH_3/ND_3 • Adduct ions support identification • Nitrogen rule works but is inverted • Negative ionization possible

spectra as compiled by the National Institute of Standards and Technology (NIST).[5]

CI is a lower energy alternative to EI for volatile analytes and the two techniques are compared in Table 3.4. In CI, the ionization is effected with reagent gas (typically ammonia or methane) in the ion chamber. The ionization process is softer and yields less fragmentation and thus a simpler mass spectrum is obtained and usually an easily identifiable molecular ion can be detected. In many modern GC-MS instruments, EI ion sources can usually be switched to CI operation very quickly, *i.e.*, they are constructed as EI/CI combination ion sources. Ionization in CI is the result of one or several competing chemical reactions and therefore the sensitivity in CI depends strongly on the experimental conditions, such as the primary electron energy, the electron current, the reagent gas, the reagent gas pressure and the ion source temperature. CI for gas-phase analysis is either positive or negative. Almost all neutral analytes can form positive ions whereas for negative CI (NCI), the analyte must be capable of producing a negative ion (stabilize a negative charge), for example by electron-capture ionization. Not all compounds can do this, hence selective NCI can be used for the analysis of compounds containing acidic groups or electronegative elements (especially halogens). In metabolomics, positive CI is the most common method, if CI is utilized.

3.2.1.2 *Electrospray Ionization*

ESI is a soft ionization technique that transfers ions from the solution to the gas phase. It is particularly useful for the analysis of large, non-volatile and chargeable molecules and it is widely utilized in metabolic applications. ESI is a leading method of choice for coupling LC or CE to MS. The limitation of ESI is that analyte should be intrinsically ionic or easily transformed to ionic in solution, via protonation, deprotonation or cation or anion attachment. Advantages of ESI are that it causes little fragmentation, it has a wide mass range and it can form multiple charged ions and, thus, shift high-mass analytes into an m/z range accessible by most mass analysers. In addition, miniaturized electrospray, namely nanoelectrospray, ionization (nanoESI) is also widely used technique especially in shot-gun MS. NanoESI has a narrower spray capillary, offering smaller droplets and nanolitre flow rates. The advantages of nanoESI are the extremely low sample consumption, toleration of higher loads of buffer salts and the possibility of using high-polarity solvents. A chip-based nanoESI sprayer on an etched silicon wafer allows automated loading by a pipetting robot and is preferred in lipidomics.[6]

3.2.1.3 *Atmospheric Pressure Chemical Ionization and Atmospheric Pressure Photoionization*

In APCI, ions are transferred into the mass analyser by use of the same atmosphere-to-vacuum interface as in ESI and, therefore, ESI sources can easily be switched to the APCI mode. In APCI, the ionization takes place by analyte collisions with charged reagent ion plasma, formed by a corona discharge needle, and therefore, neutral compounds also form ions. The nature of the APCI plasma varies widely as both the solvent and nebulizing gas contribute to the composition of the plasma, hence APCI spectra depend on the actual conditions and the ion polarity. Compared with other ionization methods, APCI suffers from comparatively low reproducibility. The greatest advantage of APCI over ESI is the ability to generate ions from neutral compounds and APCI is applied to low- to medium-polarity analytes. An alternative technique to APCI is APPI, in which a UV light source replaces the corona discharge needle and a UV light-transmitting quartz tube is used to guide the hot vapour towards the sampling orifice. Otherwise, the constructions of these two techniques are almost the same. APCI and APPI are suitable for same kinds of compounds, *i.e.*, from small to medium molecular mass compounds with low polarity.

3.2.2 Mass Analysers

Selection of the mass analyser depends greatly on the analytical demands, such as the resolution needed, mass range, concentration of analytes and budget for acquisition and operation. Different analysers are characterized by their mass range, resolution, resolving power, mass accuracy and scan rate.

3.2.2.1 Quadrupole Analysers

Quadrupole analysers have been widely used as detectors for chromatographic systems because they are cheap and light-weight and they use low ion acceleration voltages. However, quadrupole analysers have fairly low mass range and resolution. Different quadrupole analysers are available, from simple linear quadrupole mass analysers to triple quadrupoles that have hexapole or octopole collision cells. Specific quadrupoles can also be used in other ways as ion guides, collision cells and for collisional cooling. Often quadrupole analysers are combined and used for tandem mass spectrometric (MS/MS) analysis. In triple quadrupole mass spectrometers, both the first and third quadrupoles are used mass selectively, and the second quadrupole works as a collision cell. The collision cell contains low-pressure gas and collisions with analyte molecules lead to fragment ions, which are retained and directed to the third quadrupole to separate the fragments. Triple quadrupole MS can be used in four main modes: product ion scan, precursor ion scan, neutral loss scan and multiple reaction monitoring.

3.2.2.2 Quadrupole Ion Trap Analysers

The three-dimensional quadrupole ion trap (QIT) works on the same physical principles as a quadrupole mass analyser, but the ions are trapped in a potential well and are sequentially ejected. The QIT can be also used in multiple tandem mass spectrometry (MS^n). The linear quadrupole ion trap (LIT or LTQ) works on the same principles as the QIT with the difference that LIT is two-dimensional and is structurally similar to a quadrupole analyser. The LIT can be used as an individual analyser or to concentrate ions before their entry to another analyser, such as a TOF or FT-ICR system. As an advantage, the LIT can load more ions without overloading, is more sensitive, stores ions more effectively and has a faster scan rate than a traditional QIT.

3.2.2.3 Time-of-flight Analysers

The basic principle of a TOF analyser is that ions of different m/z are dispersed in time during their flight along a field-free drift path of known length. The pulsed ions are accelerated in an electric field, so that all ions receive the same kinetic energy. The velocity of ions with the same kinetic energy depends on their m/z, so that lighter ions travel faster than heavier ions. The ion mass can be calculated from the time of flight of the ion.

The resolving power of TOF analysers has been improved with the use of delayed pulsed ion extraction, which narrows the energy distribution. Also, the use of a reflector has improved the resolution by narrowing the flight time deviation. The recent development of orthogonal acceleration TOF (oaTOF) analysers has made it possible to produce pulsed ions orthogonally from a continuous ion beam. oaTOF permits combination with other non-pulsed ion sources such as ESI and offers accurate mass capabilities. The advantages of TOF instruments are that they have in principle an unlimited m/z range, fast scan rate, comparatively simple and inexpensive construction and high sensitivity and resolution.

3.2.2.4 Fourier Transform Ion Cyclotron Resonance and Orbitrap

Modern FT-ICR mass spectrometers offer ultrahigh resolving power, the highest mass accuracy, low detection limits, high mass range and the possibility of MS^n. Most FT-ICR instruments are hybrid systems with a linear quadrupole or LIT front end.

An Orbitrap analyser also delivers high resolving power and accurate mass measurements at levels competing with FT-ICR to a certain extent. FT-ICR and Orbitrap have common features, such as image current detection of ion oscillations and Fourier transformation for the conversion of the transient into the frequency domain. The main difference is that the Orbitrap is operated without a magnetic field and is available at a much lower price and installation room requirements.[7] The Orbitrap needs proper ion injection, hence a bent RF-only quadrupole (c-trap) is employed to accumulate, store and thermalize the ions by a low pressure of nitrogen prior to injection. The Orbitrap is combined with LIT in the Thermo Fisher LTQ-Orbitrap.

3.2.2.5 Ion Mobility Mass Spectrometer

Ion mobility mass spectrometry (IMS) is an analytical technique that can be used to separate gaseous ions. Separation is based on different velocities of ions when they are accelerated by a constant electric field along a drift tube with a countercurrent inert gas, typically nitrogen or helium. The average velocity of a drifting ion is related to the number of soft collisions it experiences with the inert gas within the drift tube. The drift time is characteristic of different ions and based on an ion's mass, charge, size and shape. IMS can separate isobaric ions of the same charge state resulting from their distinct velocity when they travel through the electric field of the ion mobility tube. Isobars of the same charge state can be also distinguished by their steric properties, which makes IMS a useful tool for the separation of both isomers and conformers. A hybrid instrument containing a mass-selecting quadrupole, a travelling wave ion mobility separator and an oaTOF analyser has been marketed by Waters as Synapt HDMS and Synapt G2 instruments.

3.2.2.6 Tandem Mass Spectrometry

Hybrid mass spectrometers combine different types of mass analysers and ion-guiding devices in a single hybrid instrument. In tandem mass spectrometers, the first mass analyser is used to select ions for further fragmentation and subsequent analysis of the product ions in a second mass analyser. Tandem MS can be divided into two basic instrumental concepts: tandem-in-space MS and tandem-in-time MS. Tandem-in-space refers to MS/MS instrumentation where product ion spectra are recorded using separated m/z analysers, for example QqQ and Q-TOF. Tandem-in-time MS employs a single m/z analyser such as a

QIT or FT-ICR instrument. Higher order MS^n usually requires a tandem-in-time setup, whereas tandem-in-space instrumentation is usually designed for MS/MS.

The most prominent fragmentation technique in MS/MS is CID. CID allows the fragmentation of gaseous ions by passing an ion beam through a collision cell where the collision gas (He, N_2, Ar) is set to a pressure considerably above of the pressure of surrounding high vacuum. The analytes fragment after collisions with the neutral gas molecules. A CID cell can be, for example, an RF-only quadrupole, hexapole, octopole or a travelling wave ion guide. Tandem MS with a triple quadrupole instrument usually utilizes CID in product ion, precursor ion and neutral loss scans. In a product ion scan, an ion of interest is selected in the first mass analyser (Q1). The selected ion is fragmented in the collision cell (Q2) and the fragments formed are analysed with the second mass analyser (Q3). The product ion scan is valuable for the characterization of unknown metabolites and to confirm the identity of known metabolites. In the precursor ion scan (PIS) mode, the first mass analyser is scanning and the second analyser is set to transmit a constant ion. Ions from the first mass analyser are recorded only if they produce in the collision cell a fragment with a specific m/z that the second mass analyser has been set to monitor. The PIS can be used, for example, to detect molecular species in a lipid class, since those species produce the same characteristic head group-derived fragment ion. A neutral loss scan (NLS) is also selective for a particular functional group and it is used when a common fragment for a group is uncharged. In the NLS, both Q1 and Q3 are scanned in parallel with a constant specific mass difference.

In other tandem mass spectrometer configurations, such as in Q-TOF and LTQ-Orbitrap, special modes utilizing full-scan and fragment information can be used. The mode called MS^e is used with Q-TOF to collect MS and MS/MS data simultaneously by changing between high and low collision energy within one run. Data-dependent acquisition (DDA) is also a mode of data collection where the user defines certain rules to pick precursor peaks from a survey scan, which are later subjected to MS/MS analysis.[8]

The CID in the LIT of the LTQ-Orbitrap instrument is sometimes not hard enough to achieve fragmentation of comparatively stable precursor ions. In the LTQ-Orbitrap, a C-trap can be used as a collision cell by raising the RF voltage. The nitrogen present at $\sim 1.3 \times 10^{-3}$ mbar in the C-trap is sufficient to serve as the collision gas. After several oscillations, precursor ions have undergone dissociating fragmentation and all ions have accumulated in the middle of the C-trap. This kind of collision cell is termed higher-energy C-trap dissociation (HCD). In addition to CID and HCD, other collision cells have also been used, such as electron capture dissociation (ECD) and infrared multiphoton dissociation (IRMPD) in FT-ICR MS.[9]

3.3 Identification

Identification of unknown compounds is a very challenging task. The Chemical Analysis Working Group of the Metabolomics Standards Initiative (MSI;

Table 3.5 The four levels of metabolite identification confidence defined by the Metabolomics Standards Initiative.

Level	*Confidence of identity*	*Level of evidence*
1	Confidently identified compounds	Comparison of two or more orthogonal properties with an authentic chemical standard analysed under identical analytical conditions
2	Putatively annotated compounds	Identification is based on physicochemical properties and/or spectral similarity with public/commercial spectral libraries, without reference to authentic chemical standards
3	Putatively annotated compound classes	Identification is based on characteristic physicochemical properties of a chemical class of compounds or by spectral similarity to known compounds of a chemical class
4	Unknown compounds	Although unidentified and unclassified, these metabolites can still be differentiated and quantified based on spectral data

Source: Ref. 10.

http://msi-workgroups.sourceforge.net) has defined four different levels of metabolite identification confidence, as presented in Table 3.5, and also methods on how to report metabolite identities.[10]

Definitive identification (level 1) requires the use of an authentic chemical standard and the identification should be verified by comparison of at least two properties, such as retention time or retention index, m/z or fragmentation mass spectrum, of the metabolite of interest with the same properties observed for the standard compound analysed using the identical analytical method. It should be noted that even with this level of verification, some metabolites may be difficult to distinguish. For example, many isomers have very similar chromatographic and MS characteristics. Hence for the identification of isomers, methods that allow their separation are needed. Putative identification (level 2 or 3) is usually based on one or two properties only and no authentic standards are used for the verification. The identification is usually based on comparison of the data collected in different laboratories and acquired with different analytical methods. For example, with GC-EI-MS methods, the putative identification can be based on the spectral match with the library spectra, often also combined with retention indices, and in LC-MS and CE-MS by accurately measured m/z values, sometimes combined with comparison of fragmentation spectra or retention times against experimentally or computationally derived databases.

The isotopic abundance pattern can serve as a powerful additional constraint for removing wrong elemental composition candidates. The actual ratios of the stable isotopes differ slightly for each element within a narrow range.[11] Therefore, each monoisotopic molecular ion is always accompanied by additional isotope ions and the abundance of the isotope ions is dependent on the actual elemental composition. Hence this can be used as a powerful filter in calculating unique elemental compositions from mass spectral data. It has been

Table 3.6 Number of possible molecular formulas at different levels of mass accuracy and the impact of isotopic abundance accuracy.

Molecular mass (Da)	*Without isotope abundance information*				*2% isotopic abundance accuracy*	*5% isotopic abundance accuracy*
	5 ppm	*3 ppm*	*1 ppm*	*0.1 ppm*	*3 ppm*	*5 ppm*
150	1	1	1	1	1	1
300	11	7	2	1	1	6
500	115	64	21	2	3	33
700	538	321	108	10	10	97
900	1712	1045	345	32	18	196

Source: modified from Ref. 12.

shown that high mass accuracy (1 ppm) and high resolving power alone are not sufficient for obtaining a small number of molecular formula candidates for the elucidation of molecular structure in MS determination.[12] However, using an orthogonal isotopic abundance pattern filter, it is possible to reduce substantially the number of molecular formula candidates, as demonstrated in Table 3.6.

3.3.1 GC-MS

In GC-MS-based methods, spectral libraries are a valuable tool in identification. However, just having a reasonable spectral match with a library spectrum does not guarantee that the identification is correct. Even when working with GC×GC-TOFMS, in which the separation efficiency is very high and therefore the quality of the spectra is generally good, an identification based on spectral match alone is far from sufficient. For example, in the analysis of serum extracts, typically over 1500 metabolites can be detected with GC×GC-TOFMS and around 50–60% are tentatively identified using the automated identification procedure with a spectral match better than 800.[13] On closer inspection, when the spectral match is below 900, almost 80% of these identifications are not correct, which is directly observed when retention indices of the compounds are calculated and compared with literature values. Retention indices are available for a large number of compounds and utilization of these data together with the spectra makes the identification much more reliable. Obviously, further confirmation by using authentic standards is still required. It should also be noted that a large number of compounds still cannot be identified as they are not found in the spectral libraries. The fragmentation patterns are helpful in the identification, but often the compound spectra lack intense fragments in the high-m/z region and therefore structural elucidation of unknown peaks is very challenging. The molecular mass of a compound can be determined by GC equipped with CI and accurate mass detection (*e.g.*, TOFMS); however, the mass spectral libraries for GC-CI-MS are not as extensive as for GC-EI-MS. Combination of the information obtained with the two ionization methods is a highly efficient way to identify unknown compounds.

3.3.2 LC-MS and CE-MS

In LC-MS and CE-MS analysis, identification requires the use of accurate mass measurements for the determination of elemental composition, combined with MS^n experiments to obtain information on the molecular fragments. In addition, in LC-MS and CE-MS, one metabolite may produce multiple peaks due to the presence of isotopes, adducts and neutral loss fragments. Therefore, ion annotation is crucial in order to recognize a group of ions likely to originate from the same compound. Some spectral libraries are available for LC-MS,[14,15] including the Human Metabolome Database (HMDB),[16] the METLIN Metabolite Database[17,18] and the MassBank.[19] Unfortunately, construction of universal spectral databases for API-MS is challenging owing to the poor reproducibility and high inter-instrument variability of fragmentation patterns. In the interpretation of the mass spectra from LC-MS and CE-MS studies, the first step is to ensure that the signal of interest really corresponds to a monoisotopic ion and not to a natural isotopologue ion or an adduct ion. Elemental compositions can be deduced from accurate mass measurements and used for further database queries. To aid the identification, a number of chemical databases containing huge numbers of chemical structures to aid in structural elucidation after fragmentation of the metabolite are available (*e.g.*, ChemSpider).[20] Also, several *in silico* software packages for the prediction of *in silico* spectra for compounds are available.[21]

Metabolite identifications are mainly made through mass-based searches, *i.e.*, searching the m/z value of a molecular ion of interest against the database(s), at a given tolerance range. The metabolites having molecular masses within a specified tolerance range are considered as putative identifications. For the verification of the identification, authentic standard compounds should be analysed together with the actual compounds using MS/MS.

The main problem with this approach is that even putative identification cannot be made with sufficient accuracy, owing to the existence of isomers and the limited accuracy of mass spectrometers.[22] In the worst case, one molecular ion can have more than 100 putative identifications, making the manual verification very tedious and time-consuming and thus poorly suited for the identification of large number of compounds.[2] Recently, several *in silico* methods have been developed for identification, particularly for the interpretation of tandem mass spectra. A possible structure or sub-structure of an unknown molecular ion can be obtained by comparing the spectrum with hypothetical spectra predicted through *in silico* fragmentation approaches. Prediction can be made by, *e.g.*, using a rule-based predictor, which relies on fragmentation pathways collected from the literature. Commercial tools are available for rule-based prediction, *e.g.*, ACD Fragmenter (ACD/Labs) and Mass Frontier (High-Chem). In addition, other *in silico* fragmentation tools are available, including Fragment Identificator (FiD)[23] and MetFrag.[24] They rely on a different principle, namely generation of a list of possible fragments through combinatorial disconnection of chemical bonds in the compound. The combinatorial approach does not depend on any knowledge base, so it eliminates the laborious rule collection step and avoids the potential bias derived from a limited set of fragmentation rules. However, both methods have their limitations and the identification suggested by the *in silico* fragmentation requires careful check-up.

Further experiments with a parallel technique are often needed, both to confirm the preliminary identification, if standards are not available, and for obtaining more information in the case of unknown compounds. Collection of fractions with automated systems and analyzing these fractions using, *e.g.*, high-sensitivity capillary NMR spectroscopy can be utilized in metabolite structure characterization down to the low microgram level.

3.4 Fluxomics

MS in combination with chromatography is a valuable tool for metabolic flux analysis (MFA), which in turn is a powerful methodology for the determination of metabolic pathway fluxes. In particular, the measurement of metabolic rates is of vital importance for metabolic engineers and biomedical studies. Since the estimated flux distribution reflects the several regulatory processes, MFA allows the determination of the control mechanisms in the cellular system. For example, MFA can be utilized in comparing the flux distributions of a gene knockout mutant with the wild type, showing the effect of the gene knockout.[25]

In practice, the first step in MFA is to feed a cell culture, tissue or organ of a plant a labelled organic substrate, *i.e.*, a compound that has a carbon chain labelled with a stable isotope (*e.g.*, ^{13}C), for example glucose, glutamine or atypical substrates such as propionate or succinate.[26,27] As the labelled substrate enters the metabolic network, the carbon chain is cleaved and the carbon atoms are rearranged to form new metabolites and the fate of the precursor atoms can be followed as demonstrated in Figure 3.2.[28] The metabolites can then be analysed by MS or NMR spectroscopy and the isotope labels can be detected. The isotope-enriched compounds formed in the MFA experiments are called isotopomers, which are isobaric substances with the same number of labelling isotopes (*e.g.*, ^{13}C) in different positions, and isotopologues – substances with different numbers of isotopes.

The analysis is then followed by the construction of a mathematical representation of the metabolic network for the simulation of the isotope label by assuming a certain distribution of fluxes over the network.[29] The goal in the study of the metabolic fluxes with ^{13}C MFA is to obtain as much information as possible about the amount of ^{13}C label incorporation and the distribution of ^{13}C atoms within molecules at key nodes in metabolism. However, in practice, it is not necessary to quantify isotopic labelling for all metabolites in a given metabolic network but rather to measure the ^{13}C patterns for a selected key metabolites, from which comprehensive metabolic flux maps can then be reconstructed.

Common methods for MFA analysis are GC-MS, LC-MS and NMR spectroscopy. MS-based methods are preferable to NMR spectroscopy since the sensitivity is considerably higher, allowing the analysis of smaller samples and

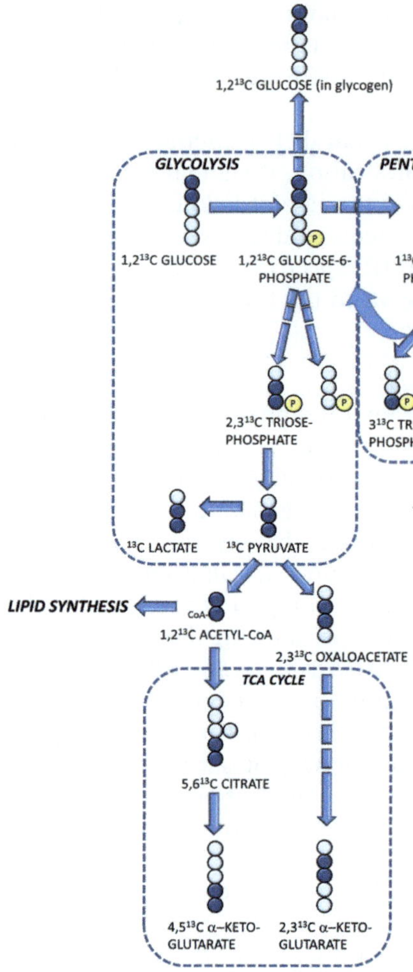

Figure 3.2 Expected mass isotopomers in glycolysis, pentose-phosphate cycle and TCA cycle from incubations with [1,2-$^{13}C_2$]glucose. When [1,2-$^{13}C_2$]glucose enters into the metabolic pathway, it is converted into glucose-6-phosphate (G6P), which can undergo glycogen synthesis, and enter either the glycolytic pathway or the pentose phosphate cycle. From glycolysis, two triose-phosphate molecules are formed, one of them with two ^{13}C and the other one without the label. Both can then form pyruvate, and therefore 50% [2,3-$^{13}C_2$]lactate. Pyruvate can also enter lipid synthesis, forming molecules with a paired number of ^{13}C atoms, or the TCA cycle, obtaining two different labeling distributions in α-ketoglutarate (which is in equilibrium with medium glutamate). When G6P enters the PPC, one ^{13}C is lost in CO_2 formation, giving ribose-5-phosphate with only one labeled carbon. This molecule can also enter the non-oxidative pentose phosphate pathway forming triose-phosphate molecules with only one ^{13}C, and all subsequent products labeled in one atom. Furthermore, when products from glycolysis of [1,2-$^{13}C_2$]glucose undergo gluconeogenesis, two different labeling patterns are expected in glucose isotopomers: [1,2-$^{13}C_2$]glucose, which is the initial isotope, and [5,6-$^{13}C_2$]glucose, formed as a result of the isotopic equilibrium between the labeled and unlabeled triose-phosphates (modified from Ref. 28).

less abundant metabolites In practice, GC-MS, LC-MS and CE-MS can be used for MFA, provided that the MS platform provides a fragmentation process allowing positional assessment of the ^{13}C labelling. Therefore, GC-EI-qMS or LC or GC with MS instrumentation equipped with CID/fragmentation capability is required. Simple GC-QMS is sufficient if the positions of the labelled carbons are known or if reference spectra of the specific metabolites with specific labels are available. If the positions of the labels are not known, GC-MS/MS is very useful in the measurement of isotopic labelling when two fragmentation steps can be utilized in stepwise cleavage of the molecules into continually smaller fragments, therefore obtaining detailed positional labelling.[30,31] In addition, with utilization of specific derivatization methods, the fragmentation can be further controlled. With respect to the building blocks of, *e.g.*, biopolymers, GC-MS has the advantage that most major biomass components (starch as glucose, proteins as amino acids and lipids as fatty acid methyl esters and glycerol) can be measured on the same system. In a similar manner, LC-MS/MS has been applied for isotopic measurements, *e.g.*, isotopic measurements of sugar phosphates,[32,33] amino acids[34] and central carbon metabolites.[35] The advantage of LC-MS over GC-MS is that no derivatization is needed; however, control of fragmentation is more difficult in LC-MS than in GC-MS.

3.5 Conclusion

MS techniques have developed rapidly during the last decade and the novel MS instruments allow highly sensitive and accurate measurements. The choice of the optimal system depends on the aim of the analysis, as typically targeted, quantitative analyses are better performed with, *e.g.*, triple quadrupole instruments owing to their better quantitative features. Non-targeted analyses require HRMS instruments, such as TOFMS, QTOFMS or Orbitrap systems. The recent developments in HRMS instrumentation now allow quantitative analyses with a relatively large dynamic range. Currently, the bottleneck is the identification of unknown compounds, which is one of the most time-consuming steps in the analysis. The Human Metabolome Project is one of the largest initiatives for the identification of metabolites, combining data from both MS and NMR measurements. The most recent version of the database (version 2.5) contains almost 8000 metabolite entries, including both water-soluble and lipid soluble metabolites and metabolites found at both high ($>1\,\mu M$) and low ($<1\,nM$) concentrations. However, the coverage of the metabolome is still not sufficient. Also, better *in silico* fragmentation models for the identification are urgently needed.

References

1. J. Binkley, Current Trends in Mass Spectrometry, *LCGC Eur.*, 2010, July available online: http://www.chromatographyonline.com/lcgc/article/articleDetail.jsp?id=677420&pageID=1&sk=&date=.

2. J. F. Xiao, B. Zhou and H. W. Ressom, *Trends Anal. Chem.*, 2012, **32**, 1.
3. http://fiehnlab.ucdavis.edu/Metabolite-Library-2007/.
4. GMD, *Golm Metabolome Database*, http://gmd.mpimp-golm.mpg.de (last accessed 8 June 2013).
5. NIST, *NIST 08 Mass Spectral Database*, National Institutes of Standards and Technology, www.nist.gov (last accessed 8 June 2013).
6. B. Fuchs and J. Schiller, *Lipids Health Dis.*, 2008, **49**, 541.
7. Q. Hu, R. J. Noll, H. Li, A. Makarov and M. G. Hardman Cooks, *J. Mass Spectrom.*, 2005, **40**, 430.
8. S. S Bird, V. R. Marur, M. J. Sniatynski, H. K. Greenberg and B. S. Kristal, *Anal. Chem.*, 2011, **83**, 6648.
9. K. Schuhmann, R. Herzog, D. Schwudke, W. Metelmann-Strupat, S. R Bornstein and A. Shevchenko, *Anal. Chem.*, 2011, **83**, 5480.
10. L. W. Sumner, A. Amberg, D. Barrett, M. H. Beale, R. Beger and C. A. Daykin, *Metabolomics*, 2007, **3**, 211.
11. J. R. De Laeter, J. K. Böhlke, P. De Bièvre, H. Hidaka, H. S. Peiser, K. J. R. Rosman and P. D. P. Taylor, *Pure Appl. Chem.*, 2003, **75**, 683.
12. T. Kind and O. Fiehn, *BMC Bioinformatics*, 2006, **7**, 234.
13. S. Castillo, I. Mattila, J. Miettinen, M. Orešič and T. Hyötyläinen, *Anal. Chem.*, 2011, **83**, 3058.
14. D. S. Wishart, C. Knox, A. C. Guo, R. Eisner, N. Young, B. Gautam, D. D. Hau, N. Psychogios, E. Dong, S. Bouatra, R. Mandal, I. Sinelnikov, J. Xia, L. Jia, J. A. Cruz, E. Lim, C. A. Sobsey, S. Shrivastava, P. Huang, P. Liu, L. Fang, J. Peng, R. Fradette, D. Cheng, D. Tzur, M. Clements, A. Lewis, A. De Souza, A. Zuniga, M. Dawe, Y. Xiong, D. Clive, R. Greiner, A. Nazyrova, R. Shaykhutdinov, L. Li, H. J. Vogel and I. Forsythe, *Nucleic Acids Res.*, 2009, **37**, D603.
15. H. Horai, M. Arita, S. Kanaya, Y. Nihei, T. Ikeda, K. Suwa, Y Ojima, K. Tanaka, S. Tanaka, K. Aoshima, Y. Oda, Y. Kakazu, M. Kusano, T. Tohge, F. Matsuda, Y. Sawada, M. Y. Hirai, H. Nakanishi, K. Ikeda, N. Akimoto, T. Maoka, H. Takahashi, T Ara, N. Sakurai, H. Suzuki, D. Shibata, S. Neumann, T. Iida, K. Funatsu, F. Matsuura, T. Soga, R. Taguchi, K. Saito and T. Nishioka, *J. Mass Spectrom.*, 2010, **45**, 703.
16. HMDB, *Human Metabolome Database*. http://www.hmdb.ca/ (last accessed 8 June 2013).
17. Scripps Center for Metabolomics, *METLIN: Metabolite and Tandem MS Database*, http://metlin.scripps.edu (last accessed 8 June 2013).
18. C. A. Smith, G. O. Maille, E. J. Want, C. Qin, S. A. Trauger, T. R. Brandon, D. E. Custodio, R. Abagyan and G. Siuzdak, *Ther. Drug Monit.*, 2005, **27**, 747.
19. MassBank Project, *MassBankHigh Quality Mass Spectral Database*, http://www.massbank.jp (last accessed 8 June 2013).
20. ChemSpider, *ChemSpider: the Free Chemical Database*, http://www.chemspider.com (last accessed 8 June 2013).

21. MetFrag, *In Silico Fragmentation for Computer Assisted Identification of Metabolite Mass Spectra*, http://msbi.ipb-halle.de/MetFrag (last accessed 8 June 2013).
22. M. Brown, D. C. Wedge, R. Goodacre, D. B. Kell, P. N. Baker, L. C. Kenny, M. A. Mamas, L. Neyses and W. B. Dunn, *Bioinformatics*, 2011, **27**, 1108.
23. M. Heinonen, A. Rantanen, T. Mielikäinen, J. Kokkonen, J. Kiuru, R. A. Ketola and J. Rousu, *Rapid Commun. Mass Spectrom.*, 2008, **22**, 3043.
24. S. Wolf, S. Schmidt, M. Muller-Hannemann and S. Neumann, BMC Bioinformatics, 2010, **11**, 148.
25. K. Shimizu, *Adv. Biochem. Eng. Biotechnol.*, 2004, **91**, 1.
26. N. Zamboni, S. M. Fendt, M. Rühl and U. Sauer, *Nat. Protocols*, 2009, **4**, 878.
27. C. M. Metallo, J. L. Walther and G. Stephanopoulos, *J Biotechnol.*, 2009, **144**, 167.
28. S. Marin, K. Chianga, S. Bassilian, W.-N. P. Lee, L. G. Boros, J. M. Fernández-Novella, J. J. Centellesa, A. Medranoc, J. E. Rodriguez-Gilc and M. Cascante, *FEBS Lett.*, 2003, **554**, 342.
29. Y. Toya, N. Kono, K. Arakawa and M. Tomita, *J. Proteome Res.*, 2011, **10**, 3313.
30. J. Choi, M. T. Grossbach and M. R. Antoniewicz, *Anal. Chem.*, 2012, **84**, 4628.
31. W. S. Ahn and M. R. Antoniewicz, *Metab. Eng.*, 2011, **13**, 598.
32. P. Kiefer, C. Nicolas, F. Letisse and J. C. Portais, *Anal. Biochem.*, 2007, **360**, 182.
33. J. C. van Dam, M. R. Eman, J. Frank, H. C. Lange, G. W. Van Dedem and J. J. Heijnen, *Anal. Chim. Acta*, 2002, **460**, 209.
34. K. Shimbo, A. Yahashi, K. Hirayama, M. Nakazawa and H. Miyano, *Anal. Chem.*, 2009, **81**, 5172.
35. M. Ruhl, B. Rupp, K. Noh, W. Wiechert, U. Sauer and N. Zamboni, *Biotechnol. Bioeng.*, 2011, **109**, 763.

CHAPTER 4

Liquid Chromatographic Techniques in Metabolomics

GEORGIOS A. THEODORIDIS,[a] FILIPPOS MICHOPOULOS,[a,b] HELEN G. GIKA,[c] ROBERT S. PLUMB[d] AND IAN D. WILSON*[d]

[a] Department of Chemistry, Aristotle University Thessaloniki, 541 24 Thessaloniki, Greece; [b] Oncology IM, AstraZeneca, Mereside, Alderley Park, Macclesfield, Cheshire SK10 4TG, UK; [c] Department of Chemical Engineering, Aristotle University Thessaloniki, 541 24 Thessaloniki, Greece; [d] Department of Surgery and Oncology, Sir Alexander Fleming Building, Imperial College, Exhibition Road, South Kensington, London SW7 2AZ, UK
*Email: i.wilson@imperial.ac.uk

4.1 Introduction

The large-scale use of liquid chromatography (LC) for untargeted metabolic profiling of the type routinely employed in metabolomic/metabonomic[1,2] applications is relatively new compared with the use of, *e.g.*, gas chromatography (GC) or nuclear magnetic resonance (NMR) (especially ^{1}H NMR) spectroscopy. Published applications using LC-MS probably only began to appear early in the twenty-first century.[3,4] Since these early applications, LC-MS-based analysis of biological samples has become increasingly popular as a result of perceived advantages such as compatibility with biological samples, *e.g.*, urine (which can often be analysed directly without the need for extraction and derivatization), high sensitivity, good spectral information content allowing analyte characterization (including identification or confirmation of identity),

RSC Chromatography Monographs No. 18
Chromatographic Methods in Metabolomics
Edited by Tuulia Hyötyläinen and Susanne Wiedmer

Published by the Royal Society of Chemistry, www.rsc.org

ease of use and widespread availability in many laboratories (compared with, *e.g.*, high-field NMR spectroscopy) and at (relatively) modest cost of instrumentation and per sample analysis. For all of these reasons, the last decade has seen an enormous expansion in the development and use of LC-MS-based techniques (see reviews 5–7). The LC systems employed have ranged from conventional high-performance liquid chromatographic (HPLC) methods using 3–5 μm particle sizes for the chromatographic separation, through more advanced 'ultra- high-pressure' liquid chromatographic (UHPLC) [or ultra-performance liquid chromatographic (UPLC)] and, more recently, supercritical fluid chromatographic (SFC) techniques, with stationary phases based on sub-2 μm particles, to capillary and miniaturized LC systems. Now, of course, it is possible to apply MS-based methods directly to the sample without first using an LC separation, with concomitant benefits to throughput and reduced analytical complexity, but experience suggests that, except for the very simplest matrices, there are significant disadvantages to this type of approach. These disadvantages centre especially on problems based on matrix interferences, in particular ion suppression or ion enhancement, which can distort the resulting metabolite profile, combined with the difficulty of distinguishing between isobaric species and structural isomers. By linking the MS instrument to an LC separation, the effects of ion suppression/enhancement can at least be moderated and at best be eliminated and isomeric/isobaric substances can be separated and determined individually (greatly aiding identification). In our opinion, these advantages more than outweigh the added complexity introduced by hyphenation in respect of, *e.g.*, the extra levels of expertise required and the increased analysis time.

As will become clear in the discussion of the various separation types and strategies that follow, there is an ever present tension between the desire to analyse the maximum number of samples in the shortest possible time and the need to obtain the most comprehensive metabolite profiles, which normally requires longer chromatographic runs. Thus, although it is possible using, *e.g.*, UPLC-MS to profile samples such as urine with very short run times[8] and very high throughput (20–30 samples per hour), metabolome coverage suffers. Longer chromatographic run times, typically 10–20 min, are more widely used for routine 'screening' in this type of global metabolite profiling, providing reasonable sample throughput combined with good chromatographic resolution and thereby minimizing, without necessarily eliminating, problems such as matrix effects and coelution of isobaric species.[9,10] Where in depth profiling is required, longer analysis times, using very high-resolution systems such as capillary LC to provide the highest chromatographic resolution available, may be indicated (see below), but clearly the throughput with such systems is severely limited. The various modes and options for LC are described below.

4.2 Liquid Chromatography for Metabolic Profiling: General Comments

Although the chromatographic separation may be performed using a variety of different formats, including HPLC, UHPLC (UPLC), capillary LC (capLC) and nano-LC, SFC, and in columns of various lengths, the modes of chromatography and the chromatographic phases employed tend to be similar. The most commonly used mode is reversed-phase (RP) chromatography using solvent gradients, from weakly to strongly eluotropic solvents, optimized for the particular matrix and its metabolic profile. However, RPLC cannot easily accommodate the whole range of analytes that might be present in a matrix such as urine or serum/plasma and, because polar compounds tend not to be well retained on the column, generally eluting with the solvent front, it is best suited to profiling 'medium' to non-polar metabolites. Where more polar substances need to be analysed then the requirement is for techniques such as hydrophilic interaction LC (HILIC), ion-pair LC (IPLC), ion-exchange LC (IELC) or aqueous normal-phase chromatography (ANPC).

Of major importance in obtaining valid data in metabolic profiling studies is the requirement for robust methodologies that provide repeatable results. In the absence of such data quality, there is no realistic possibility for true biomarker discovery and for any useful biological interpretation. At a minimum, this means that for any LC-MS system, peak retention time, signal intensity and mass accuracy must be stable across samples to facilitate their comparison. As we have described elsewhere,[9–15] the first few injections of biological samples are often associated with changes in retention times; however, these usually stabilize after 5–10 injections of the matrix as a result of the 'conditioning' of the column (a process where we assume that matrix components interact with the stationary phase to modify, *e.g.*, active sites). Thereafter, changes in retention time are generally small and the bulk of the variability results from MS source contamination. Both column lifetime and conditioning are matrix dependent, with urine requiring fewer conditioning steps and giving a longer lifetime before column replacement than, *e.g.*, serum or plasma (which usually require significantly more in the way of column conditioning). In order to monitor the performance of the LC-MS system, a quality control (QC) approach, based on the use of either a representative bulk sample of matrix or prepared from aliquots of the samples themselves, are used (a) for column conditioning and (b) to monitor the analysis *via* their repeated analysis (*e.g.*, a QC sample about every 5–10 injections) over the course of the run. The resulting data obtained from these QC samples can then be evaluated to assess system suitability, system stability and sample repeatability.[9–16]

4.3 Formats for Liquid Chromatography

4.3.1 HPLC

For metabolomic/metabonomic studies, HPLC separations are normally performed using stationary phases of 3–5 µm particle size, in columns of length 5–15 cm and internal diameter (i.d.) of 2.1–4.6 mm. In order to chromatograph the widest range of molecules in a single run, these analyses are usually performed using reversed-phase gradients. A typical HPLC system employed in our work is a 10 cm × 2.1 mm i.d. C_{18}-bonded Symmetry column containing

3.5 μm particles. To help to ensure retention stability, the temperature of the column is maintained at 40 °C in a column oven. For analysis on a column of these dimensions, a solvent flow rate of 0.6 mL min^{-1} is suitable.[17] Typical analysis times for such HPLC separations range from 10 to 30 min, depending on the solvent gradient profile, *etc.*

4.3.2 UHPLC (UPLC)

UHPLC is performed using sub-2 μm stationary phases in columns of length 5–15 cm and ~2 mm i.d.[8–10,13–15,18] It is, in principle similar to conventional HPLC, employing similar mobile phases and gradient elution programmes, but, because of the smaller particle sizes employed, UHPLC generally requires higher operating pressures. The peak widths generated by UHPLC separations are usually much narrower than those obtained *via* conventional LC separations, generally in the region of 2–5 s width at the base, providing much greater peak capacity than can be obtained by HPLC. Although the greater chromatographic efficiency provided by UHPLC compared with HPLC can enable samples to be analysed with shorter analysis times (in some cases less than 2 min per sample[8]), in practice this extra resolution has been used to improve metabolome coverage while keeping the analysis time at ~10–15 min per sample. Since the introduction of U(H)PLC, there has been a general move towards the use of this variant of LC for metabolic profiling studies. Such methods have been used for both plant- and animal-derived samples such as *Arabidopsis thaliana* extracts[19] (following treatment with iso-nitrosoacetophenone), human serum,[15] plasma[14] and urine.[10,13]

4.3.3 Narrow-bore and Capillary LC (capLC)

In addition to HPLC and UHPLC, a range of column formats are available that involve the use of either microbore columns of 0.5–1.0 mm i.d. or capillary columns of varying length formats and, although not yet in widespread use, have been employed in a number of metabolic profiling studies. For example, extracts of the plant *Arabidopsis thaliana*[20] have been analysed via C_{18}-bonded silica monolithic columns (0.2 mm i.d.), of between 30 and 90 cm in length, with MS detection. Urine from Zucker rats was analysed on a relatively short (10 cm) 320 μm i.d. column packed with a 3.5 μm C_{18}-bonded packing[21] using a reversed-phase gradient with about twice as many ions detected in the samples compared with HPLC-MS on a column of the same length with the same stationary phase. Despite using much less sample than HPLC, capLC-MS provided greatly increased sensitivity for some metabolites. Furthermore, an increased number of ions were detected, presumably owing to reduced ion suppression. Multivariate statistical analysis of the data from both types of separation was able to show diurnal variations in the samples from these animals; however, there were few discriminating ions in common between capLC-MS and HPLC-MS. An impressive, and high resolution, separation using a 50 μm i.d. fused-silica capillary, 200 cm long and filled with 3 μm porous

C_{18}-bonded particles was applied to the separation of metabolites present in cell lysates of the microorganism *Shewanella oneidnedensis*.[22.] A reversed-phase gradient was used, at a pressure of 20 000 psi, resulting in the detection of more than 5000 metabolites in about 2000 min. Figure 4.1 shows an example of this type of analysis.

More recently, fully integrated ceramic microfluidic systems have become available that combine a capillary format (*e.g.*, 300 or 150 μm i.d.) with sub-2 μm particles to give a high-performance separation system that combines low sample (and solvent) consumption with high sensitivity. The major benefit of these ceramic devices is the mechanical strength (up to 12 000 psi), which allows for the operation of the LC system at the optimal linear flow velocity. An example is shown in Figure 4.2, where a sample extracted from a dried blood spot has been profiled, using reversed-phase gradient elution. The separation was performed on a 10 cm × 150 μm i.d. capillary column packed with a 1.7 μm C_{18} stationary phase. A linear reversed-phase gradient from 5 to 95% acetonitrile–0.1%

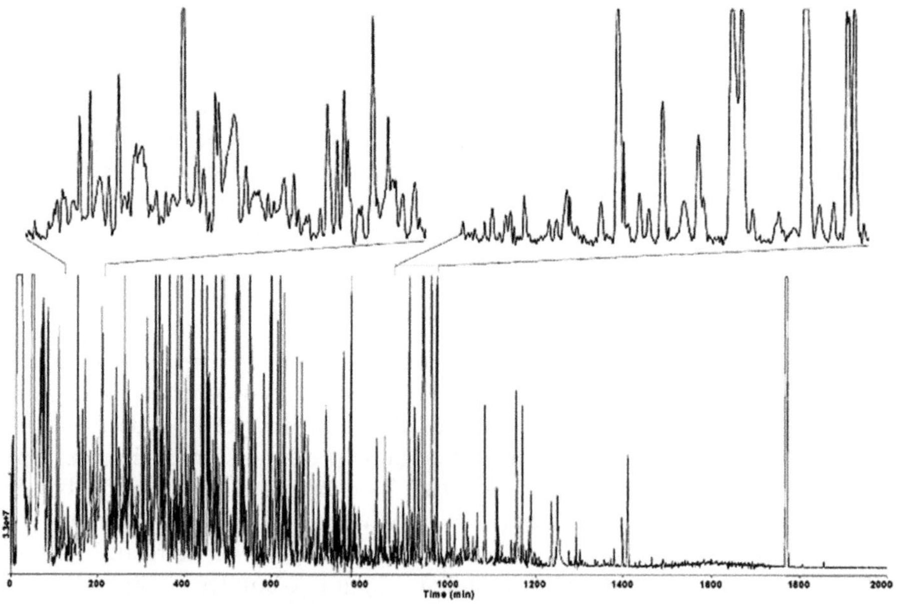

Figure 4.1 Capillary RPLC-MS at 20 kpsi on an extract of *Shewanella oneidensis*. In this single LC-MS experiment, global metabolite analysis detected >5000 metabolites. Conditions: a 20 μL volume of the *S. oneidensis* metabolomic sample (unknown concentration) was loaded on a 200 cm × 50 μm i.d. capillary containing 3 μm porous C_{18} particles. The sample was first loaded on the microSPE column and then transferred to the capillary RPLC column; an 11.4 T Fourier transform ion cyclotron resonance (FT-ICR) MS instrument was used for detection (scan speed 6 s per scan); a linear velocity of $0.12\ \mathrm{cm\ s^{-1}}$ at 20 kpsi was obtained.
Reproduced with permission from the American Chemical Society from Y. Shen, *Anal. Chem.*, 2005, **77**, 3090.

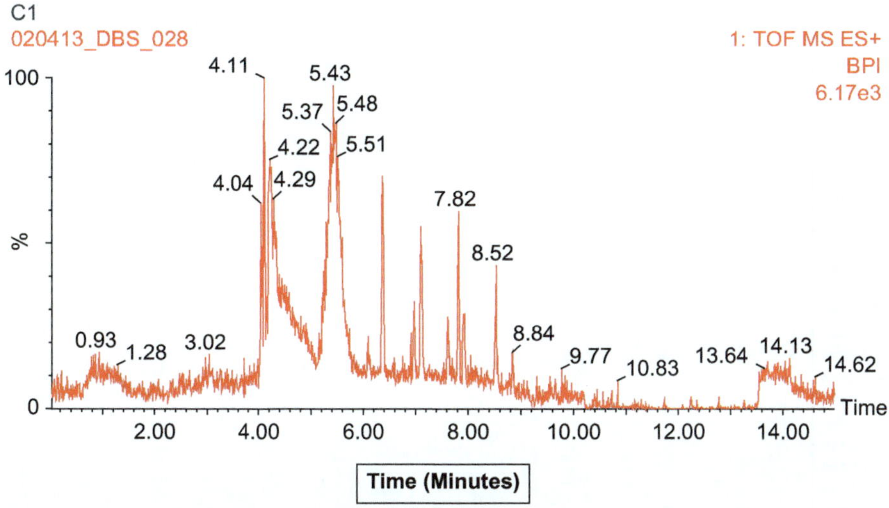

Figure 4.2 Analysis of a dried blood spot extract using microfluidic capillary LC-MS performed via a linear reversed-phase gradient, from 5 to 95% acetonitrile –0.1% aqueous formic acid, delivered at 3 μL min^{-1}, on a 10 cm × 150 μm i.d. capillary column packed with a 1.7 μm C_{18} stationary phase. MS detection was via positive ESI.

aqueous formic acid (over 10 min) was delivered at 3 μL min^{-1} with MS detection in the positive ESI mode.

4.3.4 Multidimensional LC

Neither RPLC nor HILIC alone can provide comprehensive metabolic profiles of biological fluids or tissue/cell extracts. In addition to simply profiling samples by each method separately, combining them in a single, two-dimensional, separation could be used. An example of such a combined HILIC and RP separation, via automated column switching, coupled to UV or MS detection has been described.[23] The polar compounds were separated on a TSK gel amide-80 column (50 × 2.0 mm i.d., 3 μm particle size) and the less polar compounds by RPLC on a Hypersil ODS2 C_{18} column (150 × 4.6 mm i.d., 5 mm particle size). The resulting optimized method was applied to metabolic profiling of urine from lung cancer patients and was found to be reproducible and robust. Figure 4.3 shows an LC-MS trace from the analysis of urine from a lung cancer patient. The first 22 min of the analysis represent the trace of analytes eluted from the HILIC column and the subsequent 26 min represent the effluent from the RPLC column. Using multivariate statistical analysis, the authors proposed 11 potential biomarkers of lung cancer (*i.e.*, they were upregulated in patients *versus* controls). Of these, 10 were polar and detected via HILIC and the remaining one was found *via* RPLC. The compounds included amino acids such as taurine, valine and phenylalanine and also betaine, hippuric acid, pipecolic acid and carnitine.

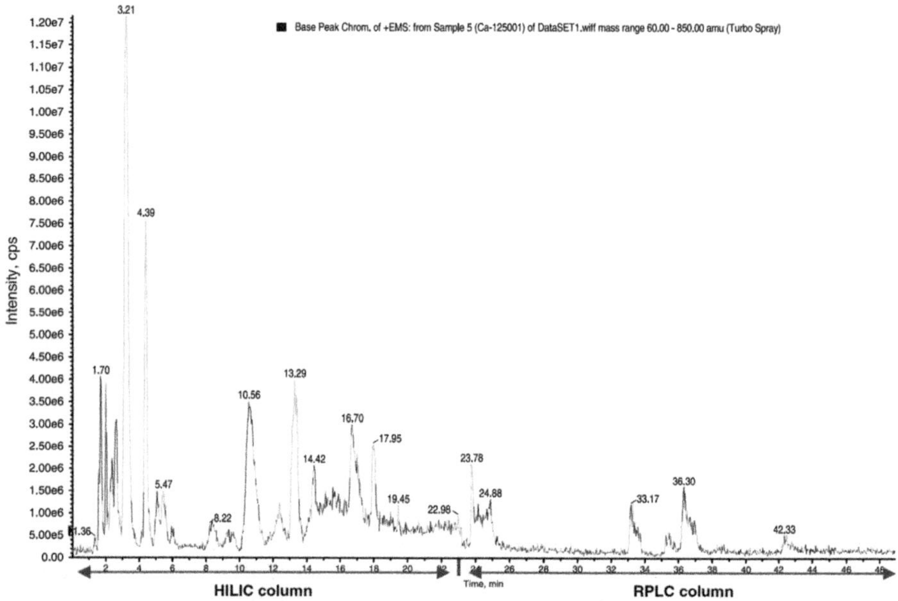

Figure 4.3 A typical chromatogram of urine analysis from a lung cancer patient performed with the HILIC/RPLC-MS system.
Reprinted with permission from Wiley from Q. Yang, X. Shi, Y. Wang, W. Wang, H. He, X. Lu and G. Xu, *J. Sep. Sci.*, 2010, **33**, 1495–1503.

4.3.5 Turbulent Flow Chromatography

So-called 'turbulent flow chromatography' (TFC) has also been investigated as a means of rapid metabonomic LC-MS analysis of plasma. The approach provides an alternative to solvent-based protein precipitation, avoiding the need for off-line sample preparation such as protein precipitation. This is possible because the use of TFC enables protein-rich samples such as blood plasma to be injected directly into to the system[24,25] as, under the conditions used in this mode of separation [high flow rates and large (25–50 μm) chromatographic particles], proteins pass through the column unretained and long-term chromatographic performance is not lost. The reason why large particles are used is that turbulent flow conditions are usually achieved at high flow rates and if small particles were to be employed this would be associated with unacceptably high back-pressures. In TFC, the analytes present in the plasma sample are first passed though a column containing these large particles on which they, but not the proteins, are retained. Analysis is effectively performed in a 2D-LC mode as, once the proteins have been removed in this way, the retained analytes are then eluted from the TFC column on to a conventional high-resolution analytical column for analysis. Preliminary studies have shown the potential of the technique,[26] but it should be noted that the data obtained by TFC analysis for human plasma shows differences in metabolite profiles compared with methanol-precipitated HPLC-MS. Differences include reduced amounts of phospholipids (about a 10-fold reduction) detected *via* TFC compared with protein-precipitated samples.

4.3.6 Supercritical Fluid Chromatography

Supercritical fluid chromatography (SFC) represents an interesting form of LC that proved to be particularly useful for separations of lipids and chiral compounds.[27] The technique was briefly popular in the latter part of the twentieth century but, probably as a result of a lack of commercially available instrumentation, never gained widespread use. In SFC, the mobile phase is usually liquid carbon dioxide, modified with an organic solvent such as acetonitrile or methanol to increase the eluotropic strength in order to elute more polar analytes. Both capillary and packed column formats for SFC have been described and linking either to MS is not particularly technically demanding. Typical stationary phases that have been used for SFC include silica, C_{18}-bonded silica and chiral stationary phases of various sorts.[27] A recent advance, which may increase the take up of SFC for applications such as metabolic profiling (see Section 4.4.6) is the introduction of packed column SFC using sub-2 μm packings. This is advantageous because the solvent properties of supercritical fluids are such that very high flow rates, which minimize analysis times, can be employed without loss of chromatographic efficiency.

4.4 Modes of Chromatography

4.4.1 Reversed-Phase Liquid Chromatography (RPLC)

As noted above, RPLC is currently the workhorse of LC-MS-based global metabolic profiling and was the first mode of LC to be widely applied to this type of analysis. In part, this is because it is generally immediately compatible with largely aqueous sample types such as urine and bile, and also protcin-precipitated plasma. Standard systems employ solvent gradient elution, of the type already described in Sections 4.3.1 and 4.3.2, with typical analysis times from 10 to 30 min (with wash and re-equilibration steps adding a further 2–5 min per run) and with start and final conditions dependent on the sample type. In the case of urine, minimal sample preparation is needed for RPLC and generally centrifugation and dilution are all that are required.[10,13] For blood-derived samples, such as plasma or serum[9,14,15] or tissues,[28–32] more extensive sample preparation is needed to remove proteins. In addition, as the balance between polar metabolites and more non-polar metabolites such as lipids is different from that for urine, the analysis of these samples typically starts at a more eluotropic solvent composition than that used for urine. Further, the need to remove strongly retained lipophilic substances (present in these samples) from the column at the end of the analysis requires stronger and more prolonged washing steps. For such sample types, methanol (or other alcohols) appear to be more efficient at eluting such contaminants than acetonitrile, possibly as a result of their ability to break hydrogen bonds. Figure 4.4

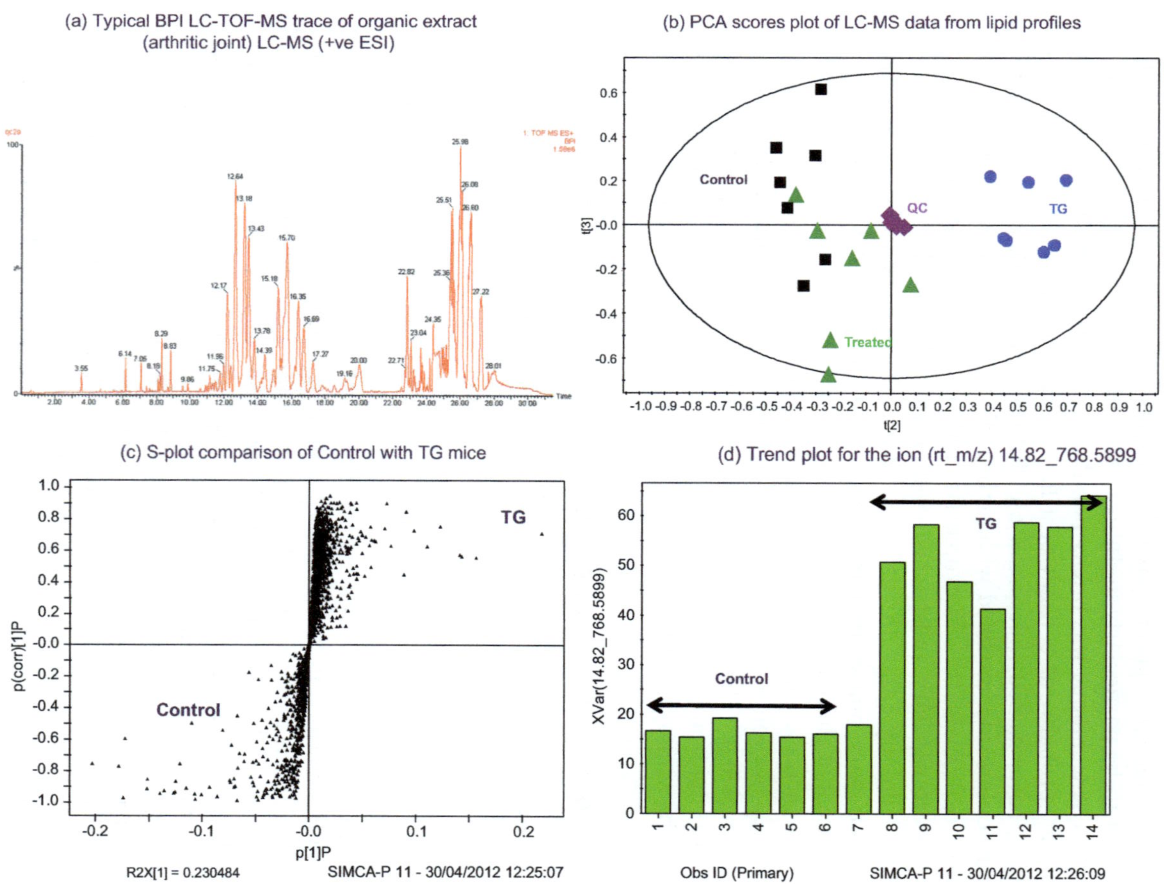

Figure 4.4 Example of the analysis and data mining for metabolomics in biomarker discovery: analysis of organic extracts from tissue from animal models (see text for a detailed explanation of the content of the panels).

provides an example of the analysis and data mining for metabolomics in biomarker discovery, in this case on the analysis of organic extracts from tissue from animal models. Figure 4.4a depicts a typical LC-TOF-MS trace in LC-MS (positive ESI mode). Following data extraction, statistical analysis was performed by principal component analysis (PCA) and the generated scores plot (Figure 4.4b) provided separation of the groups, in this case the transgenic (diseased) animals, the transgenic animals treated with a therapeutic agent and the control animals. Subsequent data mining using visualization tools such as the S-plot (Figure 4.4c) show the contribution of features to the group separation. The trend plot for a selected discriminating ion [(rt_m/z) 14.82_768.5899] is given in Figure 4.4d, showing a higher analyte signal in the transgenic (TG) animals.

For a matrix such as urine, a typical solvent system would employ 0.1% aqueous formic acid and acetonitrile (also containing 0.1% formic acid) to form the gradient, which would follow a trajectory involving a gradual increase in organic content from 0 to 20% acetonitrile over 0.5–4 min and then a relatively rapid rise to 95% acetonitrile over the next 4 min. The solvent composition is held at 95% acetonitrile for 1 min, before returning to 100% aqueous formic acid, to remove strongly retained contaminants before the next sample injection, resulting in ~10 min per analysis. Protocols for the RPLC-MS analysis of serum/plasma, urine and tissue extracts using this type of chromatography have been reported.[9,10,28]

Such acidic mobile phases are in widespread general use, but recently there have been reports of the potential to use mobile phases operating at pH >10, with MS detection in positive ESI mode, for metabonomics. The separations were performed on a 100 × 2.1 mm i.d. ACQUITY UPLC 1.7 μm BEH C_{18} column (maintained at 40 °C) at a flow rate of 500 μL min^{-1}. The solvent system employed was a linear acetonitrile–aqueous gradient over 10 min, starting at 5% and rising to 95% acetonitrile over the course of the gradient. The basic mobile phase was 0.1% ammonium hydroxide. Compared with separation using the same gradient profiles but an acidic mobile phase (0.1% formic acid), benefits include improved retention of some analytes, better chromatographic peak shape and increased signal-to-noise ratio compared with acidic solvent systems.[33]

Another alternative to conventional gradient RPLC approaches is to perform the separation with the column at elevated temperatures, employing either isothermal conditions with a solvent gradient[34] or an isocratic solvent system (in this instance purely aqueous) and a thermal gradient.[35] Both types of separation have been successfully applied to urine samples in metabonomic studies.

4.4.2 Hydrophilic Interaction Liquid Chromatography (HILIC)

When polar metabolites need to be separated, HILIC represents the current LC mode of choice. For successful and repeatable results, it is important not to exceed 50% water in the mobile phase, to employ slow gradients and to ensure

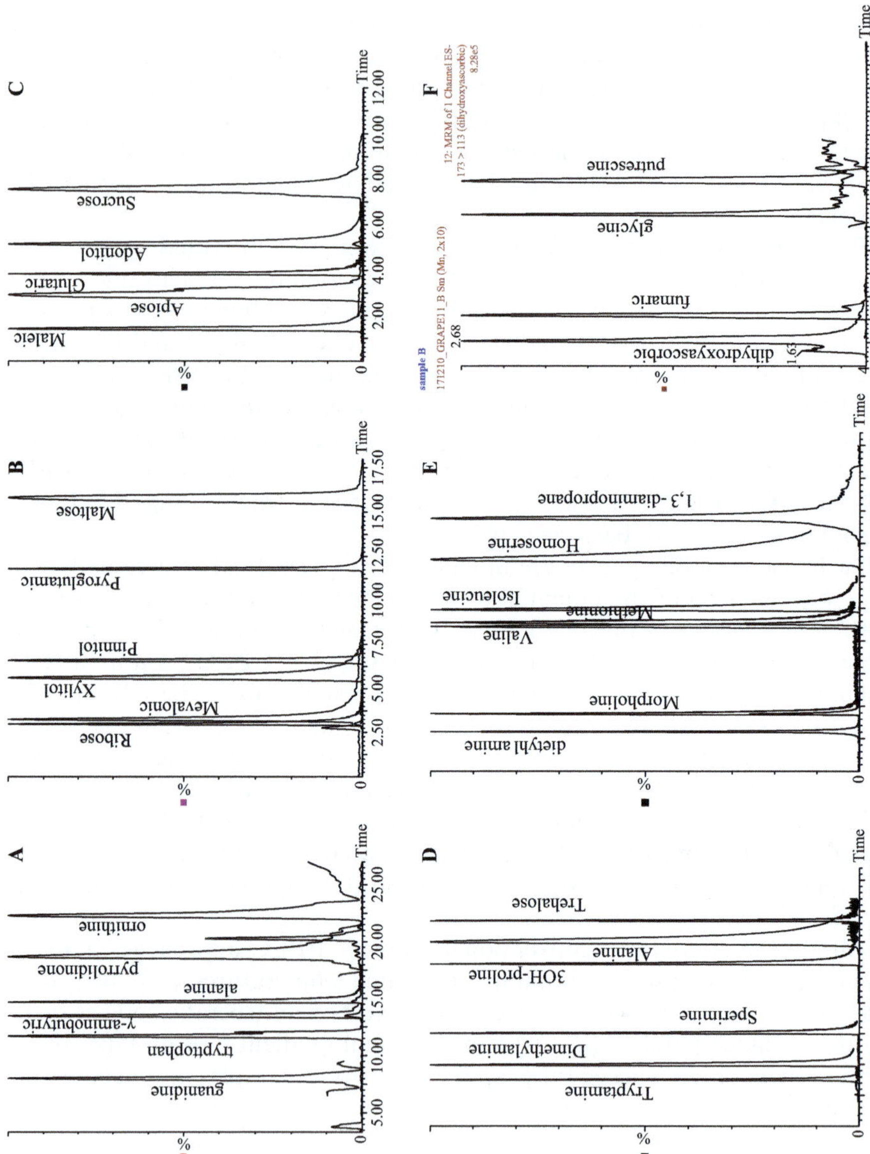
A
guanidine
tryptophan
γ-aminobutyric
alanine
pyrrolidinone
ornithine
B
Ribose
Mevalonic
Xylitol
Pinnitol
Pyroglutamic
Maltose
C
Maleic
Apiose
Glutaric
Adonitol
Sucrose
D
Tryptamine
Dimethylamine
Sperimine
3OH-proline
Alanine
Trehalose
E
diethyl amine
Morpholine
Valine
Methionine
Isoleucine
Homoserine
1,3 -diaminopropane
F
dihydroxyascorbic
fumaric
glycine
putrescine
sample B
171210_GRAPE11_B Sm (Mn, 2x10)
12: MRM of 1 Channel ES-
173 > 113 (dihydroxyascorbic)
8.28e5
Time
%

that the re-equilibration time between injections is sufficient. As indicated above, samples such as urine are composed mainly of polar compounds, many of which elute close to the solvent front in RP systems and, as shown in numerous applications,[36–40] HILIC offers the potential to improve the profiling of these molecules, providing a more comprehensive coverage of the urinary metabolome when used in combination with RPLC. In addition, because of the use of mobile phases containing a high proportion of organic solvents, such as acetonitrile, HILIC typically offers higher MS sensitivity than RP methods as a result of the increased ionization efficiency due to efficient generation of spray conditions. The first published HILIC-MS metabolite profiling study provided an application to the analysis of rat urine samples,[36] with the aim of detecting highly polar metabolites by HILIC-ESI-MS as a complementary analysis to RPLC-ESI-MS. In this work, solid-phase extraction (SPE) was used to prepare the samples with the unretained wash fraction analysed on a ZIC-HILIC column (100 × 2.1 mm i.d., 3.5 mm particle size) whereas metabolites retained by SPE were analysed by RPLC on a C_{18} column. Since then, there have been an increasing number of HPLC and UHPLC applications of HILIC-based separations (see review[38]) using a range of column types. Typical examples include ZIC-HILIC (100 × 4.6 mm i.d., 3.5 μm particle size)[39] or the Aphera NH2 polymer column (150 × 2 mm i.d., 5 μm particle size)[38]. The first UHPLC application of HILIC for global metabolic profiling[37] used ACQUITY BEH HILIC material for the separation (150 × 2.1 mm i.d., 1.7 μm particle size) and more recent UHPLC applications also used the ACQUITY HILIC phase in a toxicological application[41] or for cancer biomarker work.[42]

The recent example of a UPLC-HILIC method applied to the profiling of polar metabolites in the urine of rats exposed to the model hepatotoxin galactosamine[41] also employed an ACQUITY BEH HILIC column (100 × 2.1 mm i.d., 1.7 μm particle size). The gradient separation was performed at 40 °C with 0.1% v/v formic acid and 10 mM ammonium acetate in 95:5 acetonitrile–water as solvent A and 0.1% aqueous formic acid and 10 mM ammonium acetate in 50:50 acetonitrile–water as solvent B. The starting conditions were 99.0% A and 1% B for 1.0 min, changing linearly to 100% B over the next 11.0 min, after which the solvent composition returned to starting conditions over 0.1 min, followed by re-equilibration for 4 min prior to the next

Figure 4.5 UHPLC–MS/MS traces for 34 primary metabolites on the HILIC amide column with the developed methodology. For clarity, the 34 different ion traces are depicted in six different panes. Metabolite traces from upper left to bottom right and with increasing retention time. (a) Guanidine, tryptamine, γ-aminobutyric acid, alanine, pyrrolidinone, ornithine. (b) Ribose, mevalonic acid, xylitol, pinitol, pyroglutamic acid, maltose. (c) Maleic acid, apiose, glutaric acid, adonitol, glucose. (d) Typtamine, dimethylamine, spermine, 3-hydroxyproline, serine, trehalose. (e) Diethylamine, morpholine, valine, methionine, homoserine, 1,3-diaminopropane. (f) Dihydroascorbic acid, fumaric acid, glycine, putrescine.
Reprinted with permission from Elsevier from H. G. Gika, G. A. Theodoridis, U. Vrhovsek and F. Mattivi, *J. Chromatogr. A*, 2012, **1259**, 121.

injection. A flow rate of $0.4\,mL\,min^{-1}$ was employed. UPLC-HILIC when coupled to MS/MS can offer a powerful tool for multi-analyte quantitative determination or so-called targeted metabolomics, as shown in Figure 4.5 for a typical example of the simultaneous analysis of more than 130 primary metabolites. In this case, an ACQUITY amide column was used in HILIC conditions by using a gradient of solvent A (acetonitrile–water, 95:5 v/v, 0.1% formic acid and 0.075% NH_4OH) and solvent B (acetonitrile–water, 2:98 v/v, 0.2% formic acid and 0.1% NH_4OH). The method allowed the determination of selected metabolites from four major classes (sugars, organic acids, amino acids and amines) in a single injection, which with existing methods would require a multitude of analyses with different instrumental configurations.[43]

4.4.3 Ion-Exchange Liquid Chromatography (IELC)

Although not widely used so far for metabolome profiling, IELC-MS should offer many advantages for the analysis of polar, ionic compounds present in biofluids. In preliminary (unpublished) studies, we have used the technique to profile samples of rat urine obtained from Zucker rats [(fa/fa) and (–/–) representing obese and lean phenotypes]. The samples were analysed using an IonPac AS18 column (250 × 2 mm i.d.) (fitted with a 50 × 2 mm i.d. IonPack AG18 guard column) at ~30 °C during the analysis. Separations were performed using gradient elution with the starting conditions being 5 mM NaOH for 1 min, changing exponentially to 100 mM NaOH at 12 min at a flow rate of $0.3\,mL\,min^{-1}$. These conditions were maintained for a further 3 min before returning to the starting conditions for the next 8 min (to re-equilibrate the column) prior to next injection. Good separation between the two strains of animals was observed using PCA, although the metabolites responsible were not characterized further or identified in this preliminary study.

4.4.4 Ion-Pair Liquid Chromatography (IPLC)

The use of IPLC is not ideal for MS-based applications because of contamination of the mass spectrometer by the ion-pair reagent, which be prove extremely difficult to remove. However, for the analysis of acidic metabolites by negative ionization electrospray methods it does have some potential utility. Pragmatically, it may be necessary to dedicate an instrument to this technique to the exclusion of more general studies. To date, the IPLC studies that have been described have most often used the technique for the targeted analysis of acidic compounds. However, there is no reason, in either principle or practice, why IPLC should not be used for general metabolite profiling of acidic compounds. A recent application[44] involves the use of RP-IPLC separation on a Phenomenex Synergy Hydro-RP 2.5 μm C_{18} column (100 × 2 mm i.d.) employing a flow rate of $0.2\,mL\,min^{-1}$. The separation was achieved via a water–methanol gradient and 10 mM tributylamine and 15 mM acetic acid as the IP reagent. The starting conditions for the gradient were 97:3 water–methanol for 2.5 min, increasing the methanol content to 20% at 5 min, then, from 7.5 min, increasing to 20% methanol at 13 min, 55% at 15.5 min and finally 95% at 18.5 min. The solvent composition was maintained at 5:95 for a further 0.5 min before returning to the starting conditions at 19 min, with the next injection made following re-equilibration at 25 min. The column temperature was maintained at 25 °C. The separation was linked to an Orbitrap mass spectrometer and provided sensitive and specific detection of both unknown and known metabolites (with a limit of detection of 5 ng mL^{-1} for 80 metabolite standards). In a study on a strain of yeast engineered such that a gene of unknown function (YKL215C) was knocked out, the accumulation of oxoproline was observed, allowing the gene to be identified as oxoprolinase. In this study, a total of 137 metabolites were also monitored. These analytes covered a wide range of metabolite classes and included amino acids (*e.g.*, tryptophan, alanine and methionine), acidic metabolites such as citraconic acid, citrate/isocitrate, pyruvate and phenyllactate and phosphorylated compounds including 3-phosphoserine, glucose-6-phosphate, AMP, ADP and ATP. Tributylamine was also recently employed with a nano-scale LC-MS system[45] for the analysis of cell extracts of *Methylobacterium extorquens* and a hexylamine-based IP system[46] has been applied to the LC-MS analysis of extracts of *Lactobacillus plantarum*, *Escherichia coli* and *Bacillus subtilis*.

The example shown in Figure 4.6 (Michopoulos *et al.*, in preparation) shows a tributylamine-based UHPLC separation performed on an ACQUITY 1.8 μm

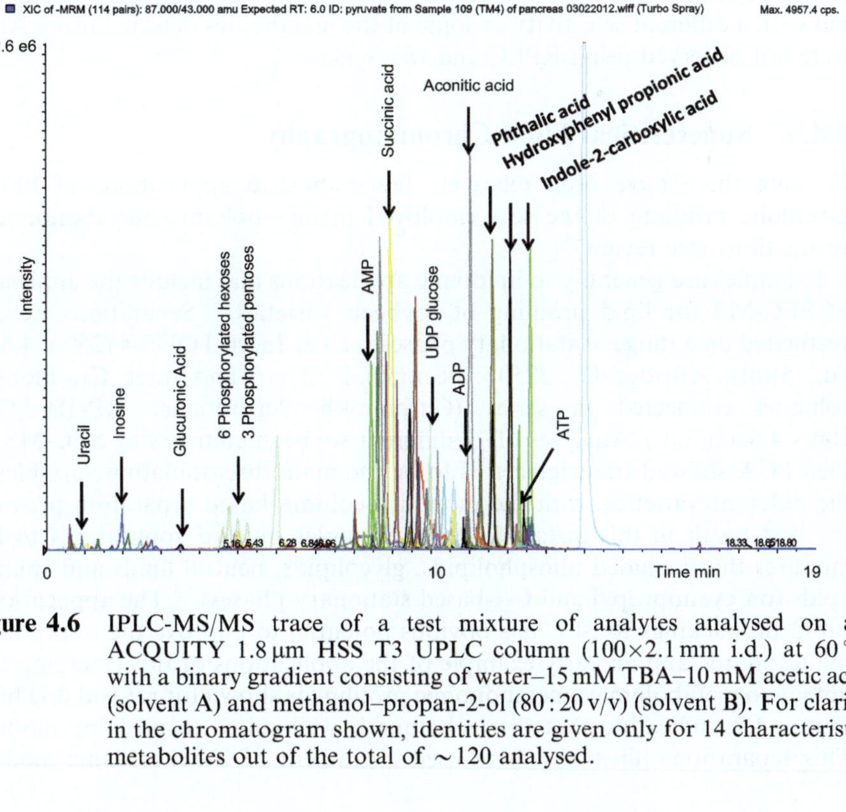

Figure 4.6 IPLC-MS/MS trace of a test mixture of analytes analysed on an ACQUITY 1.8 μm HSS T3 UPLC column (100×2.1 mm i.d.) at 60 °C with a binary gradient consisting of water–15 mM TBA–10 mM acetic acid (solvent A) and methanol–propan-2-ol (80 : 20 v/v) (solvent B). For clarity in the chromatogram shown, identities are given only for 14 characteristic metabolites out of the total of ~120 analysed.

HSS T3 UPLC column (100 × 2.1 mm i.d.) at a temperature of 60 °C and a flow rate of 0.4 mL min^{-1}. Samples were analysed for a range of acidic analytes, including organic acids, amino acids and phosphorylated sugars, using a binary solvent gradient consisting of water containing 15 mM TBA and10 mM acetic acid (solvent A) and methanol–propan-2-ol (80:20 v/v) (solvent B) using the following gradient elution profile: 0 min, 0% B; 0.5 min, 0% B; 4 min, 5% B; 6 min, 5% B; 6.5 min, 20% B; 8.5 min, 20% B; 14 min, 55% B; 15 min, 100% B; 17 min, 100% B; 18 min, 0% B; 21 min 0% B.

4.4.5 Aqueous Normal-phase Chromatography

Another methodology that may prove to be suitable for the separation of highly polar metabolites is aqueous normal-phase chromatography (ANPC) of the type that has been described on the Diamond hydride stationary phase.[47] In this example, a Cogent Diamond Hydride column (100 × 2.1 mm i.d., 4 μm particle size) was used, in combination with a conventional RP separation to profile metabolites such as nucleotides in human urine and plant extracts. The solvent systems were LC-MS compatible and consisted of gradient programme of strong solvent A (15.9 mM ammonium formate or 13.0 mM ammonium acetate) *versus* weak solvent B (90 : 10 v/v acetonitrile–water containing 15.9 mM ammonium formate or 13.0 mM ammonium acetate) at 0.4 mL min^{-1}. This study showed that ANPC provided similar features to the RPLC method, but with a different selectivity as some of the metabolites detected using ANPC were not observed using RPLC and *vice versa*.

4.4.6 Supercritical Fluid Chromatography

To date there have been relatively few published applications of SFC in metabolic profiling of the sort employed in metabolomics/metabonomic investigations (see review[48]).

Examples are generally to lipidomic applications and include the application of SFC-MS for lipid profiling of soybean varieties.[49] Separations were investigated on a range of stationary phases such as Inertsil ODS-4 (250 × 4.6 mm i.d., 5 μm), XBridge C_{18} (250 × 4.6 mm i.d., 5 μm) and three C_{18} monolith columns connected in series (Chromolith Performance RP-18e ODS, 100 × 4.6 mm i.d.). Analyses of 12 different soybean cultivars by SFC-MS and then PCA showed triacylglycerols to be the main discriminating variables for the different varieties, with the monolith column-based separation providing the best result in this instance. Other examples involve applications to lipid mixtures that included phospholipids, glycolipids, neutral lipids and sphingolipids (on cyanopropyl and C_{18}-based stationary phases).[50] The application of sub-2 μm packings to SFC has obvious potential to improve the resolution of the technique and an early example of the applications of this emerging technology to metabolomics/metabonomic profiling is shown for rat and dog bile in Figure 4.7, indicating that the technique is well suited to profiling biofluids. This separations illustrated here used a gradient of 0–5% organic modifier,

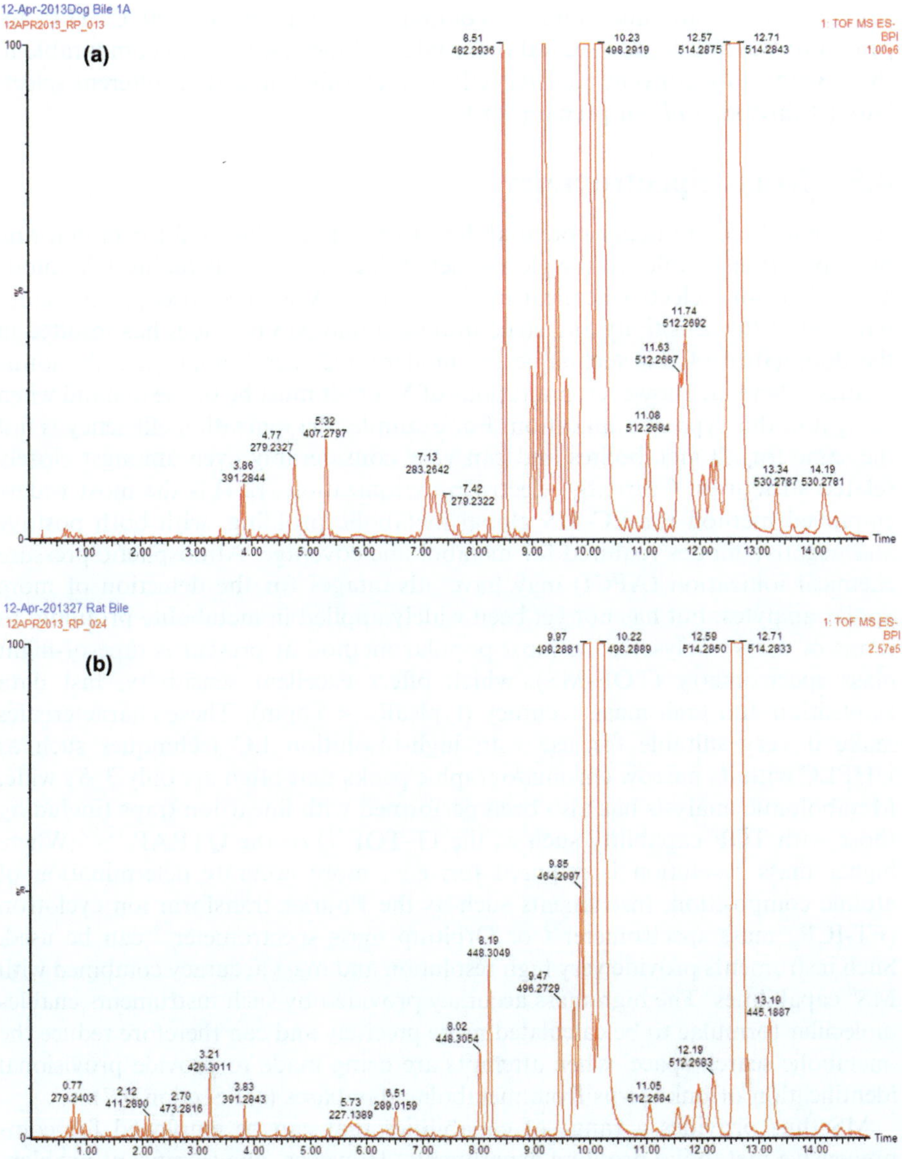

Figure 4.7 SFC separations of (a) rat and (b) dog bile on a 15 cm × 3 mm i.d. BEH column packed with sub-2 μm particles with liquid CO_2 modified with acetonitrile–20 mM aqueous ammonium formate containing 0.1% formic acid (95:5 v/v) using a gradient of 0–5% over 15 min. A flow rate of 3 mL min^{-1} at 50 °C was used.

consisting of acetonitrile–20 mM aqueous ammonium formate containing 0.1% formic acid (95:5 v/v), *versus* liquid CO_2 with the separation performed on a 15 cm × 3 mm i.d. BEH column packed with sub-2 μm particles. The flow rate employed was 3 mL min^{-1} and the column was maintained at 50 °C. SFC appeared to be robust and repeatable for bile and the results seem comparable to those with more conventional RP-UPLC methods but with a different selectivity (Rainville *et al.*, in preparation).

4.5 Mass Spectrometry

Although there are many potential detectors that can be used for monitoring the separations employed for global metabolite profiling, including UV (diode array detector), electrochemical methods and NMR spectroscopy, the combination of the sensitivity and structural data that MS provides has resulted in the dominance of this technique for modern LC metabolomic/metabonomic studies. There are, however, limitations of MS that must be borne in mind when using it in this type of application. For example, the ionization efficiency is not the same for all metabolites and can vary considerably even amongst closely related structures. Currently, electrospray ionization (ESI) is the most widely employed method for LC-MS global metabolic profiling, with both positive and negative modes required for metabolome coverage. Atmospheric pressure chemical ionization (APCI) may have advantages for the detection of more apolar analytes, but has not yet been widely applied in metabolite profiling. In terms of mass analysers, the most popular method at present is time-of-flight mass spectrometry (TOF-MS), which offers excellent sensitivity, fast data acquisition and high mass accuracy (typically <5 ppm). These characteristics make it very suitable for use with high-resolution LC techniques such as UHPLC with its narrow chromatographic peaks that often are only 3–5 s wide. Metabolomic analysis has also been performed with linear ion traps (including those with TOF capability such as the IT-TOF[51]) or the QTRAP.[12,52] Where higher mass resolution is required for, *e.g.*, more accurate determination of atomic composition, instruments such as the Fourier transform ion cyclotron (FT-ICR) mass spectrometer[53] or Orbitrap mass spectrometer[54] can be used. Such instruments provide very high resolution and mass accuracy combined with MS^n capabilities. The high mass accuracy provided by such instruments enables molecular formulae to be calculated more precisely and can therefore reduce the 'metabolic search space' when attempts are being made to provide provisional identification of unknowns from metabolite databases (see Section 4.7).

MS thus provides a range of capabilities that can be employed for comprehensive metabolic profiling experiments. However, one significant problem that researchers using LC-MS-based techniques for metabonomics need to be aware of is that there can be significant difficulties in comparing and correlating LC-MS data derived, for the same samples, from different MS instruments, even when the same separation is used. Thus, we have recently demonstrated that even the simultaneous analysis of the same chromatographic eluate on a QTOF and a QTRAP instrument for LC-MS-based metabolite profiling gave different results. While the control and test groups could easily be separated by multivariate statistical analysis (using PCA), this was based on different, instrument-dependent, markers.[52]

4.6 Quality Control and Data Analysis

Major methodological problems that can arise from the use of LC-MS-based methods result from analytical, rather than biological, variability. Thus, during the course of the analysis of a modest-sized batch of a 100 or so samples, which typically would take ~24 h to complete, there is the potential for mass accuracy, analyte response and retention times to change. All of these can cause problems in the subsequent data analysis if they become excessive. A number of approaches have been advocated to overcome this type of problem, including the use of a number of known molecules added as internal standards, the use of technical replicates or the method that we advocate, namely the use of a standard quality control (QC) sample.[9–15] The QC samples that we usually employ are prepared by making a pooled sample using small aliquots of each of the samples that form the analytical batch. These samples are then dispersed at regular intervals throughout the analytical run and the variability in the results obtained from them is used to assess the likely variability in the whole run. We have also found, through the use of these samples, the absolute need to precondition the LC-MS system with the injection of a sample-dependent number of injections of the QC sample to obtain stable analytical data, particularly retention times. Examination of the results from these conditioning injections also allows the 'system suitability' to be determined prior to committing the valuable test samples to analysis.

The data collected during an LC-MS run comprises mass, intensity and retention time information. The MS instrument typically operates in full-scan mode over the range from ~70 to ~1200 amu and a raw data file will usually contain the information as a series of full-scan mass spectra, including systematic noise, *etc.*, acquired over successive time points (each covering 2–20 ms). To deal with the hundreds of data files that comprise a UHPLC-MS run, specialist software packages are required and numerous proprietary, freeware and in-house programs have been developed for this. These programs enable processes such as noise filtering, baseline correction, centring, normalization, peak picking, peak integration and alignment to be performed automatically so that the peaks for metabolites within the chromatographic run are detected. From these data, a two-dimensional data matrix is produced which comprises a peak table listing the samples together with the detected metabolite peaks and their intensities with the three-dimensional retention time/mass/intensity LC-MS data compressed into two dimensions by combining the mass and retention time information into a single 'feature.' Open-source software programs such as MZmine,[55] MetAlign[56] and XCMS,[57] all of which are freely available on the Internet, can be used to analyse data from any manufacturer's mass spectrometer once the data have been converted into a format such as netCDF or mzXML. This can be useful when wishing to compare data obtained on different instruments, as it removes the potential to distort the results due to differences in the way in which these software programs operate. In addition, open-source software such as XCMS (which operates in R) can be customized so that, *e.g.*, the peak-picking algorithm can be made to fit better to the specific needs of certain analyses that have particularly broad, or narrow, peaks, *etc*. This is a useful feature because, although they provide an excellent platform for processing data, software programs available from mass spectrometer manufacturers (*e.g.*, MarkerLynx, MarkerView, Sieve) are generally not customized to the same extent by the user and are limited to data generated on mass spectrometers produced by the particular manufacturer. Whilst initial examination of the data is generally via multivariate analysis, to detect features that may be associated with the condition under study, we recommend both univariate statistical analysis and manual examination of the raw data to ensure that such features are real and not artefacts. For example, if a very large number of features are identified by the software, it may mean that the peak alignment has failed rather than the presence of large numbers of metabolites in the samples or that the time window selected for peak picking is too short or that instrument 'noise' is being identified as a feature. Even when the peaks are real, it must be borne in mind that a single metabolite may result in more than one feature in the data table due to, *e.g.*, the presence of isotope peaks, adduct formation with solvents or, *e.g.*, sodium ions, dimerization or fragmentation in the ion source. However, despite the above phenomena, isotope peaks can be easily recognized and de-isotoping procedures are built into most peak-picking software programs.

Data analysis is one of the most time-consuming steps of any metabolomic/metabonomic investigation, but care taken at this point is essential if the full benefits of the study are to be realized. A useful review of MS-based data analysis for this type of application has been published.[58]

4.7 Biomarker Identification and Validation

Although global metabolic profiling combined with comprehensive data analysis may highlight potential biomarkers, significant amounts of work may be required before the identity of these molecules is confirmed. Such biomarker identification still represents a major bottleneck in this type of LC-MS-based global metabolic profiling study. This is because of the huge variety of molecules that may present themselves as candidate biomarkers and the lack of readily transferable spectral and retention time libraries in LC-MS compared with, *e.g.*, those for GC-MS analysis. However, having determined, using accurate mass data for example, the probable elemental composition of the target molecules and collected as much fragmentation data as possible, tentative assignments can be made as to potential identities *via* database searching (*e.g.*, METLIN,[59] LIPID MAPS,[60] the Human Metabolome Data Base (HMDB),[61,62] ChemSpider[63]) and also comparison with spectral databases built in-house from authentic standards. Database hits can only be considered provisional until confirmed by comparison with an authentic standard (if available). In the absence of an authentic standard, then isolation and further spectroscopic analysis such as with NMR spectroscopy may also be required.

Once identity has been confirmed, the biochemical 'relevance' of these markers must be determined to ensure that they directly reflect some underlying feature of the condition under study rather than some 'collateral damage.' For example, in the case of disease or toxicity, some apparent 'biomarkers' may simply reflect the downstream effects of ill health (malnutrition, poor condition, stress, *etc.*) rather than providing discriminating biomarkers for a particular cancer or organ toxicity. Once the biochemical relevance of the putative biomarkers has been established, it is then incumbent upon the investigator to develop validated and specific methods for these compounds and then perform 'follow-up' studies using these targeted methods to confirm the utility of the metabolites for defining that particular application. Only when this has been done can the process of biomarker discovery be considered to be complete.

4.8 Conclusion

LC-MS-based methods for metabolite profiling, of the type required for metabolomic and metabonomic studies, are increasingly becoming widely available. Currently there is no one method that provides a comprehensive metabolite profile and the best attempts at this require multiple analysis using, *e.g.*, RPLC for medium to non-polar metabolites and HILIC (or some other type of LC) for polar metabolites, combined with both positive and negative ESI. Although relatively robust protocols now exist for metabolomics, difficulties remain in the identification of potential biomarkers (as opposed to their detection), and there is a need for post-discovery validation.

References

1. J. K. Nicholson, J. Connelly, J. C. Lindon and E. Holmes, *Nat. Rev. Drug Discov.*, 2002, **1**, 153.
2. J. K. Nicholson and J. C. Lindon, *Nature*, 2008, **455**, 1054.
3. R. S. Plumb, C. L. Stumpf, M. V. Gorenstein, J. M. Castro-Perez, G. J. Dear, M. Anthony, B. C. Sweatman, S. C. Connor and J. N. Haselden, *Rapid Commun. Mass Spectrom.*, 2002, **16**, 1991.
4. R. S. Plumb, J. Granger, C. Stumpf, I. D. Wilson, J. A. Evans and E. M. Lenz, *Analyst*, 2003, **128**, 819.
5. G. Theodoridis, H. G. Gika and I. D. Wilson, *Trends Anal. Chem.*, 2008, **27**, 251.
6. G. Theodoridis, H. G. Gika and I. D Wilson, *Mass Spectrom. Rev.*, 2011, **30**, 884.
7. Z. Wu, Z. Huang, R. Lehmann, C. Zho and G. Xu, *Chromatographia*, 2009, **69**, S23.
8. R. S. Plumb, J. H. Granger, C. L. Stumpf, K. Johnson, B. W. Smith, S. Gaulitz, I. D. Wilson and J. Castro-Perez, *Analyst*, 2005, **130**, 844.
9. W. B. Dunn, D. Broadhurst, P. Begley, E. Zelena, S. Francis-McIntyre, N. Anderson, M. Brown, J. D. Knowles, A. Halsall, J. N. Haselden, A. W. Nicholls, I. D. Wilson, D. B. Kell and R. Goodacre, *Nat. Protocols*, 2011, **6**, 1060.

10. E. J. Want, I. D. Wilson, H. Gika, G. Theodoridis, R. S. Plumb, J. Shockcor, E. Holmes and J. K. Nicholson, *Nat. Protocols*, 2010, **5**, 1005.
11. T. Sangster, H. Major, R. S. Plumb, A. J. Wilson and I. D. Wilson, *Analyst*, 2006, **131**, 1075.
12. H. G. Gika, G. A. Theodoridis, J. E. Wingate and I. D. Wilson, *J. Proteome Res.*, 2007, **6**, 3291.
13. H. G. Gika, E. Macpherson, G. A. Theodoridis and I. D. Wilson, *J. Chromatogr. B*, 2009, **871**, 299.
14. F. Michopoulos, L. Lai, H. Gika, G. Theodoridis and I. Wilson, *J. Proteome Res.*, 2009, **8**, 2114.
15. E. Zelena, W. B. Dunn, D. Broadhurst, S. Francis-McIntyre, K. M. Carroll, P. Begley, S. O'Hagan, J. D. Knowles, A. Halsall, I. D. Wilson and D. B. Kell, *Anal. Chem.*, 2009, **81**, 1357.
16. W. B. Dunn, I. D. Wilson, A. W. Nicholls and D. Broadhurst, *Bioanalysis*, 2012, **4**, 2249.
17. I. D. Wilson, R. Plumb, J. Granger, H. Major, R. Williams and E. M. Lenz, *J. Chromatogr. B*, 2005, **817**, 67.
18. I. D. Wilson, J. K. Nicholson, J. Castro-Perez, J. H. Granger, K. Johnson, B. W. Smith and R. S. Plumb, *J. Proteome Res.*, 2005, **4**, 591.
19. N. E. Madala, F. Tugizimana, P. A. Steenkamp, L. A. Piater and I. A. Dubery, *Chromatographia*, 2013, **76**, 279.
20. V. T. Tolstikov, A. Lommen, K. Nakanishi, A. Tanaka and O. Fiehn, *Anal. Chem.*, 2003, **75**, 6737.
21. J. Granger, R. Plumb, J. Castro-Perez and I. D Wilson, *Chromatographia*, 2005, **61**, 375.
22. Y. Shen, Y. Zhang, R. J. Moore, J. Kim, T. O. Metz, K. K. Hixon, R. Zhao, E. A. Livesay, H. R. Udseth and R. D. Smith, *Anal. Chem.*, 2005, **77**, 3090.
23. Q. Yang, X. Shi, Y. Wang, W. Wang, H. He, X. Lu and G. Xu, *J. Sep. Sci.*, 2010, **33**, 1495.
24. A. Edge, in *A Handbook of Bioanalytical Separations*, ed. I. D. Wilson, Elsevier, Amsterdam, 2003, 91.
25. P. Turnpenny, D. Fraier, C. Chassaing and J. Duckworth, *J. Chromatogr. B*, 2007, **856**, 131.
26. F. Michopoulos, A. M. Edge, G. Theodoridis and I. D. Wilson, *J. Sep. Sci.*, 2010, **33**, 1472.
27. J. R. Dean (ed.), *Applications of Supercritical Fluids in Industrial Analysis*, Blackie Academic and Professional, Glasgow, 1993.
28. E. J. Want, P. Masson, F. Michopoulos, I. D. Wilson, G. Theodoridis, R. S. Plumb, J. Shockcor, E. Holmes and J. K. Nicholson, *Nat. Protocols*, 2013, **8**, 17.
29. P. Masson, A. C. Alves, T. M. D. Ebbels, J. K. Nicholson and E. J. Want, *Anal. Chem.*, 2010, **82**, 7779.
30. P. Masson, K. Spagou, J. K. Nicholson and E. J. Want, *Anal Chem.*, 2011, **83**, 382.

31. N. Loftus, A. Barnes, S. Ashton, F. Michopoulos, G. Theodoridis, I. Wilson, C. Ji and N. Kaplowitz, *J. Proteome Res.*, 2011, **10**, 705.
32. H. Wu, A. D. Southam, A. Hines and M. R. Viant, *Anal. BioChem.*, 2008, **372**, 204.
33. P. D. Rainville, N. W. Smith, D. Cowan, J. K. Nicholson, J. P. Shockcor, St-J. Skilton and R. S. Plumb, *Bioanalysis*, 2012, **4**, 2833.
34. H. Idborg, L. Zamani, P.-O. Edlund, I. Schuppe-Koistinen and S. P. Jacobsson, *J. Chromatogr. B*, 2005, **828**, 9.
35. R. S. Plumb, P. Rainville, B. W. Smith, K. A. Johnson, J. Castro-Perez, I. D. Wilson and J. K. Nicholson, *Anal. Chem.*, 2006, **78**, 7278.
36. H. G. Gika, G. Theodoridis, J. Extance, A. M. Edge and I. D. Wilson, *J. Chromatogr. B*, 2008, **871**, 279.
37. H. G. Gika, G. A. Theodoridis and I. D. Wilson, *J. Sep. Sci.*, 2008, **31**, 1598.
38. K. Spagou, H. Tsoukali, N. Raikos, H. Gika, I. D. Wilson and G. Theodoridis, *J. Sep. Sci.*, 2010, **33**, 716.
39. S. Cubbon, T. Bradbury, J. Wilson and J. Thomas-Oates, *Anal. Chem.*, 2007, **79**, 8911.
40. K. Kim, P. Aronov, S. O. Zakharkin, D. Anderson, B. Perroud, I. M. Thompson and R. H. Weiss, *Mol. Cell. Proteomics*, 2009, **8**, 558.
41. K. Spagou, I. D. Wilson, P. Masson, G. Theodoridis, N. Raikos, M. Coen, E. Holmes, J. C. Lindon, R. S. Plumb, J. K. Nicholson and E. J. Want, *Anal. Chem.*, 2011, **83**, 382.
42. X. Cai, J. Dong, L. Zou, X. Xue, X. Zhang and X. Liang, *Chromatographia*, 2011, **74**, 391.
43. H. G. Gika, G. A. Theodoridis, U. Vrhovsek and F. Mattivi, *J. Chromatogr. A*, 2012, **1259**, 121.
44. W. Lu, M. F. Clasquin, E. Melamud, D. Amador-Noguez, A. A. Caudy and J. D. Rabinowitz, *Anal. Chem.*, 2010, **82**, 3212.
45. P. Kiefer, N. Delmotte and J. A. Vorholt, *Anal. Chem.*, 2010, **83**, 850.
46. L. Coulier, R. Bas, S. Jespersen, E. Verheij, E. van der Werf and T. Henkemeier, *Anal. Chem.*, 2006, **78**, 6573.
47. M. T. Matyska, J. J. Pesek, J. Dudley, M. Zamzami and S. M. Fischer, *J. Sep. Sci.*, 2010, **33**, 930.
48. T. Bamba, J. W. Lee, A. Matsubara and E. Fukusaki, *J. Chromatogr. A*, 2012, **1250**, 212.
49. J. W. Lee, T. Uchikata, A. Matsubara, T. Nakamura, E. Fukusaki and T. Bamba, *J. Biosci. Bioeng.*, 2012, **113**, 262.
50. T. Bamba, N. Shimonishi, A. Matsubara, K. Hirata, Y. Nakazawa, A. Kobayashi and E. Fukusaki, *J. Biosci. Bioeng.*, 2008, **105**, 460.
51. N. Loftus, K. Miseki, J. Iida, H. G. Gika, G. Theodoridis and I. D. Wilson, *Rapid Commun. Mass Spectrom.*, 2008, **22**, 2547.
52. H. G. Gika, G. A. Theodoridis, M. Earll, R. W. Snyder, S. J. Sumner and I. D. Wilson, *Anal. Chem.*, 2010, **82**, 8226.

53. S. Forcisi, F. Moritz, B. Kanawati, D. Tziotis, R. Lehmann and P. Schmitt-Kopplin, *J. Chromatogr. A*, 2013, **1292**, 51.
54. T. Zhang, D. J. Creek, M. P. Barrett, G. Blackburn and D. G. Watson, *Anal. Chem.*, 2012, **84**, 1994.
55. M. Katajamaa, J. Miettinenand and M. Orešič, *Bioinformatics*, 2006, **22**, 634.
56. A. Lommen, *Anal. Chem.*, 2011, **81**, 3079.
57. C. A. Smith, E. J. Want, G. O'Maille, R. Abagyan and G. Siuzdak, *Anal. Chem.*, 2006, **78**, 779.
58. M. Katajamaa and M. Oresic, *J. Chromatogr. A*, 2007, **1158**, 318.
59. Scripps Center for Metabolomics, *METLIN: Metabolite and Tandem MS Database*, http://metlin.scripps.edu/ (last accessed 10 June 2013).
60. Lipidomics Gateway, *LIPID MAPS: Advancement and Innovation in Lipidomics*, http://www.lipidmaps.org/ (last accessed 10 June 2013).
61. D. S. Wishart, C. Knox, A. C. Guo, R. Eisner, N. Young, B. Gautam, D. D. Hau, N. Psychogios, E. Dong, S. Bouatra, R. Mandal, I. Sinelnikov, J. Xia, L. Jia, J. A. Cruz, E. Lim, C. A. Sobsey, S. Shrivastava, P. Huang, P. Liu, L. Fang, J. Peng, R. Fradette, D. Cheng, D. Tzur, M. Clements, A. Lewis, A. De Souza, A. Zuniga, M. Dawe, Y. Xiong, D. Clive, R. Greiner, A. Nazyrova, R. Shaykhutdinov, L. Li, H. J. Vogel and L Forsythe, *Nucleic Acids Res.*, 2009, **37**, D603.
62. Genome Alberta, *Human Metabolome Data Base (HMDB)*, http://www.hmdb.ca/ (last accessed 10 June 2013).
63. Royal Society of Chemistry, *ChemSpider*, http://www.chemspider.com/ (last accessed 10 June 2013).

CHAPTER 5

Gas Chromatographic Techniques in Metabolomics

CHRISTIAN J. WACHSMUTH, FRANZISKA C. VOGL, PETER J. OEFNER AND KATJA DETTMER*

Institute of Functional Genomics, University of Regensburg, Josef-Engert-Strasse 9, 93053 Regensburg, Germany
*Email: katja.dettmer@klinik.uni-regensburg.de

5.1 Introduction

Gas chromatography coupled to mass spectrometry (GC-MS) is an indispensable tool in metabolomics. It is perfectly suited for the analysis of small volatile metabolites. A basic prerequisite for GC analysis is the transfer of the analytes into the gas phase without thermal decomposition or rearrangement. This is often facilitated by precolumn chemical derivatization, as many metabolites are polar compounds containing functional groups that impede their GC analysis in native form.

Capillary GC is capable of resolving very complex mixtures and MS is a universal detection technique providing highly selective and sensitive detection of metabolites along with their identification. Furthermore, mass spectral deconvolution can be applied to distinguish partly coeluting compounds. Capillary GC can be easily interfaced with an MS instrument as carrier gas flow rates of about 1 mL min^{-1} are compatible with modern MS vacuum systems. Moreover, analytes have already been transferred into the gas phase before reaching the ion source of the MS. Ionization is commonly performed under high-vacuum conditions using electron ionization (EI).

RSC Chromatography Monographs No. 18
Chromatographic Methods in Metabolomics
Edited by Tuulia Hyötyläinen and Susanne Wiedmer

Published by the Royal Society of Chemistry, www.rsc.org

GC-MS is ideally suited both for metabolite profiling and for metabolic fingerprinting. For metabolite profiling, *i.e.*, the quantitative analysis of pre-selected metabolites, GC-MS methods are tailored for the separation and sensitive detection of the target analytes. Reference substances in known concentrations and an appropriate set of internal standards are used to construct calibration curves to quantify the target analytes in biological specimens. Alternatively, isotope dilution analysis can be employed for quantification. Stable isotope-labelled analogues of the target analytes are spiked into the samples in known concentration and quantification is performed by comparing the peak area of the target analyte with that of the labelled standard.

For metabolic fingerprinting, the complete GC-MS chromatogram is exported and analysed. One can either search for signals that change significantly between sample groups or use the chromatograms to classify samples. This does not require knowledge of the identity of the signals, often referred to as features or markers. Metabolite identification is then restricted to those features that yielded significant differences in signal intensity between sample groups.

Aiming at the simultaneous detection of the highest possible number of metabolites, sample preparation is typically kept to the minimum in metabolite fingerprinting in order to prevent analyte losses. In targeted studies, on the other hand, sample preparation is tailored to the specific analytes and, thus, typically more extensive to reduce the chemical complexity of the samples prior to analysis. In order to minimize variability introduced during sample preparation, standardized protocols for quenching, extraction and derivatization steps are essential and should be followed as proposed by Sumner *et al.* in 2007.[1] Still, it is almost impossible to avoid analyte losses during sample preparation owing to the high chemical diversity of metabolites. Therefore, stable isotope-labelled internal standards (ISs) are commonly used to account for analyte losses.

This chapter presents the principles of GC-MS methodologies, including novel approaches, such as comprehensive two-dimensional GC-MS [GC×GC-time-of-flight (TOF) MS] and GC coupled to high-resolution MS. The main emphasis is placed on metabolite identification strategies.

5.2 Sample Preparation

5.2.1 Analysis of Volatile Metabolites

Headspace (HS) GC-MS, which exploits differences in volatility between analytes and the sample matrix, is a convenient means of on-line separation of volatile analytes from non-volatile compounds. Developments and trends in HS-GC-MS were reviewed by Snow and Slack in 2002.[2]

In the case of static HS-GC, the sample (liquid or solid) is placed in a gas-tight vial and an aliquot from the headspace above the sample is taken by means of a gas-tight syringe and injected into the GC column. Volatile metabolites go into the headspace of the vial according to their temperature-dependent partition coefficient K, which is the ratio of the concentrations of analyte in the sample phase to that in the gas phase. Partitioning of analytes into the gas phase can be increased by several means, such as incubating the sample vial at higher temperatures, adding inorganic salts or increasing the sample volume within the vial. Sampling is usually performed after equilibrium between the gas phase and the sample phase has been reached. Nevertheless, sampling can be performed under non-equilibrium conditions, but the concentration of the analyte in the gas phase is lower and a higher variability can arise. In 1999, Wahl *et al.* employed HS-GC for the analysis of volatile metabolites in urine samples, resulting in the identification of 34 volatile metabolites.[3]

In the case of extremely low analyte concentrations, dynamic HS-GC can be used. Dynamic HS-GC is an exhaustive extraction step, whereas static HS-GC is based on the establishment of equilibrium. Dynamic HS-GC, which is also known as 'purge and trap,' was first described in 1973.[4] The sample is continuously purged with carrier gas to strip out volatile analytes, which are subsequently trapped on a solid sorbent. Finally, the analytes are desorbed from the trap using a thermal desorption unit, which is in-line with the gas chromatograph.

HS-GC can be combined with an enrichment step such as solid-phase microextraction (SPME). SPME was first introduced by Zhang and Pawliszyn in 1993[5] and is more and more frequently used to study the volatile fraction of the metabolome. HS-SPME-GC was extensively reviewed by Mills and Walker in 2000 for a range of matrices,[6] such as blood and urine. SPME uses a fused-silica fibre that is coated with a stationary phase. The fibre is housed in a hollow needle for protection. It can be immersed in a liquid sample or placed in the headspace of the sample. Adsorbed analytes are subsequently desorbed in the hot GC injector. A simple scheme of HS-SPME-GC sampling is shown in Figure 5.1. Extraction is based not only on equilibrium distribution between the sample and the headspace but also the headspace and the stationary phase on the fibre. The proportion of metabolites in the sample and headspace compartment is affected by the phase ratios, temperature and analyte–matrix interactions. This requires calibration for quantitative analysis and both the extraction time and temperature have to be carefully optimized beforehand. A number of fibre coatings with different polarity are available, enabling selectivity tuning of the method. Commercial coatings include polydimethylsiloxane, poly(ethylene glycol), polyacrylate, polydimethylsiloxane–divinylbenzene, carboxen–polydimethylsiloxane and divinylbenzene–carboxen–polydimethylsiloxane.

HS-SPME provides a simple and inexpensive way of separating volatile compounds from non-volatile matrix compounds. Furthermore, derivatization of analytes for HS-SPME-GC-MS can be performed.[6] This is accomplished either by adding the derivatization reagent to the sample matrix, derivatization on the SPME fibre by coating it with the reagent or by derivatization in the GC injection port.[6] Liebich *et al.* used trimethyloxonium tetrafluoroborate as a methylating agent to derivatize organic acids directly in urine samples followed by immersion SPME extraction using a polyacrylate phase and GC analysis.[7] HS-SPME-GC-MS can be integrated into an automated workflow and,

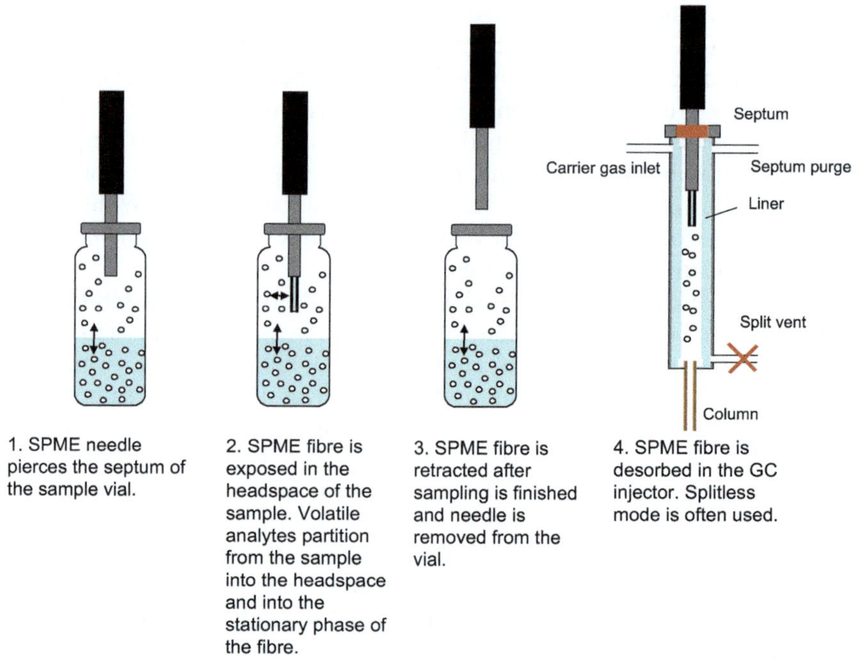

Figure 5.1 Simple scheme of HS-SPME sampling for GC analysis.

therefore, is widely used in the analysis of foodstuffs[8,9] and homogenized leaf samples.[10] SPME has also been employed successfully in combination with GC×GC-TOFMS for the analysis of volatiles in apples in both headspace and immersion modes[11] and for the analysis of wine.[12,13]

5.2.2 Derivatization of Semi- and Non-volatile Metabolites

Major prerequisites for GC-MS analysis include a sufficient vapour pressure and thermal stability of metabolites. Most metabolites, such as amino acids, organic acids, sugars and alcohols, possess polar functional groups, which have to be derivatized prior to GC analysis. Apart from increasing the volatility and thermal stability, derivatization of the analytes can improve the chromatographic properties. It can provide additional structural information and enhance sensitivity and selectivity for quantitative measurements. However, derivatization prolongs the overall analysis time and adds variance to the analysis. It can also hamper the identification of metabolites if derivatives are not contained in mass spectral libraries. Overall, derivatization should be simple, fast, automatable, efficient and reproducible, and spectral libraries of the derivatives are needed. Halket and Zaikin reported extensively on derivatives for MS in a series of reviews, which in each case dealt with a single derivatization reaction, *e.g.*, silylation.[14] Among others, silylation, acylation, alkylation/esterification, the formation of cyclic derivatives and generation of mixed derivatives of polyfunctional compounds have been performed for GC-MS. Liu *et al.* in 2001 employed acylation for HS-SPME-GC-MS analysis of amphetamines in hair.[15] Esterification to form fatty acid methyl esters (FAMEs) is the preferred method for the analysis of fatty acids and was used by Gharaibeh and Voorhees in 1996 to classify bacteria based on their fatty acid profile.[16] Derivatization reagents containing fluorine atoms are particularly suited for the determination of trace levels of amino acids[17] and aromatic amines[18] by GC coupled to electron capture negative ion MS.

In 2011, Koek *et al.*[19] reviewed a large number of metabolomics studies with respect to the derivatization strategy pursued. The majority of GC(-MS) methods employed silylation with or without prior oximation. Oximation reagents, such as hydroxylamines and alkoxyamines, react with aldehyde and keto groups. They react with open-chain reducing sugars, thereby preventing ring formation of reducing sugars. The derivatization results in the formation of two stereoisomers (*syn* and *anti* forms), which can be separated chromatographically (*e.g.*, see the glucose signal in Figure 5.2). Further, decarboxylation of α-ketocarboxylic acids and keto–enol tautomerism are hindered. *N*-Methyl-*N*-trimethylsilyltrifluoroacetamide (MSTFA) and *N*,*O*-bis(trimethylsilyl)trifluoroacetamide (BSTFA) are typically used reagents for silylation. They

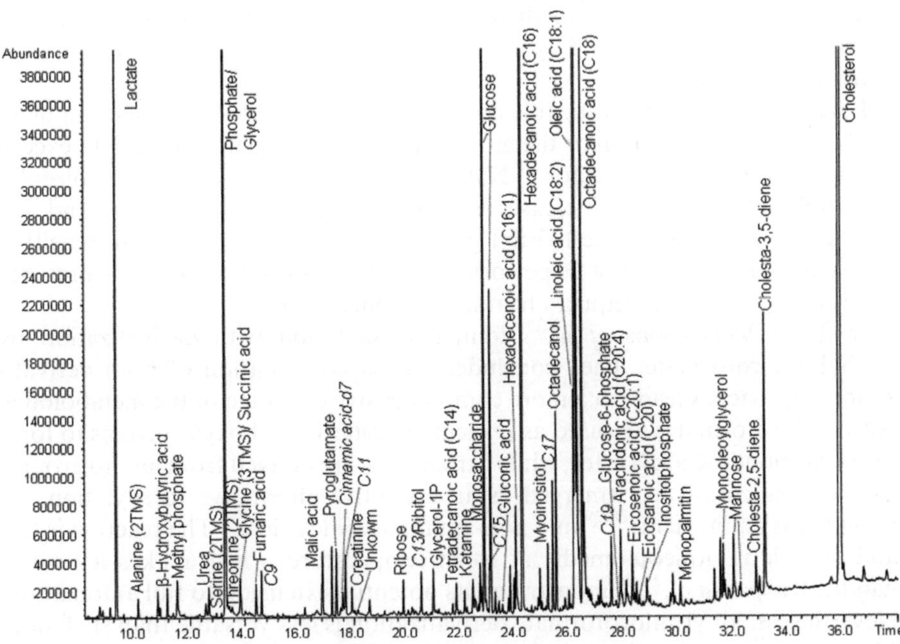

Figure 5.2 GC-qMS (quadrupole MS) of a liver tissue sample. Liver tissue (20 mg) from a male BALB/c mouse fed a non-alcoholic steatohepatitis (NASH)-inducing diet (15% lard, 1.25% cholesterol, 0.5% cholate) was homogenized and extracted using methanol–chloroform, and the extract was dried, derivatized (methoximation/silylation) and analysed by GC-qMS. Signals marked in italics refer to internal standards.

replace active hydrogen atoms in functional groups, such as –COOH, –OH, –NH and –SH, with a trimethylsilyl (TMS) group. In the case of silylation reagents that contain chlorine atoms, small amounts of a catalyst, *e.g.*, 1% of trimethylchlorosilane, are added. Silyl derivatives are prone to hydrolysis in the presence of traces of water. Consequently, anhydrous conditions have to be employed. Bulkier silylation reagents, such as *N*-methyl-*N*-(*tert*-butyldimethylsilyl)trifluoroacetamide (MTBSTFA), have been shown to be less susceptible to hydrolysis.[20] Moreover, *tert*-butyldimethylsilyl derivatives show characteristic $[M - 57]^+$ ions in EI-MS, which may benefit the identification of unknown compounds,[21] but harsher reaction conditions are required.

Oximation followed by silylation is often employed in non-targeted metabolomics[22,23] because of the ready derivatization of many different functional groups and the availability of large spectral libraries.[19] However, degradation or rearrangement reactions have been reported, *e.g.*, arginine is decomposed to form ornithine[24] and glutamate rearranges to pyroglutamate. Partial silylation yielding multiple peaks for amino acids is a well-known problem, which can be addressed by either weighting the multiple peaks correctly[25] or by summing the responses of all derivatives.[26]

Modern GC instruments equipped with advanced sample robots (*e.g.*, from CTC or Gerstel) permit the sequential automated derivatization of sample analytes immediately prior to their analysis instead of off-line *en bloc* derivatization, thereby minimizing hydrolysis and variability in the ratio of partially derivatized metabolites.

Large numbers of derivatizable groups may lead to indefinite retention of an analyte on the GC column or the molecular mass of the derivative might exceed the mass range of the detector.[14] Native lactose, for example, has a molecular mass of 342 Da, which increases to 947 Da after methoximation and the introduction of eight trimethylsilyl groups. Furthermore, silylation is neither simple nor fast as it cannot be performed in aqueous solution or in the presence of protic solvents and requires heating for some time.

In 2011, Villas-Bôas *et al.*[27] compared silylation with derivatization by methyl chloroformate. They concluded that a combination of both derivatization approaches leads to a more comprehensive coverage of the metabolome. Alkyl chloroformates are used as reagents to esterify carboxylic groups to form the corresponding alkyl ester, while carbamates are derived from amino groups. This derivatization has emerged as an attractive alternative to silylation, especially for amino acids[28,29] and non-amino organic acids.[30] The reaction is fast and feasible in aqueous media at room temperature, which makes it easy to handle. Kaspar *et al.*[28] demonstrated its potential extension to full automation. Nevertheless, alkyl chloroformate derivatization is less versatile than silylation. Moreover, comprehensive mass spectral libraries are still lacking and need to be developed, which renders methoximation/silylation overall the method of choice in GC-MS-based metabolomics.

Interestingly, in contrast to the study by Koek *et al.* in 2006,[31] Villas-Bôas *et al.*[27] reported a poor analytical performance of TMS derivatives. This underscores the importance of strictly enforced standard operating procedures such as the regular replacement of GC inlet liners and trimming of the inlet end of the capillary column, and also the addition of several stable isotope-labelled ISs to monitor the performance of extraction and derivatization.[31] ISs labelled with ^{13}C, ^{15}N or ^{2}H exhibit physicochemical properties almost identical with those of their unlabelled analogues. Aside from their use in the calculation of retention indices and the calibration of the mass scale in high-resolution MS, they are routinely employed to ensure accurate and reproducible metabolite quantification.[22,23] In their plant metabolomics studies, Gullberg *et al.*[32] included 11 stable isotope-labelled ISs representing different compound classes and optimized their extraction and derivatization protocols using a design of experiment approach. Koek *et al.*[31] distinguished three performance classes of silylated metabolites and recommended the inclusion of representative compounds from each class, preferably labelled with stable isotopes, for method optimization and validation. However, stable isotope-labelled ISs are expensive and their availability is limited. In addition, they can complicate several data processing steps, such as deconvolution, peak picking and integration, especially in the case of GC×GC-MS.[33] Lien *et al.*[34] employed deuterated MSTFA as a derivatizing reagent (MSTFA-d_9) for a mixture of standard compounds to overcome the limited number of available stable isotope-labelled ISs. This deuterated standard mixture was then spiked into MSTFA-derivatized samples prior to GC-MS/MS analysis. However, this strategy does not allow recognition of loss of metabolites during sample preparation and derivatization. Alternative approaches for overcoming the lack of commercially available stable isotope-labelled ISs include the use of MCF-d_3/MeOH-d_3 to generate stable isotope-labelled methyl chloroformate (MCF) derivatives[35] and their *in vivo* biosynthesis by feeding microorganisms isotopically labelled substrates.[36] However, such approaches will cover only parts of more complex metabolomes such as those of plants and mammals.

5.3 Principles of GC and MS

The following textbooks provide an in-depth account of the fundamentals of GC-MS techniques and readers are advised to consult them for details beyond the scope of this chapter:

- H.-J. Hübschmann, *Handbook of GC/MS: Fundamentals and Applications*, Wiley-VCH, Weinheim, 2008.
- M. McMaster, *GC/MS: a Practical User's Guide*, Wiley, Hoboken, NJ, 2011.
- J. H. Gross, *Mass Spectrometry: a Textbook*, Springer, Heidelberg, 2011.
- J. T. Watson and O. D. Sparkman, *Introduction to Mass Spectrometry: Instrumentation, Applications and Strategies for Data Interpretation*, Wiley, Chichester, 2007.

GC coupled to MS is a major tool in metabolomics research owing to the high chromatographic resolution of GC and the high selectivity and sensitivity of MS. MS is ideally suited for the identification of metabolites and provides an additional dimension for the separation of compounds with similar chromatographic retention behaviour.

For GC separation, analytes have to be transferred into the gas phase, which is typically achieved by a hot split/splitless injector or by a programmed-temperature vaporizer (PTV). With the latter, the sample can be injected into a cold injector, which is then rapidly heated using a defined temperature programme. Analytes are evaporated according to their vapour pressure, reducing temperature stress for thermally labile compounds. Hot injection in the split/splitless mode is easy to handle and segregation of analytes and matrix is achieved, as non-volatile matrix components remain in the injection port. To prevent sample overload, the amount of sample transferred on to the column can be controlled by the split ratio.

Nowadays, GC separation is commonly performed using capillary columns. These are open tubular columns of length 1–100 m and <1 mm i.d. The stationary phase is coated on the inner wall of the column. It can be either a highly viscous liquid (WCOT – wall-coated open-tubular column), with analyte separation attained by partitioning of the analytes into the liquid (GLC – gas–liquid chromatography) or a solid adsorbent (PLOT – porous layer open-tubular column) for adsorption-based separation (GSC – gas–solid chromatography). GSC is mainly used for highly volatile analytes such as permanent gases. In metabolomics, GLC prevails. A variety of stationary phases (SPs) are available. Non-polar SPs provide separation of analytes as a function of differences in vapour pressure, whereas polar SPs allow the exploitation of additional polar interactions. Common SPs are polysiloxanes, which are modified with both non-polar methyl groups (PDMS – polydimethylsiloxane) and polar phenyl- or cyanopropyl groups to tune the polarity of the phase. Polyethylene glycol (Carbowax) is employed in the separation of very polar analytes. The recently introduced ionic liquid SPs are an attractive alternative to polysiloxane SPs. Ionic liquids are a class of organic solvents with a very low melting point that include organic cations and either inorganic or organic counterions. Their unique properties allow GC separations over extended temperature ranges and result in longer column lifetimes due to higher resistance to moisture and oxygen. Furthermore, they offer unique selectivity for polar compounds and they can be considered as truly orthogonal to PDMS phases. However, owing to the huge chemical diversity of metabolites, there is no single ideal column chemistry capable of covering of entire metabolomes. Rather, multiple SP chemistries tailored to different classes of metabolites or their derivatives have to be employed sequentially to achieve comprehensive coverage.

In addition to SP type and film thickness, it is the column temperature that affects retention of analytes the most. The logarithm of the retention factor k ($\ln k$) is inversely proportional to the column temperature, T_C. Therefore, in most applications, the temperature of the column oven is ramped to analyse compounds that cover a wide (>100 °C) boiling point range. When setting up a temperature programme, the maximum operational temperature of the column has to be taken into account, otherwise degradation of the stationary phase can occur. To a minor extent, degradation of the stationary phase is a normal process known as column bleeding, which is more pronounced for thick-film or polar SPs. It is mainly visible as an elevated baseline at higher column temperatures, which is caused by the formation of linear and cyclic polysiloxanes in the case of siloxane-based stationary phases. In particular, oxygen in the carrier gas in combination with higher column temperatures will increase column bleeding. This background noise adversely affects the quality of mass spectra and limits of detection. Nowadays, most manufacturers offer low-bleed columns that are suitable for GC-MS.

Capillary GC lends itself to interfacing with MS, as analytes have already been transferred into the gas phase and the carrier gas flow rates are compatible with modern vacuum systems. Coupling of GC and MS is typically accomplished by means of electron ionization (EI), which is a hard ionization technique under high vacuum. Electrons are emitted from a heated wire filament, usually at an electron energy of 70 eV. In an initial step, radical cations are generated from gas-phase molecules. Subsequent fragmentation reactions, losses of neutral groups and rearrangement of secondary ions occur because the electron energies far exceed the ionization energies of organic molecules. This yields predictable ion patterns, which can be interpreted by the experienced scientist. Moreover, EI is highly reproducible and, therefore, huge commercial spectral libraries, currently containing ~213 000 unique compounds (NIST 11 Mass Spectral Library), have been built for the routine identification of analytes. However, structural isomers often show similar EI spectra, thus precluding unambiguous compound identification in the absence of additional information such as differences in chromatographic retention. Further, both the still incomplete coverage of natural metabolites in spectral libraries and the frequent absence of molecular ions in EI mass spectra continue to constitute major bottlenecks in the identification of group-discriminating features. Therefore, soft ionization techniques that preserve the quasi-molecular ions have become prominent in recent years, such as chemical ionization (CI) and atmospheric pressure chemical ionization (APCI). Operation in the CI mode takes place under low vacuum in the presence of a reagent gas, *e.g.*, methane, isobutane or ammonia. Electrons emitted from a heated wire filament initially ionize the reagent gas, which is present in great excess. Subsequently, the analytes are ionized by ion–molecule reactions. In the APCI source, on the other hand, plasma is formed from nitrogen gas in the high electric field generated at the tip of a corona-discharge needle operated at atmospheric pressure. In the presence of traces of water, formation of hydronium-ion water clusters occurs, which finally donate protons to target molecules that have higher gas-phase basicity than water to produce protonated molecular ions in the positive ionization mode.[37] The benefits of APCI for metabolite identification are shown in Figure 5.3, which depicts mass spectra of fructose-1MeOx-5TMS acquired with GC-EI-qMS and GC-APCI-TOFMS, respectively. Information on the entire molecule is only preserved in the latter case. However, commercial spectral libraries are not available yet for soft ionization techniques.

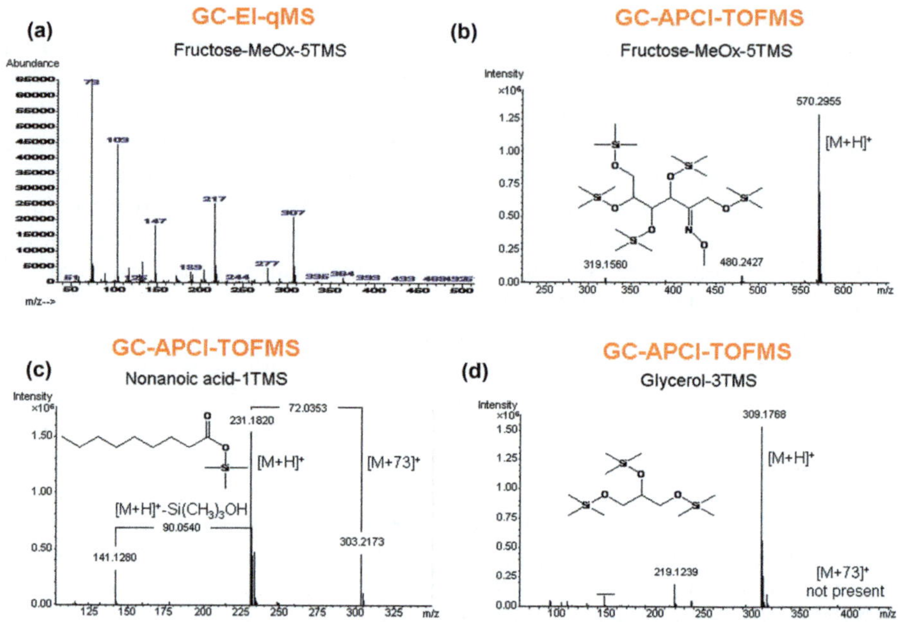

Figure 5.3 GC-MS traces from standards (62.5 μM) using hard and soft ionization techniques as described elsewhere.[23] Identification of fructose-1MeOx-5TMS requires spectral libraries for GC-EI-qMS (a) or can be achieved starting from the quasi-molecular ions in the case of GC-APCI-TOFMS (b), where structure-dependent adduct formation may serve as a plausibility check as exemplified for (c) nonanoic acid-1TMS and (d) glycerol-3TMS, with latter showing no adduct formation as it lacks a carboxylic group (MeOx = group introduced by methoximation).

Recently, Wachsmuth *et al.*[23] reported that APCI was superior to CI with regard to metabolite identification of phosphorylated compounds and sugars, which either fail to form or form only to a minor extent quasi-molecular ions in the CI mode. In contrast, the $[M+H]^+$ ions almost always represented the most abundant mass signal in the APCI spectra of 43 compounds from various chemical classes and core metabolic pathways.[23] Wachsmuth *et al.*[23] further reported on a silylation adduct, which was formed only in the case of compounds with a carboxylic acid group, and thus, helped to strengthen a putative identification, *e.g.*, nonanoic acid-1TMS (Figure 5.1c) showed adduct formation as opposed to glycerol-3TMS (Figure 5.1d), which lacks a carboxylic group. However, APCI still suffers from poor reproducibility and awaits further improvement.

The appropriate choice of mass analyser depends on the aim of the study. Mainly quadrupole (qMS) and time-of-flight mass spectrometry (TOFMS) are applied to metabolomics. Quadrupole mass analysers enjoy widespread use, mostly because they are less expensive, but also robust and easy to handle. However, they are limited in mass range and scan speed and in most cases restricted to nominal mass resolution. Therefore, quadrupoles are commonly used in routine analysis, either in full-scan mode or selected ion monitoring (SIM) mode, with the latter achieving lower limits of quantification due to a longer dwell time on preselected masses. Even more powerful is the sequential combination of three quadrupoles, which offers highly selective and sensitive quantitative analysis. In addition, structural information on the analytes is provided as an asset by tandem mass spectrometry (MS/MS). TOFMS instruments equipped with time-to-digital (TDC) converters provide not only low limits of quantification but also wide dynamic ranges of up to four orders of magnitude. Fast-acquisition TOFMS instruments, which acquire up to 500 spectra per second, are a prerequisite for the successful resolution and integration of the very narrow chromatographic peaks obtained by fast GC or GC×GC. Typically, however, they provide only nominal mass resolution. On the other hand, TOFMS instruments that are operated at acquisition rates of 20–50 spectra per second provide enhanced mass spectral resolution and accuracy of ion masses, which facilitates the identification of unknown compounds. However, TOFMS instruments are sensitive to voltage or temperature fluctuations requiring both external and internal (re-)calibration of the mass scale on a regular basis. Even better suited for the identification of unknowns are hybrid qTOFMS instruments that combine quadrupoles with a TOF mass analyser. This type of MS instrument permits the determination of parent and fragment ions along with their isotopic pattern with high mass accuracy.

5.4 GC×GC-MS

A number of different measures are employed to characterize chromatographic separation power and column efficiency. The most widely used measure of column efficiency is the plate count (N). However, it is only valid for a given model substance under isothermal conditions. In 1967, Giddings proposed the concept of peak capacity.[38] Peak capacity is defined as the maximum number of peaks that can be separated on a given column in a defined retention time window with a fixed resolution. For temperature-programmed GC, peak capacity is calculated by dividing the separation time $t_l - t_f$ (t_l = retention time of the last peak, t_f = retention time of first peak equalling the column void time) by the average base peak width (4σ criterion) assuming that the peak width remains constant over the separation time. For example, a separation time of 30 min and a peak width of 9 s would result in a peak capacity of 200. However, the peak capacity is a theoretical value assuming that the peaks are evenly distributed across the chromatogram. Davis and Giddings showed that peak resolution is impaired if the number of analytes exceeds 37% of the peak capacity.[39] They further demonstrated that the peak capacity must be 100 times higher than the number of analytes to resolve 98% of randomly distributed analytes. That means to resolve 200 metabolites, a peak capacity of 20 000 is required, resulting in a hypothetical separation time (peak width 9 s) of about 3000 min. Therefore, despite high chromatographic resolution, GC separations of complex metabolite mixtures will still suffer from coelution and peak overlap. Figure 5.2 shows an example GC-MS analysis of a methoximated and silylated murine liver tissue extract.

One means of increasing chromatographic resolution and peak capacity is multidimensional gas chromatography (MDGC), whereby coeluting peaks are transferred on to a second column of different selectivity (polarity).[40] For MDGC (GC-GC), two columns of different polarity but conventional dimensions (*e.g.*, 30 m×0.25 mm i.d.) are coupled by a valve (Deans switch) to transfer selected bands from the first column on to the second column. GC-GC provides an additive increase of the peak capacity, which depends on the number of cuts performed ($n_{total} = n_1 + n_2 + n_3 \ldots$). However, only a few cuts can be performed per run. The leap from GC-GC to comprehensive two-dimensional GC (GC×GC) was realized by Liu and Phillips in 1991,[41] who analysed all components of a sample in two dimensions within one run. For a true comprehensive two-dimensional separation, the following criteria must be met:[42]

- The complete sample is analysed in two dimensions.
- The separation in the two dimensions should be orthogonal, *i.e.*, of different selectivity.
- The resolution obtained in the first dimension is maintained.
- All sample components are detected.

This results in a multiplicative increase in theoretical peak capacity ($n_{total} = n_1 \times n_2$).

A GC×GC system comprises two columns with orthogonal separation characteristics positioned in series with an interjacent transfer device called a modulator.[43] Typically, an apolar column is used in the first dimension to separate analytes based on their vapour pressure, and a polar column is used in the second dimension to exploit polar interactions, *e.g.*, hydrogen bridge linkages or dipole–dipole interactions, of analytes with the stationary phase.[44] However, inverse column setups have also been reported.[45,46] To preserve the resolution achieved in the first column, a modulator is used to focus, fraction and transfer eluting analytes on to the secondary column. First dimension peak widths are typically in the range 6–25 s. As 3–4 modulations are required per peak to maintain first-dimension resolution, modulation times range from 2 to 8 s.[47] This necessitates short, narrow second-dimension columns to ensure a complete separation of the analytes of one fraction before the next is transferred on to the column. Otherwise, analytes from different modulation cycles can coelute in the second dimension, which is known as wrap-around.[48]

Modulators can be grouped into thermal and flow modulators. The thermal sweeper was one of the first modulators. The eluate from the primary column is trapped on a short thick-film capillary, which is placed between the two columns. By means of a rotating slotted heater, the analytes are mobilized and swept on to the secondary column.[48,49] A further development was the longitudinally modulating cryogenic system (LMCS). The LMCS uses a moving modulator, which traps and focuses the analytes at the head of the secondary column using expanding liquid carbon dioxide. Thereafter, the analyte fractions are released by removing the modulator and reheating the column section using the column oven temperature.[49,50] Today, stationary cryogenic jets are typically employed for focusing the analytes eluting from the first dimension.[48] The analytes are released either by the oven temperature or by hot jets. Such cryogenic modulators are available as dual-stage setups with two cold jets or four jets, the latter comprising two cold jets, using cold N_2 or expanding liquid or gaseous CO_2, and two hot jets, using hot air. With the dual-stage setup one cold jet is always on to prevent analyte breakthrough during hot mobilization.

Differential flow modulation represents the second type of modulators. It uses a six-port valve and a secondary flow controller for the modulation process. There are two configurations of this modulation valve, one for loading the sample loop with an analyte fraction from the primary column and the other for injecting this fraction on to the second column. The secondary flow controller regulates the gas flow of the secondary column.[51]

Modulation generates narrow peaks with widths of 50–200 ms in the second dimension. This peak focusing results in lower limits of detection and quantification compared with one-dimensional GC.[23]

Fast acquisition detectors are required to map second-dimension peaks adequately. For example, a peak width of 50 ms requires a data acquisition rate of 300 data points per second in order to acquire 15 data points across a peak. Modern generation flame ionization detection (FID) and electron-capture detection (ECD) are suitable in addition to MS, which is the preferred detection method in metabolomics. Fast acquisition TOFMS offers data acquisition rates of up to 500 Hz. Alternatively, for limited scan ranges, fast scanning quadrupole mass spectrometers may be used.[52]

GC×GC-MS data are commonly displayed as a two-dimensional contour plot (Figure 5.4b). Retention times of the first (*x*-axis) and second dimension (*y*-axis) span a two-dimensional plane, while the third dimension depicts the detector response in colour code rings. For the generation of a contour plot, the two-dimensional chromatogram is cut (Figure 5.4a) into single chromatograms, the lengths of which are defined by the modulation time. These individual chromatograms representing the separation on the second column are then stacked on each other to reconstruct the separation in the first dimension.[53] GC×GC-TOFMS chromatograms can be used for metabolic fingerprinting by exporting all detected features[22,53,54] or for the quantification of selected metabolites.[23,33] Applications of GC×GC-TOFMS to both metabolite fingerprinting and profiling and data analysis strategies were reviewed recently.[55]

Comprehensive GC×GC-TOFMS certainly constitutes a valuable tool for metabolomics owing to the multiplicative increase in peak capacity and enhanced detection sensitivity. However, for its full power to be exploited, further improvements are urgently needed. These include improved algorithms for spectral deconvolution, alignment and integration, and also stable isotope-labelled metabolite mixtures that provide a more comprehensive coverage of metabolomes of interest. A further aspect that warrants attention, although not limited to GC×GC-TOFMS, concerns batch-to-batch effects that are the

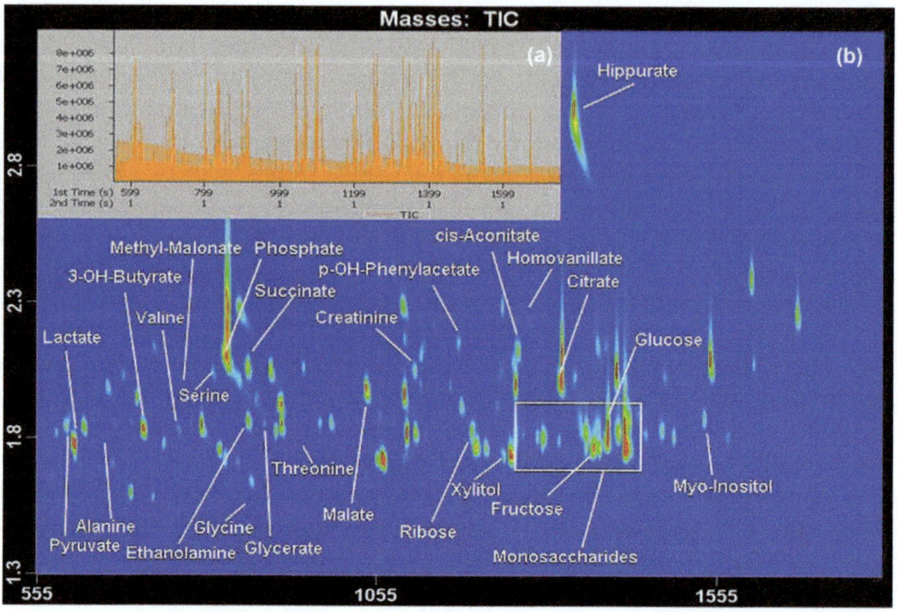

Figure 5.4 Analysis of a human urine specimen by GC×GC-TOFMS as described elsewhere.[22] Urease treatment and two-step derivatization had been applied prior to the analysis. Data analysis was performed with Leco ChromaTOF software version 4.44 as described.[22] Identification of annotated metabolites in (b) was based on spectral matching to the NIST 05 library using a cut-off score of 700 and comparing retention indices with those of authentic standards. (a) The unprocessed gas chromatogram includes in-line second-dimension chromatograms, from which the 3D contour plot is created by displaying the retention time of the first dimension on the *x*-axis and the retention time of the second dimension on the y-axis, and intensity is given as colour code.

result of analytical variables such as column ageing, shifts in retention time and/or decreasing detector response, which can severely impact biological conclusions drawn from the data.

5.5 Identification of Metabolites

Recent years have seen great progress in the identification of features, which is still considered one of the greatest bottlenecks in metabolomics (see http://fiehnlab.ucdavis.edu/staff/kind/Metabolomics-Survey-2009/). For targeted metabolomics, only very few peaks are of interest, so-called 'known unknowns' that have been either described in the literature or included in databases, but have not been reported for the organism and/or type of sample at hand. Their identification is typically accomplished by means of commercial standards. In semi-targeted or untargeted metabolomics, the use of reference compounds for the identification of features becomes cumbersome as the number of features recorded in a single run may easily extend into the hundreds. Therefore, routines for automatic feature annotation and identification of so-called 'unknown unknowns' are required. The rapid progress in MS technologies and resulting improvements in detection sensitivity have revealed a vast number of such compounds that still await positive identification. This holds especially true for secondary metabolites of plants and fungi, whose metabolomes have been estimated to contain tens of thousands of individual metabolites.[56] However, these numbers are only rough estimates.[57]

Sumner *et al.* in 2007[1] proposed four levels of metabolite identification depending on the amount of information collected on a feature of interest. They distinguished between positively (level 1) and putatively (level 2) identified compounds, compounds that are putatively assigned to a class of compounds (level 3) and unknown compounds (level 4). For positive identification (level 1), matching two independent parameters, such as retention index and mass spectrum, relative to an authentic standard is required, which emphasizes the need for chromatography as an additional dimension to MS. Several excellent reviews have appeared in recent years on advances in metabolite identification by means of chromatography and MS.[58–61]

5.5.1 Gas Chromatography

Separation of analytes in a time dimension constitutes added value to the identification process. GC-MS retention times are very stable and, therefore, of great benefit to identification. Typically, the retention time of a compound is given relative to a reference compound within the chromatogram to exclude variations due to slight instrument changes between consecutive analyses and to make retention times more comparable to those from other laboratories. In 1958, Kováts introduced the concept of calculating retention indices (RIs)[62] based on a series of alkanes as reference compounds for isothermal GC. This concept was later modified and applied to temperature-programmed GC by Van den Dool and Kraatz.[63] Temperature-programmed RIs are often referred to as linear retention indices. Retention indices are a normalized measure of retention time. They are calculated relative to compounds from a homologous series of alkanes or fatty acids that are added to each sample. In temperature-programmed GC, members of a homologous series show a linear increase in retention time with increasing number of carbon atoms for a given stationary phase. By definition, members of the homologous series are assigned a retention index by multiplying the respective carbon number by 100. The linear retention index I of an analyte x is calculated using the following equation:

$$I_x = 100n + 100\left[\frac{t_{R(x)} - t_{R(n)}}{t_{R(n+1)} - t_{R(n)}}\right]$$

where $t_{R(x)}$ = retention time of the analyte, $t_{R(n)}$ = retention time of the reference compound eluting before the analyte, $t_{R(n+1)}$ = retention time of the reference compound eluting after the analyte and n = carbon number of the reference compound eluting before the analyte.

Traditionally, *n*-alkanes have been used as RI markers,[64,65] although fatty acid methyl esters (FAMEs)[66] or any other homologous series can be used. Selection of the reference system depends on the column polarity, *e.g.*, *n*-alkanes are poorly dissolved in polar stationary phases, and the detector used, *e.g.*, *n*-alkanes do not ionize using an APCI source for GC-MS.

Retention indices are partly reported in MS databases, such as NIST, HMDB,[67] Golm Database[68] and Fiehn Metabolome Database.[66] Physicochemical properties of not yet annotated metabolites are collected in the Golm Database as mass spectral tags (MSTs), which were introduced recently[69,70] in analogy with expressed DNA sequence tags (ESTs). MSTs now await their structural elucidation.

For databases to be useful, they need to be updated regularly. Insufficient resources for database curation still limit the utility of HMDB and the Golm Database. Hence NIST is still the most widely used database. The latest version NIST 11 contains EI mass spectra for over 212 000 unique compounds and RI values are provided for more than 38 000 compounds. However, the majority of the entries do not refer to metabolites.

In the absence of RI values for metabolites of interest, they may be predicted. In 2011, Kumari *et al.*[71] introduced a workflow for re-identifying 29 trimethylsilylated metabolites from accurate mass GC-CI-TOFMS data. Elemental formulae were derived from the quasi-molecular ions in an initial step, followed by *in silico* derivatization and prediction of RIs using the NIST RI algorithm with a correction factor for TMS groups.[72] In a subsequent step, accurate mass EI spectra were predicted for remaining structures and scored against experimental spectra. As a result, the correct structure was obtained in 73% within the top five hits of the proposed candidates.[71] Earlier, Mihaleva *et al.*[73] made use of multiple linear regression (MLR) and support vector regression (SVR) models for RI prediction, which cut the hit lists obtained from matching experimental spectra of the NIST 05 MS library in half.

The concept of RIs has also been transferred to GC×GC,[74] but it is not used routinely. Nevertheless, GC×GC facilitates the identification of compounds belonging to different chemical classes as they occupy different regions within the 2D separation space. Hence structure–retention relationships serve as a plausibility check for the identification of unknown signals in the chromatogram. Quantitative structure–retention relationship (QSRR) models have also been developed to correlate molecular structures with their retention times.[75] In summary, retention times are essential in compound identification, especially in the case of isomers, which cannot be distinguished from their mass spectra alone.

5.5.2 Mass Spectrometry

Mass spectrometry is the method of choice in the identification of unknown metabolites. Matching EI fragment patterns to spectral libraries has traditionally been performed in MS-based metabolite identification. A second strategy involves the use of soft ionization techniques in conjunction with high-resolution MS. Elemental formulae are derived from the accurate mass and isotopic pattern of the quasi-molecular ions in an initial step, groups introduced by derivatization are removed and curated formulae can be fed into databases for a putative identification.

Successful identification of known unknowns based on mass spectral libraries requires mass spectra of high quality, powerful spectral matching algorithms to reveal significant degrees of spectral convergence and, particularly, comprehensive and reliable reference spectral libraries. Successful library identification depends strongly on the quality of the input spectrum. Spectral quality often suffers from poor deconvolution, ion source overload and/or varying acquisition parameters. Mass spectral deconvolution has to be performed in the case of mixed mass spectra originating from unresolved peaks. Local peak maxima are analysed for all ion traces and ions that share a peak maximum (retention time) are combined in one mass spectrum. This process gives a 'pure' mass spectrum that can be matched against reference spectra. Of course, deconvolution is only successful if overlapping peaks display a slight difference in peak maxima and if sufficient data points across the peak are available to distinguish these slight differences. Undoubtedly, deconvolution software packages, such as AMDIS supplied with the NIST library, ChromaTOF (LECO) and AnalyzerPro (SpectralWorks), constitute an added value for proper analysis of highly complex samples. However, deconvolution may generate outright false or impure spectra. In 2008, Lu *et al.*[76] evaluated the performance of AMDIS, ChromaTOF and AnalyzerPro using a mixture of 36 endogenous compounds, which exhibited varying chemical structures and concentrations. Numerous failures of AMDIS and ChromaTOF occurred, which resulted in incorrect identifications,[76] emphasizing the need for further improvements or novel approaches.

For spectral matching, several algorithms such as INCOS, probability-based matching (PBM) and dot-product scoring have been described.[77] In the case of INCOS and PBM, the number of mass/intensity pairs of library spectra considered for matching is restricted, which might result in failure to retrieve as first hit the correct library entry for a searched full spectrum of an unknown. The dot-product algorithm, in contrast, is first carried out on condensed spectra before finally considering entire spectra for definite metabolite identification. Stein and Scott in 1994[78] found dot-product scoring to be superior (75% accuracy) to four other algorithms (Euclidean distance, absolute value distance, PBM and Hertz *et al.* similarity index) based on matching test spectra against the NIST library, which contained 12 592 spectra of about 8000 unique compounds at that time.

The library search delivers a match factor, which is a measure of the agreement between search and library spectra. A NIST search delivers both a match factor (direct match) and a reverse match factor. The reverse search algorithm compares the library spectra against the search spectrum and non-matching peaks in the submitted spectrum are ignored. Match factors above 900 indicate an excellent match, with 999 being a perfect match. Match factors above 800 are still considered good, but match factors below 700 have to be treated with caution. In general, top candidates from the hit list should be checked for plausibility by considering information on retention indices or reference standards should be run for verification.

It is also possible to create custom libraries with commercial standards, as accomplished by Smart *et al.* for methyl chloroformate derivatives.[79] *In silico* mass spectral fragmentation routines can be used if reference spectra are not available. In 2009, Schymanski *et al.*[80] tested three commercial programs for *in silico* generation of mass fragments and concluded that the available software was not yet practical.

Alternatively, efficient and sensitive identification of metabolites may be accomplished by soft ionization techniques, accurate mass measurement and tandem MS. Although commonly implemented in LC-MS, exploitation of these features of modern MS equipment in GC-MS is a more recent development. McEwen and McKay in 2005[81] and Schiewek *et al.* in 2008[82] developed independently atmospheric pressure ion sources suited for operation in LC and GC modes. Several metabolomics studies using either GC-APCI-TOFMS[23,83,84] or GC-APCI-qTOFMS[85] have reported metabolites that had escaped previously identification using EI libraries.

GC-CI-TOFMS has also been employed for metabolite identification.[71,86] Abate *et al.*[86] reported that higher ion intensities improved accurate mass and isotopic abundance measurements and, therefore, allowed the determination of elemental formulae with higher confidence. In these GC-CI-TOFMS studies, high-mass tuning [86] and the careful selection of the reagent gas[71] proved critical in obtaining higher ion intensities.

Recent years have seen a flurry of technological advances in MS. Most relevant to metabolite identification are improvements in mass accuracy and resolution along with the capability to perform MS/MS experiments. Accurate mass measurements require both external and internal mass calibrations to correct for temperature or voltage drifts during TOFMS and qTOFMS analyses. For external calibration, the APCI source can be replaced with an ESI source and calibration can be performed using an ESI tune mix. Alternatively, FAMEs can be infused from the top of the APCI source at the beginning of each analytical run. However, post-acquisition recalibration of mass spectra is not as straightforward as in LC-MS. Internal standards that are added to each sample prior to analysis can be used for internal calibration as described by Wachsmuth *et al.*[23] Briefly, an average mass spectrum is created from regions within each chromatogram that correspond to the internal standards nonanoic acid, [U-^{13}C]fumarate, [$^{2}H_7$]-*trans*-cinnamate, undecanoic acid, [U-^{13}C]glucose and [U-^{13}C]lactose. The mass scale is then recalibrated according to the theoretical masses of fragment ions, quasi-molecular ions and adducts that originated from the standards. In total, 10 masses over a mass range from m/z 138.093 to 960.507 were used.[23] Nevertheless, high mass accuracy rarely suffices to achieve an unambiguous assignment of a single molecular formula to an unknown, especially in higher mass regions, where the number of elemental combinations for a given mass and the defined mass error is many times higher.

Therefore, isotopic pattern matching is required as an orthogonal filter to remove false candidate formulae.

Kind and Fiehn in 2006[87] demonstrated *via* querying over 1.6 million elemental formulae that a mass spectrum with 3 ppm mass accuracy and 2% error for isotopic abundance patterns outperforms a hypothetical mass spectrometer with 0.1 ppm mass accuracy that does not take isotope information into account.

Identification of unknowns still requires manual intervention. The analyst has to review the corresponding spectrum for proper selection of the quasi-molecular ion used to calculate an elemental formula. Accidentally subjecting masses of fragments or adducts to the elemental formula calculator may result in significant errors or an inconclusive database search. Once an elemental formula has been calculated, atoms introduced by derivatization have to be removed prior to database queries. Figure 5.5 summarizes the workflow for the identification of unknown metabolites in high-resolution GC-MS metabolite fingerprints of two *Escherichia coli* strains, which made use of the accurate mass measurements of GC-APCI-TOFMS to identify metabolites that differed significantly in abundance between the two strains.[23] As exemplified for *N*-acetylneuraminate-1MeOx-6TMS, putative annotation *via* calculation of elemental formulae and subsequent database search initially failed (step 3 on the left-hand side) because the feature extraction routine and statistical analysis (step 2) pointed to a fragment mass instead of the mass of the quasi-molecular ions. Manual review of the corresponding mass spectra finally led to positive identification (step 3 on the right-hand side). Nevertheless, several ion species within a single mass spectra can also be helpful, as reported by Wachsmuth *et al.*,[23] who observed the structure-dependent formation of a silylation adduct, which proved useful in excluding false putative database annotations.[23]

Tools for calculating elemental formulae from accurate mass measurements, *e.g.*, the SmartFormula tool from Bruker Daltonics, consider chemical and heuristic rules as proposed by Kind and Fiehn in 2007[88] in addition to comparing the isotopic patterns, which is assessed by the mSigma value (between 0 in the case of perfect match and 1000 in the case of no match). These rules include, among others, restrictions on the number and ratio of elements, and also LEWIS and SENIOR rules, and their application resulted in correct metabolite annotations with 98% probability for 6000 compounds in target databases.[88]

For calculating elemental formulae, the search space can be reduced by assuming a lower experimental mass error. However, the correct formula may be missed, whereas if a larger mass error is tolerated, hit lists become more extensive and are more likely to include hits that are closer to the experimental data than the correct formula. SmartFormula3D additionally relies on the accurate mass and isotope pattern of the fragments from a parent ion in MS/MS experiments. Accordingly, parent ions that cannot be related to the fragments are excluded, thus restricting the search space drastically up to a single remaining elemental formula in some cases.

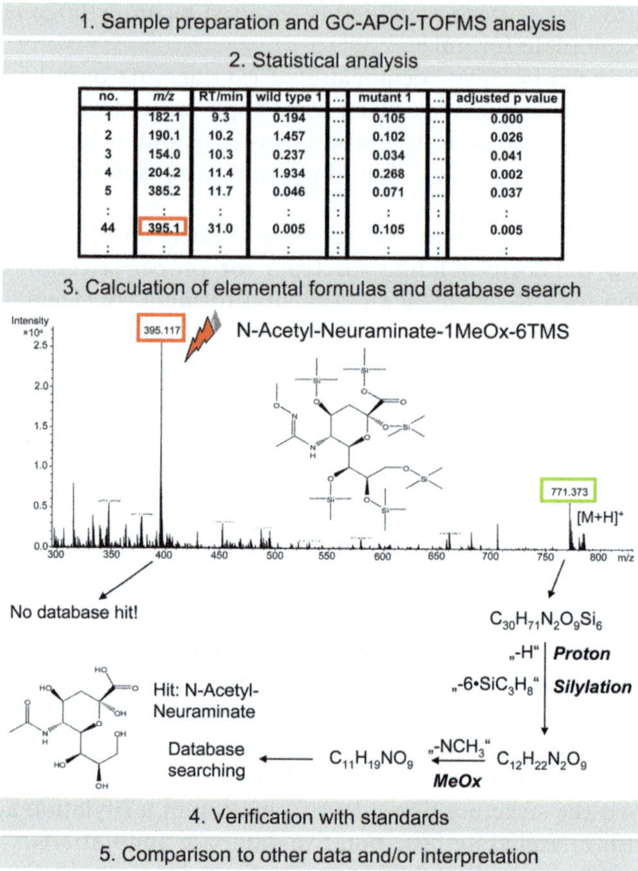

Figure 5.5 High-resolution GC-APCI-TOFMS workflow for the identification of unknown metabolites derived from a study on two *E. coli* strains.[23] The example of *N*-acetylneuraminate-1MeOx-6TMS emphasizes the need for manual review of mass spectra for the correct assignment of metabolites to features that statistical analysis had found to distinguish the two strains.

According to Sumner *et al.*,[1] two independent/orthogonal parameters, *e.g.*, retention index/mass spectrum or accurate mass in combination with MS/MS information, relative to a commercial standard are required to identify compounds positively. Whenever standards are not available, confidence in a putative identification can be increased by the use of isotope labelling to distinguish background ions from ions that originate from a true biological source. Furthermore, with the commercial availability of fraction collectors for GC, compound isolation for metabolite identification by NMR spectroscopy has become feasible. However, the stability of derivatized metabolites is critical while collecting fractions by cold trapping, and a huge discrepancy exists between amounts that are subjected to GC-MS analysis and those required for NMR analysis. Once a single or several metabolites have been identified, proper biological or medical interpretation is promoted by metabolic pathway databases such as the Kyoto Encyclopedia of Genes and Genomes (KEGG) database.[89]

5.5.3 Perspectives

Recent years have seen major developments in technologies, methodologies and software for the identification of unknown metabolites. Advances in detection sensitivity, resolution and accuracy of MS equipment along with soft ionization techniques that preserve the quasi-molecular ion have opened up new routes for feature annotation. Nevertheless, especially in comparison with the proteomics community,[90,91] validated standard operating procedures for the annotation of novel metabolites and open access to a central depository of metabolite information are still lacking. Moreover, integration of the various algorithms developed in recent years to facilitate metabolite identification[71,92] into a single suite is needed. The introduction of International Chemical Identifier (InChI) strings for consistent naming of identified metabolites in addition to supporting global data exchange formats, *e.g.*, the netCDF format, for sharing mass spectra and associated information constitute initial steps to cope with these problems. Still, greater community efforts are necessary to catch up with other 'omics' disciplines.

Acknowledgement

Funding by DFG (KFO 262) is gratefully acknowledged.

Authorship

C.J.W. drafted the entire book chapter except for the section on GC×GC-MS, which was written by F.C.V.; P.J.O. revised the manuscript; K.D. designed the outline of the chapter, contributed figures and revised the manuscript.

References

1. L. W. Sumner, *et al.*, Proposed minimum reporting standards for chemical analysis, *Metabolomics*, 2007, **3**(3), 211–221.
2. N. H. Snow and G. C. Slack, Head-space analysis in modern gas chromatography, *Trends Anal. Chem.*, 2002, **21**(9–10), 608–617.
3. H. G. Wahl, *et al.*, Analysis of volatile organic compounds in human urine by headspace gas chromatography–mass spectrometry with a multipurpose sampler, *J. Chromatogr. A*, 1999, **847**(1–2), 117–125.
4. A. Zlatkis, H. Lichtenstien and A. Tishbee, Concentration and analysis of trace volatile organics in gases and biological fluids with a new solid adsorbent, *Chromatographia*, 1973, **6**(2), 67–70.
5. Z. Zhang and J. Pawliszyn, Headspace solid-phase microextraction, *Anal. Chem.*, 1993, **65**, 1843–1852.

6. G. A. Mills and V. Walker, Headspace solid-phase microextraction procedures for gas chromatographic analysis of biological fluids and materials, *J. Chromatogr. A*, 2000, **902**(1), 267–287.
7. H. M. Liebich, E. Gesele and J. Woll, Urinary organic acid screening by solid-phase microextraction of the methyl esters, *J. Chromatogr. B*, 1998, **713**(2), 427–432.
8. E. Aprea, *et al.*, Metabolite profiling on apple volatile content based on solid phase microextraction and gas chromatography time of flight mass spectrometry, *J. Chromatogr. A*, 2011, **1218**(28), 4517–4524.
9. Y. Tikunov, *et al.*, A novel approach for nontargeted data analysis for metabolomics. Large-scale profiling of tomato fruit volatiles, *Plant Physiol.*, 2005, **139**(3), 1125–1137.
10. G. Weingart, *et al.*, Establishment and application of a metabolomics workflow for identification and profiling of volatiles from leaves of *Vitis vinifera* by HS-SPME-GC-MS, *Phytochem. Anal.*, 2012, **23**(4), 345–358.
11. S. Risticevic, J. R. DeEll and J. Pawliszyn, Solid phase microextraction coupled with comprehensive two-dimensional gas chromatography–time-of-flight mass spectrometry for high-resolution metabolite profiling in apples: implementation of structured separations for optimization of sample preparation procedure in complex samples, *J. Chromatogr. A*, 2012, **1251**, 208–218.
12. J. E. Welke, *et al.*, Characterization of the volatile profile of Brazilian Merlot wines through comprehensive two dimensional gas chromatography time-of-flight mass spectrometric detection, *J. Chromatogr. A*, 2012, **1226**, 124–139.
13. A. L. Robinson, *et al.*, Development of a sensitive non-targeted method for characterizing the wine volatile profile using headspace solid-phase microextraction comprehensive two-dimensional gas chromatography time-of-flight mass spectrometry, *J. Chromatogr. A*, 2011, **1218**(3), 504–517.
14. J. M. Halket and V. G. Zaikin, Derivatization in mass spectrometry – 1. Silylation, *Eur. J. Mass Spectrom.*, 2003, **9**, 1–21.
15. J. Liu, *et al.*, New method of derivatization and headspace solid-phase microextraction for gas chromatographic–mass spectrometric analysis of amphetamines in hair, *J. Chromatogr. B*, 2001, **758**(1), 95–101.
16. A. A. Gharaibeh and K. J. Voorhees, Characterization of lipid fatty acids in whole-cell microorganisms using *in situ* supercritical fluid derivatization/extraction and gas chromatography/mass spectrometry, *Anal. Chem.*, 1996, **68**(17), 2805–2810.
17. J. T. Simpson, *et al.*, Analysis of amino acids in biological fluids by pentafluorobenzyl chloroformate derivatization and detection by electron capture negative ionization mass spectrometry, *Anal. Biochem.*, 1996, **233**(1), 58–66.
18. M. Longo and A. Cavallaro, Determination of aromatic amines at trace levels by derivatization with heptafluorobutyric anhydride and gas chromatography–electron-capture negative-ion chemical ionization mass spectrometry, *J. Chromatogr. A*, 1996, **753**, 91–100.

19. M. M. Koek, *et al.*, Quantitative metabolomics based on gas chromatography mass spectrometry: status and perspectives, *Metabolomics*, 2011, **7**(3), 307–328.
20. C. Birkemeyer, A. Kolasa and J. Kopka, Comprehensive chemical derivatization for gas chromatography–mass spectrometry-based multi-targeted profiling of the major phytohormones, *J. Chromatogr. A*, 2003, **993**(1–2), 89–102.
21. O. Fiehn, *et al.*, Identification of uncommon plant metabolites based on calculation of elemental compositions using gas chromatography and quadrupole mass spectrometry, *Anal. Chem.*, 2000, **72**(15), 3573–3580.
22. M. F. Almstetter, *et al.*, Integrative normalization and comparative analysis for metabolic fingerprinting by comprehensive two-dimensional gas chromatography–time-of-flight mass spectrometry, *Anal. Chem.*, 2009, **81**, 5731–5739.
23. C. J. Wachsmuth, *et al.*, Performance evaluation of gas chromatography–atmospheric pressure chemical ionization–time-of-flight mass spectrometry for metabolic fingerprinting and profiling, *Anal. Chem.*, 2011, **83**(19), 7514–7522.
24. J. M. Halket, *et al.*, Chemical derivatization and mass spectral libraries in metabolic profiling by GC/MS and LC/MS/MS, *J. Exp. Bot.*, 2005, **56**(410), 219–243.
25. H. H. Kanani and M. I. Klapa, Data correction strategy for metabolomics analysis using gas chromatography–mass spectrometry, *Metab. Eng.*, 2007, **9**(1), 39–51.
26. G. Noctor, *et al.*, A comparative study of amino acid measurement in leaf extracts by gas chromatography–time of flight-mass spectrometry and high performance liquid chromatography with fluorescence detection, *Metabolomics*, 2007, **3**, 161–174.
27. S. G. Villas-Bôas, *et al.*, Alkylation or silylation for analysis of amino and non-amino organic acids by GC-MS?. *Metabolites*, 2011, **1**(1), 3–20.
28. H. Kaspar, *et al.*, Automated GC-MS analysis of free amino acids in biological fluids, *J. Chromatogr. B*, 2008, **870**(2), 222–232.
29. M. C. Waldhier, *et al.*, Comparison of derivatization and chromatographic methods for GC-MS analysis of amino acid enantiomers in physiological samples, *J. Chromatogr. B*, 2010, **878**(15–16), 1103–1112.
30. S. G. Villas-Bôas, *et al.*, Simultaneous analysis of amino and nonamino organic acids as methyl chloroformate derivatives using gas chromatography-mass spectrometry, *Anal. Biochem.*, 2003, **322**(1), 134–138.
31. M. M. Koek, *et al.*, Microbial metabolomics with gas chromatography/mass spectrometry, *Anal. Chem.*, 2006, **78**(4), 1272–1281.
32. J. Gullberg, *et al.*, Design of experiments: an efficient strategy to identify factors influencing extraction and derivatization of *Arabidopsis thaliana* samples in metabolomic studies with gas chromatography/mass spectrometry, *Anal. Biochem.*, 2004, **331**(2), 283–295.

33. M. C. Waldhier, *et al.*, Improved enantiomer resolution and quantification of free D-amino acids in serum and urine by comprehensive two-dimensional gas chromatography–time-of-flight mass spectrometry, *J. Chromatogr. A*, 2011, **1218**(28), 4537–4544.
34. S. K. Lien, H. F. Kvitvang and P. Bruheim, Utilization of a deuterated derivatization agent to synthesize internal standards for gas chromatography–tandem mass spectrometry quantification of silylated metabolites, *J. Chromatogr. A*, 2012, **1247**, 118–124.
35. H. F. Kvitvang, *et al.*, Highly sensitive GC/MS/MS method for quantitation of amino and nonamino organic acids, *Anal. Chem.*, 2011, **83**(7), 2705–2711.
36. C. Birkemeyer, *et al.*, Metabolome analysis: the potential of *in vivo* labelling with stable isotopes for metabolite profiling, *Trends Biotechnol.*, 2005, **23**(1), 28–33.
37. E. C. Horning, *et al.*, New picogram detection system based on a mass spectrometer with an external ionization source at atmospheric pressure, *Anal. Chem.*, 1973, **45**, 936–943.
38. J. C. Giddings, Maximum number of components resolvable by gel filtration and other elution chromatographic methods, *Anal. Chem.*, 1967, **39**(8), 1027–1028.
39. J. M. Davis and J. C. Giddings, Statistical theory of component overlap in multicomponent chromatograms, *Anal. Chem.*, 1983, **55**(3), 418–424.
40. L. F. de Alencastro, D. Grandjean and J. Tarradellas, Application of multidimensional (heart-cut) gas chromatography to the analysis of complex mixtures of organic pollutants in environmental samples, *Chimia*, 2003, **57**(9), 499–504.
41. Z. Liu and J. B. Phillips, Comprehensive 2-dimensional gas-chromatography using an on-column thermal modulator interface, *J. Chromatogr. Sci.*, 1991, **29**(6), 227–231.
42. J. C. Giddings, Concepts and comparisons in multidimensional chromatography, *J. High Resolut. Chromatogr.*, 1987, **10**, 319–323.
43. L. Mondello, *et al.*, Comprehensive gas chromatography coupled to mass spectrometry for the separation of pesticides in a very complex matrix, *Anal. Bioanal. Chem.*, 2007, **389**(6), 1755–1763.
44. J. Beens, R. Tijssen and J. Blomberg, Prediction of comprehensive two-dimensional gas chromatographic separations: a theoretical and practical exercise, *J. Chromatogr. A*, 1998, **822**(2), 233–251.
45. M. M. Koek, *et al.*, Higher mass loadability in comprehensive two-dimensional gas chromatography–mass spectrometry for improved analytical performance in metabolomics analysis, *J. Chromatogr. A*, 2008, **1186**(1–2), 420–429.
46. K. A. Kouremenos, J. Pitt and P. J. Marriott, Metabolic profiling of infant urine using comprehensive two-dimensional gas chromatography: application to the diagnosis of organic acidurias and biomarker discovery, *J. Chromatogr. A*, 2010, **1217**(1), 104–111.

47. J. V. Seeley, Theoretical study of incomplete sampling of the first dimension in comprehensive two-dimensional chromatography, *J. Chromatogr. A*, 2002, **962**(1–2), 21–27.
48. J. Dallüge, J. Beens and U. A. Brinkman, Comprehensive two-dimensional gas chromatography: a powerful and versatile analytical tool, *J. Chromatogr. A*, 2003, **1000**(1–2), 69–108.
49. P. J. Marriott, *et al.*, Comparison of thermal sweeper and cryogenic modulator technology for comprehensive gas chromatography, *J. High Resolut. Chromatogr.*, 2000, **23**(3), 253–258.
50. R. M. Kinghorn and P. J. Marriott, High speed cryogenic modulation – a technology enabling comprehensive multidimensional gas chromatography, *J. High Resolut. Chromatogr.*, 1999, **22**(4), 235–238.
51. J. V. Seeley, F. Kramp and C. J. Hicks, Comprehensive two-dimensional gas chromatography via differential flow modulation, *Anal. Chem.*, 2000, **72**(18), 4346–4352.
52. P. Korytar, *et al.*, Quadrupole mass spectrometer operating in the electron-capture negative ion mode as detector for comprehensive two-dimensional gas chromatography, *J. Chromatogr. A*, 2005, **1067**(1–2), 255–264.
53. M. F. Almstetter, P. J. Oefner and K. Dettmer, Metabolic fingerprinting using comprehensive two-dimensional gas chromatography–time-of-flight mass spectrometry, *Methods Mol. Biol.*, 2012, **815**, 399–411.
54. M. F. Almstetter, *et al.*, Comparison of two algorithmic data processing strategies for metabolic fingerprinting by comprehensive two-dimensional gas chromatography–time-of-flight mass spectrometry, *J. Chromatogr. A*, 2011, **1218**(39), 7031–7038.
55. M. F. Almstetter, P. J. Oefner and K. Dettmer, Comprehensive two-dimensional gas chromatography in metabolomics, *Anal. Bioanal. Chem.*, 2012, **402**(6), 1993–2013.
56. O. Fiehn, Metabolomics – the link between genotypes and phenotypes, *Plant Mol. Biol.*, 2002, **48**(1–2), 155–171.
57. T. Kind, M. Scholz and O. Fiehn, How large is the metabolome? A critical analysis of data exchange practices in chemistry, *PLoS One*, 2009, **4**(5), e5440.
58. D. S. Wishart, Computational strategies for metabolite identification in metabolomics, *Bioanalysis*, 2009, **1**(9), 1579–1596.
59. D. S. Wishart, Advances in metabolite identification, *Bioanalysis*, 2011, **3**(15), 1769–1782.
60. W. B. Dunn, *et al.*, Mass appeal: metabolite identification in mass spectrometry-focused untargeted metabolomics, *Metabolomics*, 2013, **9**(1 Suppl.), 44–66.
61. T. Kind and O. Fiehn, Advances in structure elucidation of small molecules using mass spectrometry, *Bioanal. Rev.*, 2010, **2**(1–4), 23–60.
62. E. Kováts, Gas-chromatographische Charakterisierung organischer Verbindungen. Teil 1: Retentionsindices aliphatischer Halogenide, Alkohole, Aldehyde und Ketone, *Helv. Chim. Acta*, 1958, **41**(7), 1915–1932.

63. H. Van den Dool and P. D. Kratz, A generalization of the retention index system including linear temperature programmed gas–liquid partition chromatography, *J. Chromatogr. A*, 1963, **11**, 463–471.
64. E. Malvoisin, *et al.*, Determination of Kovats retention indices with a capillary column and electron-capture detection: application to the assay of the enzymatic conversion of 3,4-epoxy-1-butene into diepoxybutane, *J. Chromatogr.*, 1979, **186**, 81–87.
65. J. Lisec, *et al.*, Gas chromatography mass spectrometry-based metabolite profiling in plants, *Nat. Protocols*, 2006, **1**(1), 387–396.
66. T. Kind, *et al.*, FiehnLib: mass spectral and retention index libraries for metabolomics based on quadrupole and time-of-flight gas chromatography/mass spectrometry, *Anal. Chem.*, 2009, **81**(24), 10038–10048.
67. D. S. Wishart, *et al.*, HMDB: a knowledgebase for the human metabolome. *Nucleic Acids Res.*, 2009, **37**(Database issue), D603–D610.
68. J. Kopka, *et al.*, GMD@CSB.DB: the Golm Metabolome Database, *Bioinformatics*, 2005, **21**(8), 1635–1638.
69. G. G. Desbrosses, J. Kopka and M. K. Udvardi, *Lotus japonicus* metabolic profiling. Development of gas chromatography–mass spectrometry resources for the study of plant–microbe interactions, *Plant Physiol.*, 2005, **137**(4), 1302–1318.
70. J. Kopka, Current challenges and developments in GC-MS based metabolite profiling technology, *J. Biotechnol.*, 2006, **124**(1), 312–322.
71. S. Kumari, *et al.*, Applying *in-silico* retention index and mass spectra matching for identification of unknown metabolites in accurate mass GC-TOF mass spectrometry, *Anal. Chem.*, 2011, **83**(15), 5895–5902.
72. S. E. Stein, *et al.*, Estimation of Kovats retention indices using group contributions, *J. Chem. Inf. Model.*, 2007, **47**(3), 975–980.
73. V. V. Mihaleva, *et al.*, Automated procedure for candidate compound selection in GC-MS metabolomics based on prediction of Kovats retention index, *Bioinformatics*, 2009, **25**(6), 787–794.
74. S. Bieri and P. J. Marriott, Generating multiple independent retention index data in dual-secondary column comprehensive two-dimensional gas chromatography, *Anal. Chem.*, 2006, **78**(23), 8089–8097.
75. K. Héberger, Quantitative structure–(chromatographic) retention relationships, *J. Chromatogr. A*, 2007, **1158**(1–2), 273–305.
76. H. Lu, *et al.*, Comparative evaluation of software for deconvolution of metabolomics data based on GC-TOF-MS, *Trends Anal. Chem.*, 2008, **27**(3), 215–227.
77. O. D. Sparkman, Evaluating electron ionization mass spectral library search results, *J. Am Soc. Mass Spectrom.*,**7**(4), 313–318.
78. S. E. Stein and D. R. Scott, Optimization and testing of mass spectral library search algorithms for compound identification, *J. Am Soc. Mass Spectrom.*, 1994, **5**(9), 859–866.
79. K. F. Smart, *et al.*, Analytical platform for metabolome analysis of microbial cells using methyl chloroformate derivatization followed by gas chromatography–mass spectrometry, *Nat. Protocols*, 2010, **5**(10), 1709–1729.

80. E. L. Schymanski, M. Meringer and W. Brack, Matching structures to mass spectra using fragmentation patterns: are the results as good as they look?, *Anal. Chem.*, 2009, **81**(9), 3608–3617.
81. C. N. McEwen and R. G. McKay, A combination atmospheric pressure LC/MS:GC/MS ion source: advantages of dual AP-LC/MS:GC/MS instrumentation, *J. Am. Soc. Mass Spectrom.*, 2005, **16**(11), 1730–1738.
82. R. Schiewek, *et al.*, Development of a multipurpose ion source for LC-MS and GC-API MS, *Anal. Bioanal. Chem.*, 2008, **392**(1–2), 87–96.
83. A. Carrasco-Pancorbo, *et al.*, Gas chromatography/atmospheric pressure chemical ionization–time of flight mass spectrometry: analytical validation and applicability to metabolic profiling, *Anal. Chem.*, 2009, **81**(24), 10071–10079.
84. T. Pacchiarotta, *et al.*, Evaluation of GC-APCI/MS and GC-FID as a complementary platform, *J. Biomol. Tech.*, 2010, **21**(4), 205–213.
85. G. Zurek, *et al.*, Metabolic profiling of a *Corynebacterium glutamicum* deltaprpD2 by GC-APCI high resolution Q-TOF analysis, *J. Biomol. Tech.*, 2011, **22**(Suppl.), S61–S62.
86. S. Abate, *et al.*, Determination of elemental compositions by gas chromatography/time-of-flight mass spectrometry using chemical and electron ionization, *Rapid Commun. Mass Spectrom.*, 2010, **24**(8), 1172–1180.
87. T. Kind and O. Fiehn, Metabolomic database annotations via query of elemental compositions: mass accuracy is insufficient even at less than 1 ppm. BMC, *Bioinformatics*, 2006, **7**, 234–243.
88. T. Kind and O. Fiehn, Seven golden rules for heuristic filtering of molecular formulae obtained by accurate mass spectrometry, *BMC Bioinformatics*, 2007, **8**, 105–124.
89. M. Kanehisa, *et al.*, KEGG for linking genomes to life and the environment. Nucleic Acids Res., 2008, **36**(Database issue), D480–4.
90. B. Hoekman, *et al.*, msCompare: a framework for quantitative analysis of label-free LC-MS data for comparative candidate biomarker studies, Mol. Cell. Proteomics, 2012, **11**(6), M111.015974.
91. P. A. Rudnick, *et al.*, Performance metrics for liquid chromatography–tandem mass spectrometry systems in proteomics analyses, *Mol. Cell. Proteomics*, 2010, **9**(2), 225–241.
92. M. Rojas-Cherto, *et al.*, Metabolite identification using automated comparison of high-resolution multistage mass spectral trees, *Anal. Chem.*, 2012, **84**(13), 5524–5534.

CHAPTER 6

Capillary Electromigration Techniques in Metabolomics

AKIYOSHI HIRAYAMA AND TOMOYOSHI SOGA*

Institute for Advanced Biosciences, Keio University, Tsuruoka, Yamagata 997-0052, Japan
*Email: soga@sfc.keio.ac.jp

6.1 Introduction

Metabolomics, together with genomics, transcriptomics and proteomics, constitute the main "omics" technologies in systems biology. Metabolomics (or metabolome analysis) is defined as the measurement of the levels of all metabolites (typically less than 1 kDa) in a biological system. It has become an important tool for gaining functional insight into cell biology. Most metabolites are not organism specific and there are estimated to be fewer of them than genes and proteins.[1] Spatial and temporal information about specific changes in metabolite concentrations complements gene expression and proteome studies and provides insight into metabolic phenotypes.

The ultimate goal of metabolomics is a single platform that can be used for the comprehensive analysis of all metabolites in a biological sample. However, this technology is not currently available. This is partly because of the wide variations in the chemical and physical properties of a metabolome and the fact that concentrations extend over an estimated seven to nine orders of magnitude (from pmol to mmol).[2] In addition, variations in extraction efficiencies for different classes of metabolites also complicate absolute quantification in metabolomics. Therefore, the term metabolomics is widely used to cover approaches that investigate subsets of the metabolome. These approaches include

RSC Chromatography Monographs No. 18
Chromatographic Methods in Metabolomics
Edited by Tuulia Hyötyläinen and Susanne Wiedmer

Published by the Royal Society of Chemistry, www.rsc.org

metabolic profiling, metabolic fingerprinting and metabolic footprinting.[3,4] Metabolic profiling involves the identification and quantitation of the set of metabolites in a selected biochemical pathway or chemical class. This also includes metabolite target analysis, which focuses only on a particular metabolite or metabolite class. Metabolic fingerprinting involves global, rapid and high-throughput analysis of crude samples or sample extracts to observe signature patterns without necessarily identifying or quantifying individual metabolites. Metabolic footprinting includes fingerprinting analysis of extracellular metabolites in cell culture media as a reflection of metabolite excretion or uptake by cells.[4]

To date, metabolomics has been performed using nuclear magnetic resonance (NMR) spectroscopy, mass spectrometry (MS) or MS with chromatography- or electrophoresis-based separation methods, such as gas chromatography (GC), liquid chromatography (LC) and capillary electrophoresis (CE). NMR spectroscopy is advantageous because it is non-destructive and features high throughput, high reproducibility and simple sample preparation.[5] However, it has low sensitivity and therefore requires a large sample. By comparison, MS has high sensitivity and selectivity and can discriminate some compound classes depending on the ionization mode. NMR spectroscopy and MS can be used as stand-alone techniques, but their combination with upstream separation methodologies generally enhances their performance for complex samples.

Many of the metabolites in biological systems are charged and capillary electrophoresis combined with mass spectrometry (CE-MS) is a new, powerful analytical tool for these species. In contrast to GC and LC, which operate based on differential interaction with a stationary phase, CE separates analytes based on their charge-to-mass ratios. In CE-MS, CE provides rapid analysis and efficient separation and MS provides excellent selectivity and sensitivity. The most attractive feature of CE is that it requires only a small sample (a few nanolitres at most). Although poor concentration sensitivity has often been cited as a disadvantage of CE with absorbance-related detectors, this does not pose a significant problem when CE is coupled with MS for detection. Additionally, in-capillary sample concentration can be performed to boost sensitivity,[6] without the need for dedicated instrumental modifications. Hence CE-MS represents a viable platform for metabolomics.

Many reviews have been published on CE-MS-based metabolomics. Monton and Soga described technical considerations and selected examples from different fields of application.[6] Ramautar and co-workers demonstrated the applicability of CE-MS in metabolomics research through examples of the analysis of bacterial extracts, plant extracts, urine, plasma and cerebrospinal fluid samples.[7,8] Barbas *et al.* reviewed the general advantages and disadvantages of CE for non-targeted fingerprinting of biological samples.[2]

This chapter provides a general description of the application of CE-MS in metabolomics, including different CE modes, MS instruments, interfaces and microfabricated devices (chip CE). The strategies for data processing and metabolite identification are described and analytical strategies of CE-MS for targeted and non-targeted metabolomics are discussed.

6.2 Technical Considerations

6.2.1 CE

CE is a versatile technique that can be used to separate a wide range of analytes, such as small inorganic ions,[9] large proteins[10] and even intact bacteria.[11] Different additives (*e.g.*, micelles, chiral selectors or complexing agents) can be incorporated in the run buffer to achieve a variety of separation modes. In CE-MS-based metabolomics, capillary zone electrophoresis (CZE) is the main CE separation mode because the buffer component is restricted when coupling with MS *via* electrospray ionization (ESI). Separation in CZE is based on the different migration velocities of analytes in an electric field based on their charge to size ratios.

The potential application of CZE in metabolomics has been demonstrated using various types of biological samples.[12–16] A CE-MS method using CZE separation was developed for the comprehensive and quantitative analysis of charged metabolites from a metabolic extract of *Bacillus subtilis*.[17] Using CE-MS conditions for cationic metabolite, anionic metabolite and nucleotides analysi, 1692 metabolites were detected from the *B. subtilis* extracts (Figure 6.1).

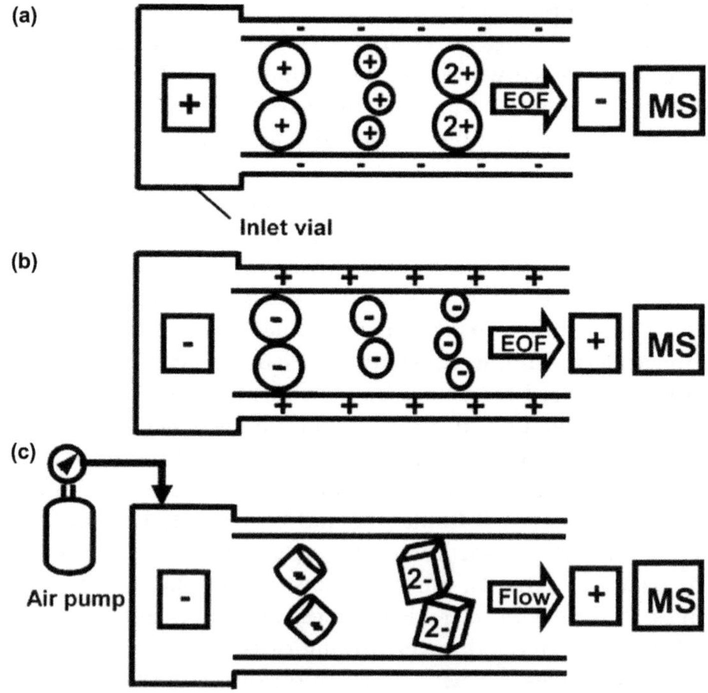

Figure 6.1 Schematic diagram of the various CE-MS methods. (a) Cationic metabolites, (b) anionic metabolites and (c) nucleotides and CoA compounds. Reprinted with permission from Ref. 17. Copyright 2003 American Chemical Society.

Micellar electrokinetic chromatography (MEKC) is a powerful mode for separating neutral compounds based on partitioning to charged micelles and it has been used for metabolic profiling. Barbas *et al.* applied two CE separation modes, CZE and cyclodextrin-modified MEKC combined with UV detection, for the analysis of urinary metabolites from a rat model of type I diabetes (streptozotocin rat).[18] CZE separations were performed with reverse polarity using a polyacrylamide-coated capillary. In addition, cyclodextrin-modified MEKC separations were performed with normal polarity using a fused-silica capillary and sodium dodecyl sulfate (SDS) and sulfated β-cyclodextrin as buffer additive. The samples from control and diabetic animals receiving either a placebo treatment or an antioxidant nutraceutical could be identified with this method. Hou *et al.* reported an application of a novel polymeric chiral surfactant, polysodium *N*-undecenoyl-L,L-leucylvalinate, as a pseudo-stationary phase for the separation and determination of warfarin enantiomers in human plasma with MEKC-MS.[19] However, MEKC is known to contaminate the ion source, suppress analyte ionization and decrease the MS response. To prevent loss of sensitivity, MEKC modifications have been adopted, including the use of volatile[20] and high molecular weight micelles[21] and the partial filling technique,[22] in which separation occurs only in a particular section of the capillary that is filled with micellar solution.

Recently, capillary electrochromatography (CEC), a hybrid separation technique that combines the high separation efficiency of CE with the stationary phase selectivity of LC, has been developed.[23–25] Separation in CEC is based on both interaction with the stationary phase and electrophoretic mobility.[26] To date, CEC-MS has not been applied in metabolomics, although Johannesson *et al.* screened biomarkers in plasma from patients with gangrenous and phlegmonous appendicitis using both CE-MS and CEC-MS.[27]

CE buffers are generally aqueous, but non-aqueous systems are also used, particularly for analytes that are insoluble or sparingly soluble in water. Non-aqueous systems allow the study of hydrophilic interactions, such as hydrogen bonding, dipole–dipole and ionic interactions, which are thermodynamically strengthened in hydrophobic environments.[28] When coupled with MS *via* ESI, non-aqueous CE circumvents buffer compatibility problems and enhances the sample ionization process, which results in improved detection limits compared with aqueous buffer systems.[29] Vuorensola *et al.* compared both aqueous and alcoholic non-aqueous systems for the separation of catecholamines by CE and CE-MS and found that the separation was more efficient in non-aqueous media than in water[30] (Figure 6.2).

Separation of biological samples by CE using fused-silica capillaries is strongly affected by solute adsorption on the capillary wall. In CE-MS, a constant and appreciable electroosmotic flow (EOF) is often essential to achieve adequate and reproducible spray conditions. However, it is impossible to remove all possible matrix components adsorbed on the capillary during sample preparation, especially in metabolic fingerprinting, which requires minimal sample pretreatment. To reduce these problems, different coatings

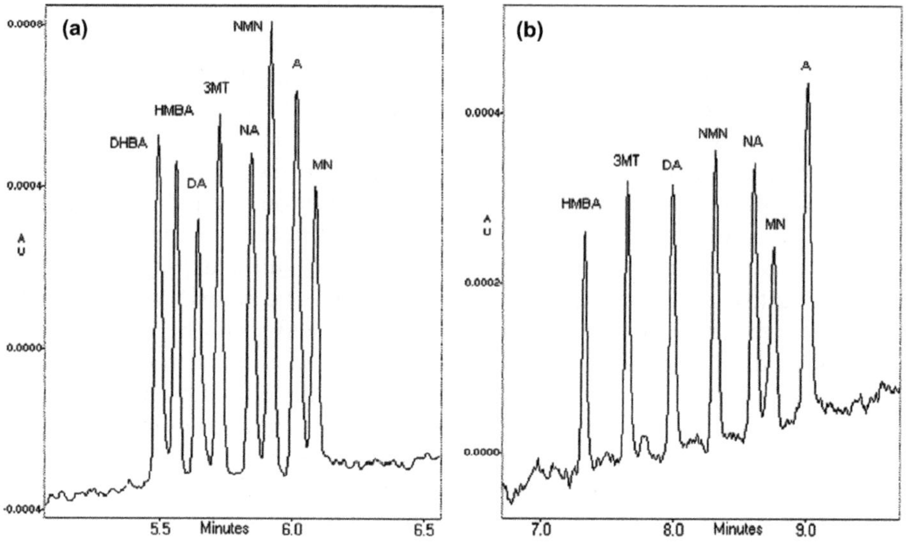

Figure 6.2 Catecholamines in optimized aqueous (a) and ethanolic (b) separations. (a) 20 mmol L^{-1} ammonium acetate in water–acetic acid (99:1 v/v), capillary 50/57 cm, separation at +20 kV, UV detection at 214 nm. (b) 15 mmol L^{-1} ammonium acetate in ethanol–acetic acid (98.5:1.5 v/v), capillary 30/37 cm, separation at +30 kV, UV detection at 214 nm. DHBA, 3,4-dihydroxybenzylamine hydrobromide; HMBA, 4-hydroxy-3-methoxybenzylamine hydrochloride; DA, dopamine; 3MT, 3-methoxytyramine; NA, [±]-noradrenaline [+]-hydrogentartrate; NMN, DL-normetanephrine; A, adrenaline; MN, DL-metanephrine.
Reprinted with permission of Elsevier from Ref. 30.

have been developed for CE-MS metabolomics and these coatings were recently reviewed by Huhn *et al.*[31]

The two major coating approaches used in CE-MS are dynamic coating and static coating. Dynamic coatings are generated by addition of coating agents to the BGE during separation and these agents compete with the analyte for adsorption sites on the capillary wall. This approach is simple, but the coating additives create background noise, ion suppression and MS contamination. Static coatings physically adsorb on the capillary wall through forces such as electrostatic and hydrophobic interactions or chemically attach to the capillary wall through covalent binding.

Capillaries with static coatings are sometimes used in CE-MS; for example, Ramauter *et al.* prepared capillaries coated with a triple layer of Polybrene–dextran sulfate–Polybrene (PB–DS–PB) for metabolic profiling of human urine.[32] The surface of the capillary was positively charged, which could reverse the EOF towards the anode and cationic compounds would migrate later than in EOF at low pH. The relative standard deviations (RSDs) for the migration times and peak areas of the test compounds, including amino acids, organic acids, nucleosides, catecholamines, small peptides and amines, were less than 1% and 10%, respectively. Using this method, about 500 molecular features were detected in human urine, which is almost twice as many as observed with the previous method from the same groups.[33]

6.2.2 MS Instrument

Almost all types of mass analysers have been coupled to CE, including single quadrupole, ion trap (IT), time-of-flight (TOF), triple quadrupole, quadrupole time-of-flight (Q-TOF) and Fourier transform ion cyclotron resonance (FT-ICR) instruments. Among these mass analysers, single quadrupole MS is used most often because of its relatively low cost. However, it has a number of major disadvantages, including slow scanning speed and low resolution and sensitivity in the scan mode. IT-MS is also used extensively and enables MS/MS experiments to be performed without the need for multiple analysers. Triple quadrupole MS can provide high sensitivity, even for complex matrices, using the multiple reaction monitoring mode, and this allows high selectivity in monitoring a specific fragment ion of a selected precursor mass.[34] This kind of MS instrument is suitable for targeted metabolome analysis.

In the case of non-targeted metabolome analysis, higher performance MS instruments, which are capable of very fast scanning speeds, are required to describe adequately very narrow peaks resulting from CE separations. In addition, the MS system should also offer high mass accuracy and high resolution so that peaks arising from closely migrating compounds with similar nominal masses can be resolved.

TOFMS can provide high resolution, high mass accuracy with errors generally below 5 ppm and high sensitivity.[26,35] This instrument also allows the determination of the elemental compositions of compounds from their isotopic patterns, which provides an additional analytical dimension for compound identification.[36] For these reasons, TOFMS instruments have been widely used in non-targeted metabolome analysis. In the same way, Q-TOFMS can provide accurate masses for the product ions, which increases the selectivity for compounds in complex matrices. This makes it possible to identify the unknown compounds from the elemental compositions of their fragments.

FT-ICR-MS allows for ultra-high mass resolution, excellent sensitivity and high mass accuracy of less than 1 ppm.[37] Baidoo *et al.* used CE-FT-ICR-MS analysis to improve the overall detection of cationic metabolites in a bacterial extract and obtained good mass resolution and sub-ppm mass accuracy.[38] However, because of its high cost, application of CE-FT-ICR has been limited.

6.2.3 Interface

The most common MS interface for CE-MS in metabolomics is ESI. ESI enables molecules in the liquid phase to be transferred directly into ions in the gas phase,[39] and it can be easily adapted for on-line coupling of MS with CE. CE separates mostly charged compounds, which makes ESI a good ionization technique for polar and ionic compounds. Matrix-assisted laser desorption/ionization (MALDI) is a popular technique for proteomics, but it has yet to be applied to metabolomics because of strong interference from matrix ions in the detection of low molecular-weight compounds.[40]

A CE-ESI-MS system has to complete the CE electrical circuit for analyte separation, while simultaneously providing an electrical potential to the spray tip. The coupling of CE to ESI-MS is performed using sheath-flow or sheathless interfaces. The sheath-flow interface is the most widely used for CE-MS because it can be easily and reproducibly constructed. In this configuration, the separation capillary is inserted in a slightly larger diameter stainless-steel tube in a coaxial setting and a sheath liquid is mixed with effluent at the capillary outlet. Furthermore, the nebulizer gas, which is supplied *via* a third coaxial tube, assists with stable spray formation and solvent evaporation.[41] A aqueous–organic mixture is generally employed as the sheath liquid and its composition and flow rate are very important. For instance, Vuorensola *et al.* tested three organic solvents as sheath liquids, namely methanol, acetonitrile and 2-propanol, mixed with water (50:50 v/v), for catecholamine analysis.[42] In this experiment, methanol–water (80:20 v/v) with 0.5% acetic acid gave the best signal intensity and spray stability. The effect of volatile salt and methanol in the sheath liquid on amino acid analysis was evaluated and 5 mmol L^{-1} ammonium acetate in methanol–water (50% v/v) was found to be the optimum sheath liquid.[43]

The stainless-steel needle that is used commercially for the sheath-flow interface affects anionic metabolome analysis in the negative-ion mode.[44] As shown in Figure 6.3, a stainless-steel spray needle placed at the anode tends to oxidize and corrode because of electrolysis and iron oxides (rust) precipitate and block the capillary outlet. Furthermore, many anionic metabolites form complexes with metal ions (*e.g.*, Fe^{2+}, Ni^{2+}) from the corroded stainless-steel needle and this complex formation considerably reduces the detection sensitivity. Use of a platinum ESI needle can prevent these problems and provide better sensitivity for anionic metabolome analysis (Figure 6.4).

To ensure a stable ESI flow, the sheath liquid is usually introduced at a higher flow rate (*e.g.*, a few microlitres per minute) than the CE flow rate.[39] This reduces the sensitivity because of dilution of the compounds of interest at the mixing point in sheath-flow CE-MS. This problem can be avoided with a sheathless interface.

In the sheathless interface configuration, electrical contact is achieved by several different methods that accomplish CE and complete the ESI electrical circuit simultaneously. For example, various conductive materials have been investigated as coatings for the outer surface of the capillary,[45–49] and these materials help establish the electrical contact. For example, Kele *et al.* constructed a sheathless CE-ESI-MS interface by combining a gold-coated nanospray borosilicate capillary with a fused-silica capillary and successfully applied the interface to peptide analysis.[46] Good results were obtained with this method, but the capillary is not durable and lasts only a few hours to a few days because the coatings are mechanically unstable.[50]

In another approach, a platinum wire was inserted into the capillary to make a stable electric contact. Cao and Moini constructed a sheathless ESI interface

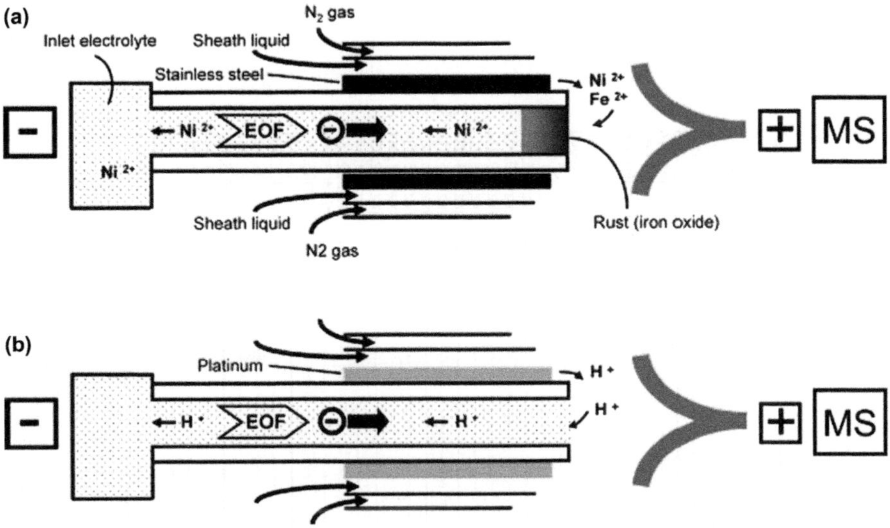

Figure 6.3 Schematic of anion analysis by CE-MS in the negative ion mode using a CE-ESI-MS sprayer with (a) a stainless-steel needle and (b) a platinum needle. (a) Metal ions such as Fe^{2+} and Ni^{2+} are generated electrochemically from the stainless-steel needle at the anode (pH 6.9) and moved into the separation capillary. $Fe^{2+}$is oxidized (forming rust) and blocks the capillary outlet because of electrolysis (pH 8.5), while Ni^{2+} migrates towards the capillary inlet (cathode). Anions exhibiting high chelating properties with metal ions form complexes with Ni^{2+} or iron oxides. (b) In the platinum needle, water oxidation ($2H_2O \rightarrow 4H^+ + O_2\uparrow + 4e^-$) occurs at the anode and the anionic metabolites do not encounter metal ions.
Reprinted with permission from Ref. 44. Copyright 2009 American Chemical Society.

by this method[51] and obtained a detection limit of ~4 fmol for detection of myoglobin. However, these interfaces are difficult to construct and may cause bubble formation and turbulent flow in the capillary.

A split-flow technique has been developed in which a small hole or a locally porous section is formed near the capillary outlet and a small percentage of the buffer exits through this hole/section and makes contact with an external electrode *via* a buffer reservoir.[52,53] The advantage of a split-flow interface is that the CE capillary is partially disconnected from the ESI emitter, so that a problem emitter can be replaced easily, which increases the lifetime of the capillary. However, most of these techniques require special equipment and skilled operators, which limit their application in CE-MS.

In 2007, Moini introduced a novel sheathless interface consisting of a porous capillary,[54] which allowed electrical connection through a porous tip at the capillary outlet. The porous tip was formed by etching 3–4 cm of the capillary outlet with a solution of 49% hydrofluoric acid. Electrical contact between the porous capillary and ESI was achieved *via* a static conductive liquid in the ESI

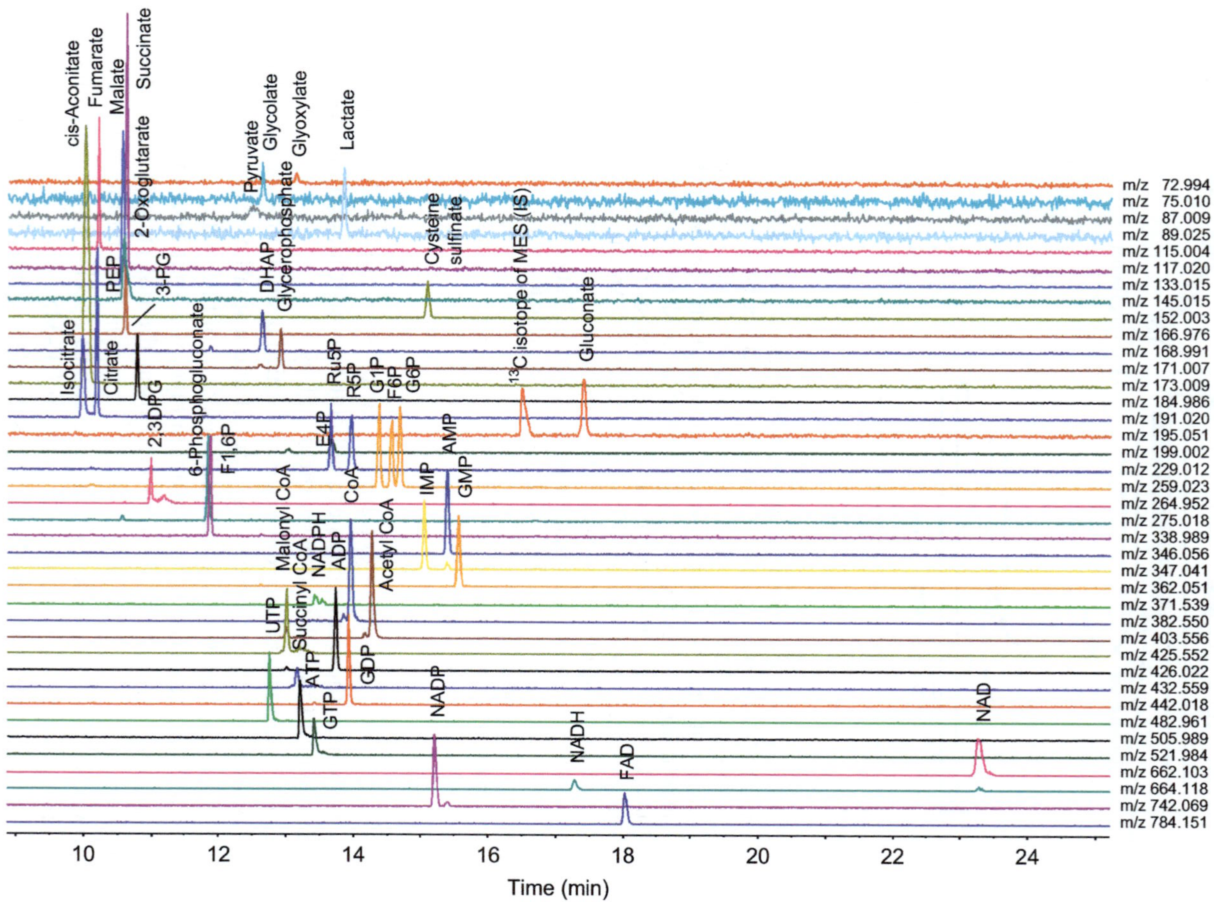

Figure 6.4 Selected CE-TOFMS ion electropherograms for a standard mixture of anionic metabolites from the glycolysis, pentose phosphate and tricarboxylic acid pathways obtained using a CE-ESI-MS sprayer attached to a platinum needle. Reprinted with permission from Ref. 44. Copyright 2009 American Chemical Society.

needle. Based on this concept, a high-sensitivity porous sprayer (HSPS) sheathless interface for CE-MS was developed by Beckman Coulter (Figure 6.5). The HSPS has been used for the analysis of peptide mixtures,[55,56] intact proteins[57] and phosphorus-containing amino acid-type herbicides.[58]

Based on the principles of the electrospray process, that is, protonation or deprotonation of solutes or formation of charged adducts, ESI is particularly suitable for the analysis of polar compounds,[59] but is rather inefficient for non-polar compounds. It also has a low tolerance for salts and is susceptible to matrix effects. Furthermore, volatile CE buffers are required for CE-MS with an ESI interface and these buffers often do not provide the same separation performance as standard buffers (*e.g.*, phosphate and borate) for CE/UV detection. In addition, the CE separation will typically be better at higher ionic strength, whereas high ionic strength is a clear disadvantage in ESI-MS where it will impair ion production of the compounds of interest.[60]

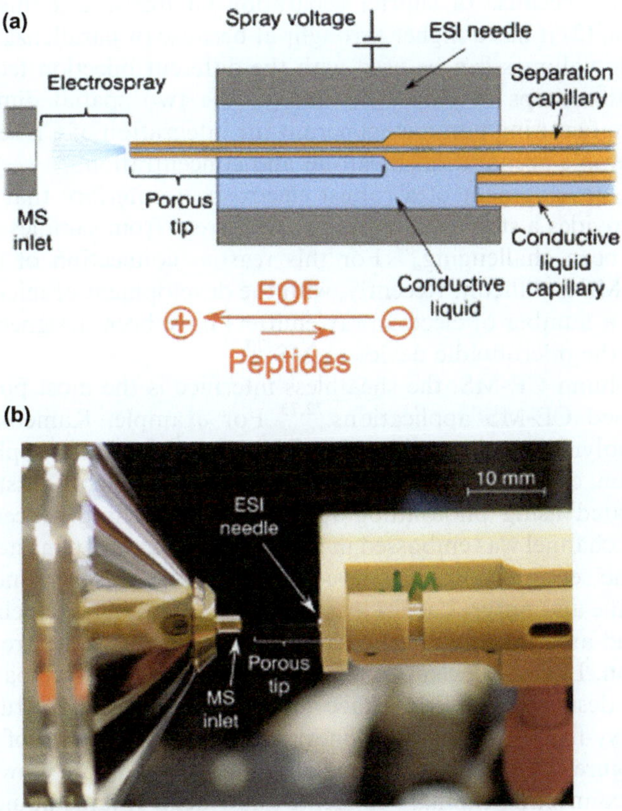

Figure 6.5 (a) Schematic diagram and (b) photograph of the prototype high-sensitivity porous sprayer interface.
Reprinted with permission from Ref. 56. Copyright 2011 American Chemical Society.

Atmospheric pressure chemical ionization (APCI) is another ionization technique, in which analyte ionization occurs through gas-phase ion–molecule reactions.[61] Because it is a harder ionization process than ESI, it is more suitable for neutral and/or less polar compounds[62] than apolar and ionic compounds, which cannot be detected following APCI. APCI-MS signals are not affected by the composition of the CE buffer, even when it contains a non-volatile salt and SDS.[63] Atmospheric pressure photoionization has also been shown to be relatively non-discriminant towards non-polar compounds and reasonably tolerant of matrix additives.[64] These techniques have been rapidly adopted in LC-MS, but only a few applications have been reported for CE-MS so far.[60,65–67]

6.2.4 Chip CE

The design and fabrication of miniaturized devices is important in CE-MS. The major advantages of chip CE compared with standard CE are as follows: (1) it is much faster because of shorter separation channels, and this reduces the analysis time; (2) it has a higher throughput because of parallelization; (3) very small sample volumes can be used with the different injection techniques and miniaturization steps in chip CE; and (4) the two spatial dimensions and various manufacturing parameters permit the integration of CE separation on-chip with sample preparation, clean-up and concentration.[68]

However, development of a robust electrospray interface that reliably and routinely provides a stable electrospray ion current from samples separated on a chip has been challenging.[69] For this reason, connection of a chip-based device with MS is difficult. Recently, with the development of microfabrication technology, a number of electrospray emitters have been designed to optimize coupling of the microfluidic device to MS.[70,71]

Unlike column CE-MS, the sheathless interface is the most popular design for chip-based CE-MS applications.[72,73] For example, Kameoka *et al.* developed a polymeric microfluidic chip for CE-ESI-MS and applied it in the determination of carnitine and its acyl esters.[74] A silicon master chip was microfabricated using photolithographic and dry etching processes and the microfluidic channel was embossed in plastic from the silicon master. Using this chip, baseline separation was achieved for three selected carnitines (carnitine, acetylcarnitine and butyrylcarnitine) in less than 10 s using a CE channel length of 3.5 cm and an applied voltage of $2\,\mathrm{kV\,cm^{-1}}$ without surface treatment.

In addition, Li *et al.* developed a chip-CE-ESI-MS interface based on a low sheath flow design.[75] A CE interface was constructed using a tungsten wire-assisted epoxy-fixing method to connect a capillary to the end of the chip-CE channel (Figure 6.6). With this capillary, the sheath liquid flowed coaxially around the connecting capillary to create a low dead volume liquid junction at the interface between the connecting capillary and the ESI emitter. This device has several advantages compared with sheathless chip CE-MS, including flexibility of buffer selection, low dilution factor and ease of operation. This device was evaluated for analysis of histamine antagonists and peptides. The

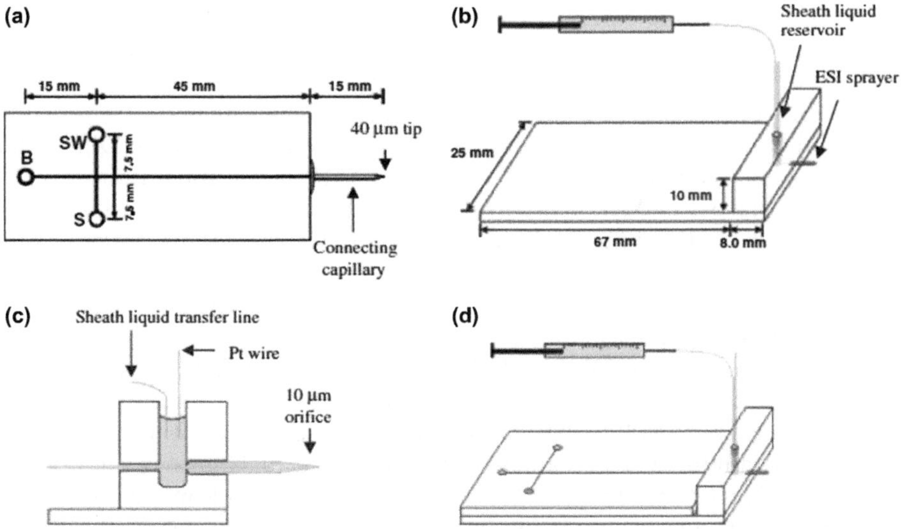

Figure 6.6 Schematic diagrams of the chip-CE low sheath flow ESI-MS device. (a) Layout of the poly(methyl methacrylate) (PMMA)-CE chip with a connecting capillary (B, buffer reservoir; S, sample reservoir; SW, sample waste reservoir). (b) The low sheath flow interface. (c) An enlarged diagram of the low sheath flow interface. The sprayer tip was tapered to ~10 µm. (d) The CE microchip slides into the low sheath flow interface. Reprinted with permission of Wiley-VCH from Ref. 75.

histamine antagonists were separated within 1.5 min and the RSDs of the peak height and area were 5.6 and 6.2%, respectively.

Chip CE techniques permit very fast and high throughput analysis and have great potential in metabolomics. However, this system is expected to be more suitable for targeted than non-targeted metabolomic analysis, because its separation efficiency may not be sufficient for comprehensive analysis. Moreover, a straightforward method for coupling chip CE with MS is yet to be developed.

6.3 Data Treatment

6.3.1 Data Processing

CE-MS is a versatile system, but a major challenge is variability of the migration times of analytes. This run-to-run variability in EOF occurs mainly because of changes in the capillary wall condition that result in greater migration time variations compared with other separation methods, such as GC and LC. Even in a single run, fluctuations of the capillary electrical conditions can cause migration time shifts.[76] For these reasons, accurate and versatile migration time correction for a large variety of metabolites is necessary in CE-based metabolomics.

Recently, several software systems, including MZmine, XCMS and MetAlign, have been developed and used in metabolomics studies.[77–80] However, most of the software currently available is used in LC-MS studies and is not optimized for CE-MS-derived data. The peak shapes in CE-MS are more diverse than those derived from chromatographic techniques such as LC-MS and GC-MS, which makes peak detection particularly challenging.[76]

To address these issues, MathDAMP (Mathematica package for Differential Analysis of Metabolite Profiles)[81,82] and JDAMP (Java application for Differential Analysis of Metabolite Profiles)[76] software have been developed and can be used for the differential analysis of CE-MS metabolite profiles. MasterHands (Metabolome Analysis and Screening Tool for Easy and Rapid HANDling of Sample data), which can be used for comprehensive peak picking of CE-TOFMS data, has also been developed.[83]

The data analysis workflow includes noise filtering, baseline correction, peak detection and integration of the peak area from sliced electropherograms (electropherogram width $= 0.02\, m/z$). The accurate m/z value for each peak detected within the time domain is calculated with Gaussian curve fitting of the mass spectrum on the m/z domain peak. Peaks from multiple measurements are aligned by dynamic programming-based techniques and the detected peaks are regarded as those with m/z differences of less than 20 ppm that are derived from the same electropherograms.

6.3.2 Metabolite Identification

A major obstacle in metabolomics is the identification of many unknown metabolites in complex biological samples. In CE-MS, metabolites are generally identified by matching their migration times and mass or tandem mass spectra with those of authentic metabolite standards. However, there are only a few thousand commercially available analytical standards because of a lack of demand and the inherent instability of many metabolites. These standards and endogenous metabolites contained in databases represent only a small proportion of all endogenous metabolites.

For instance, of the 1692 metabolites mentioned in Section 6.2.1, only 150 were positively identified and an additional 83 were assigned based on expected charge states and isotopic distributions.[17] For unknown compounds, the use of high-mass accuracy analysers (*e.g.*, TOF) allows the assignment of empirical formulae, while tandem MS permits structural identification *via* interpretation of their fragmentation patterns. However, the sheer scale of metabolomics demands novel identification methodologies, including computational approaches.

To facilitate metabolite identification, Sugimoto *et al.* developed a technique for predicting the identities and migration times of cations in CE-MS using an ensemble of artificial neural networks.[84] When evaluated against all metabolites listed in the Kyoto Encyclopedia of Genes and Genomes (KEGG) ligand database, the correct compound among the top three candidates could be predicted in 78.0% of the cases. More recently, they also reported a method to identify unknown peaks based on the predicted migration time (t_m) and accurate m/z values.[85] They developed a predictive model using 375 standard

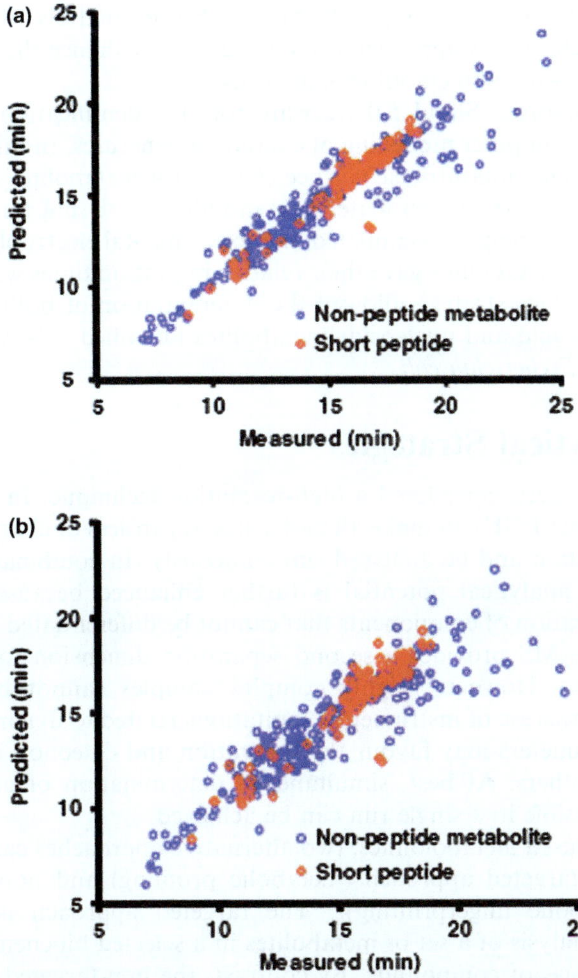

Figure 6.7 Correlation between predicted and measured normalized t_m values. The plots in (a) were obtained using the complete data sets and the plots in (b) were obtained with the CV procedure. The red and blue signals depict the results for peptides and non-peptide small molecules, respectively. Reprinted with permission of Wiley-VCH from Ref. 85.

cationic metabolites and support vector regression. The model yielded good correlations between the predicted and measured t_m ($R = 0.952$ and 0.905 using complete and cross-validation data sets, respectively) (Figure 6.7). Using the trained model, the authors subsequently predicted the t_m values for 2938 metabolites available from public databases and assigned tentative identities to noise-filtered features in human urine samples. While 38.9% of the peaks were assigned metabolite names by matching with the standard library alone, the proportion increased to 52.2% using this method. This approach resulted in reasonably confident metabolite assignment without requiring additional systematic and time-consuming MS/MS or NMR analysis of the detected features. Therefore, this approach can considerably enhance the value of non-targeted CE-MS-based metabolomic analysis.

Lee *et al.* reported Simul 5.0 freeware for the identification of unknown metabolites by computer modelling of candidate structures. In this method, the molecular volumes and intrinsic valence charges for metabolites are calculated and subsequently used to estimate their mobilities.[86] Using Simul 5.0, comparisons between computer-simulated and experimental electropherograms for several cationic metabolites gave their relative migration times with an average error of <2%. This strategy allowed the differentiation of both isomeric and isobaric amino acid and nucleoside metabolites identified as key nutrients for the growth of *Escherichia coli.*

6.4 Analytical Strategies

CE has always been considered a high-resolution technique. In principle, the presence of a fast EOF can make all molecules, regardless of charge, migrate in the same direction and be analysed simultaneously. In combination with MS detection, the analytical potential is further enhanced because CE permits temporal separation of components that cannot be differentiated by mass (*e.g.*, isomers), while MS provides a second separation dimension for metabolites that co-migrate. However, highly complex samples cannot be completely characterized because of instrumental limitations related to dynamic range and analytical parameters may favour the separation and detection of some compounds over others. At best, simultaneous determination of as many components as possible in a single run can be achieved.

In CE-MS-based metabolomics, two alternative approaches can be adopted, including the targeted approach (metabolic profiling) and non-targeted approach (metabolic fingerprinting).[4] The targeted approach is focused on quantitative analysis of a set of metabolites in a selected biochemical pathway or a specific class of compounds. By contrast, the non-targeted approach involves unbiased, global screening to classify samples based on metabolite patterns or 'fingerprints' that change in response to disease or environmental or genetic perturbations and the ultimate goal is to identify discriminating metabolites.

6.4.1 Targeted Approach

The best way to quantify metabolites accurately is to use the isotopically labeled standards. However, these standards are expensive and not suitable for routine use. Instead, several internal standards, each representing a class of compounds (*e.g.*, carboxylic acids), are often used for the quantification of metabolites.[7] However, the addition of many internal standards sometimes results in ion suppression. Although not as accurate as these methods, peak area comparison using internal standardization techniques is relatively simple and generally acceptable for targeted analysis.

Metabolic pathways, including glycolysis, the pentose phosphate pathway and the tricarboxylic acid cycle, play important roles in energy metabolism and the intermediates involved in these pathways are highly polar and charged compounds. CE-TOFMS has been used for comprehensive and quantitative analyses of endogenous metabolites in tumour and grossly normal tissues obtained from 16 colon and 12 stomach cancer patients.[87] CE-TOFMS in three different modes for cation, anion and nucleotide analyses detected 738 (normal) and 877 (tumour) peaks in colon tissue and 1007 (normal) and 1142 (tumour) peaks in stomach tissue on average, after eliminating redundant peaks, such as noise, fragments and adduct ions. Among these, 94 peaks in the colon tissue and 95 peaks in the stomach tissue were identified and quantified with metabolite standards by matching the closest m/z values and normalized migration times (Figure 6.8). Because these metabolites are involved in glycolysis, the pentose phosphate pathway, the tricarboxylic acid and urea cycles and amino acid and nucleotide metabolisms, the results identified several cancer-specific metabolic traits. Notably, low glucose concentrations and high lactate and glycolytic intermediate concentrations were found in both the colon and stomach tumour tissues, which indicates that enhanced glycolysis occurs in tumour tissues (Warburg effect).

The multiple reaction monitoring approach of MS detection has been successfully used by Harada *et al.* to identify and quantify anionic metabolites by CE-ESI-triple quadrupole MS/MS.[88] The use of a sulfonated capillary permitted the simultaneous analysis of sugar phosphates, organic acids, nucleotides and coenzyme A compounds using only a single CE-MS setting. Under the optimized conditions, the detection limits of 53 anionic metabolites were between 0.04 $\mu mol\,L^{-1}$ (shikimate) and 8.8 $\mu mol\,L^{-1}$ (fructose 2,6-bisphosphate). The RSDs of the migration times and peak areas ($n = 6$) for the most anionic metabolites were 0.5% and <10.0%, respectively. The developed method was successfully applied to determine anionic metabolites in a cell culture of *Catharanthus roseus*. Accumulation of some metabolites, including shikimate, malate and sedoheptulose 7-phosphate, by activation of methyl jasmonate was observed in the study. The results showed that shikimate, tricarboxylic acid and pentose phosphate pathways were activated. This method will be useful for the detailed analysis of primary metabolism dynamics.

Desiderio *et al.* optimized the CE-IT-MS method for the analysis of arginine, monomethylarginine, symmetric dimethylarginine (SDMA) and asymmetric dimethylarginine (ADMA) in human plasma.[89] The use of field-amplified sample injection greatly increased the method detection sensitivity and allowed the quantitation of ADMA and SDMA in human plasma with a minimal sample pretreatment procedure of rapid acetonitrile protein precipitation. Several parameters that affect sample ionization and CE separation were carefully investigated. Under the optimized conditions, complete separation of arginine, monomethylarginine, ADMA and SDMA was obtained using 1.5 $mol\,L^{-1}$ formic acid as the run buffer with a run time of <15 min and high specificity in MS/MS detection. The method was applied to the quantitation of arginine and its methylated metabolites in plasma from healthy volunteers for

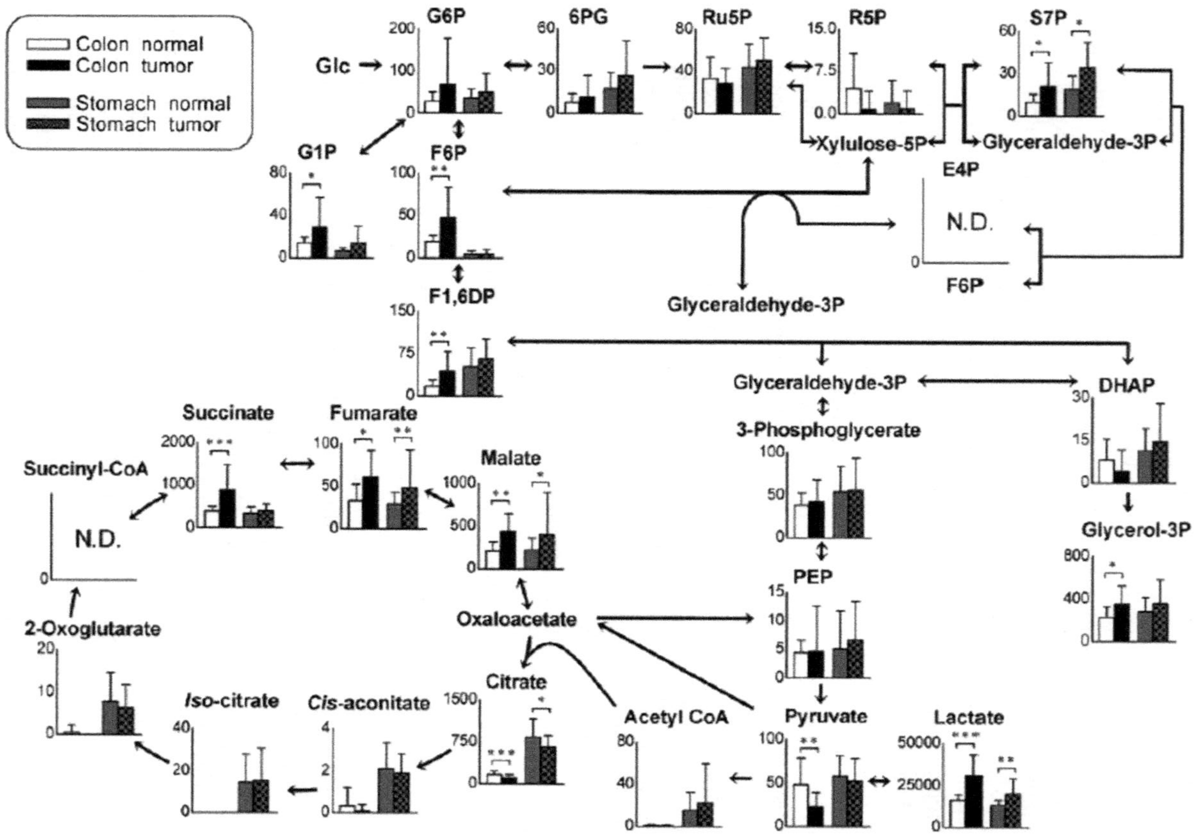

Figure 6.8 Quantified levels of metabolites involved in central carbon metabolism. Metabolite concentrations for colon and stomach tissues superimposed on a metabolic pathway map that includes glycolysis and the pentose phosphate and tricarboxylic acid pathways. Reprinted with permission of the American Association for Cancer Research from Ref. 87.

the evaluation of the endogenous levels of these compounds. The concentrations of the metabolites determined using CE-IT-MS agreed with their previously reported levels in plasma.

6.4.2 Non-targeted Approach

In non-targeted analysis, a large number of metabolites in a sample are analysed without bias under high-throughput conditions. Several sets of conditions that have been optimized for different groups of compounds can be used for the same sample. This approach could be applied in areas such as the detection of new disease-related biomarkers in clinical proteomics and genome-wide studies.[90] TOFMS is widely used for non-targeted analysis because sample complexity requires high mass accuracy and high resolution for the resolution of closely migrating compounds with similar nominal masses.

A metabolome differential display method based on CE-TOFMS was used for metabolic profiling of liver and serum metabolites from mice following acetaminophen-induced hepatotoxicity.[91] In mouse liver extracts, 1859 peaks were detected and changes in the metabolite profiles were observed. The ophthalmate biosynthesis pathway was activated and serum ophthalmate was shown to be a sensitive indicator of hepatic glutathione depletion. Therefore, it could be used as a new biomarker for oxidative stress.

Sugimoto *et al.* used CE-TOFMS for the metabolic analysis of saliva samples obtained from 215 individuals (69 oral, 18 pancreatic and 30 breast cancer patients, 11 periodontal disease patients and 87 healthy controls).[83] They identified 57 principal metabolites that could be used to predict accurately the probability of being affected by each disease. Significant correlations were found between the known patient characteristics and the discriminating metabolites, which suggested that cancer-specific signatures are embedded in saliva metabolites. Multiple logistic regression models yielded high areas under the receiver-operating characteristic curves (AUCs) to discriminate healthy controls from each disease. The AUCs were 0.865 for oral cancer, 0.973 for breast cancer, 0.993 for pancreatic cancer and 0.969 for periodontal diseases. In conclusion, this study demonstrated that CE-TOFMS can readily and effectively be applied for the detection of oral, breast and pancreatic cancers.

Ramautar *et al.* showed the advantages of using non-covalently coated capillaries with multiple layers of charged polymers for the global metabolic profiling of rat urine.[92] Bilayer [Polybrene–poly(vinyl sulfonate) (PB–PVS)] and triple-layer (PB–DS–PB) coatings were evaluated under acidic (pH 2.0) and alkaline (pH 9.0) separation conditions, which separated basic and acidic compounds. The analysis of cationic compounds with the PB–DS–PB CE method at low pH (*i.e.*, compounds detected after EOF) provided a larger separation window and number of separated peaks in urine than the analysis with the PB–PVS CE method at low pH (*i.e.*, compounds detected before EOF) (Figure 6.9). In this study, about 600 molecular features were detected in rat urine by the PB–DS–PB CE-MS method whereas about 300 features were found with the PB–PVS CE-MS method. Hence the selection of a CE method

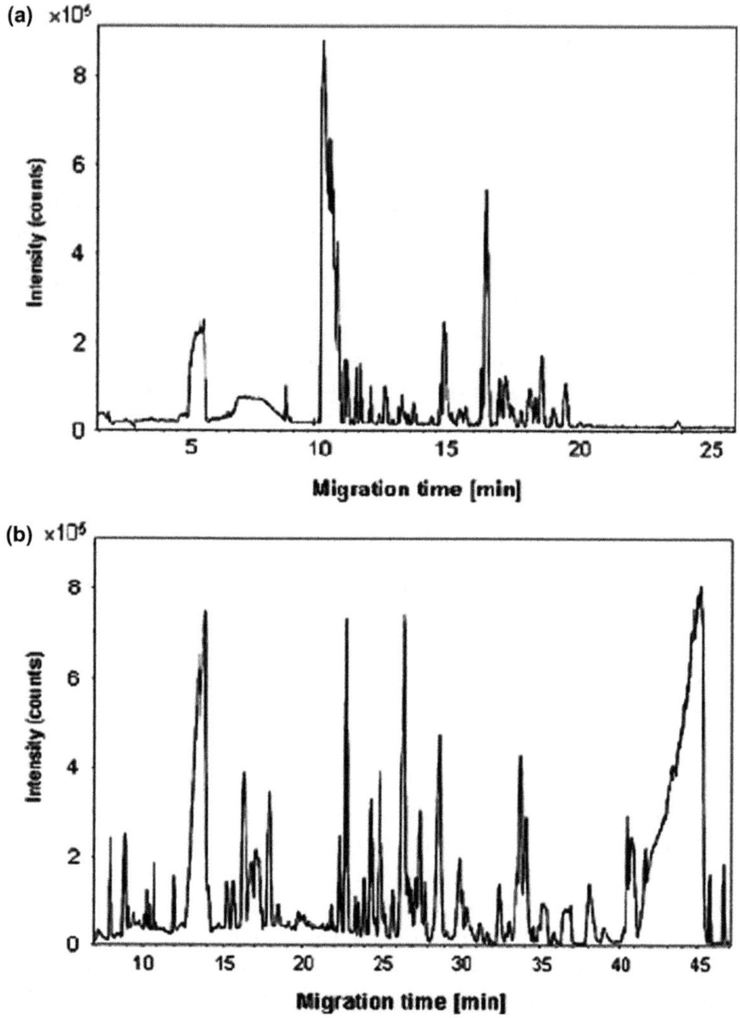

Figure 6.9 Base peak electropherograms obtained during CE-MS analysis of rat urine using (a) a Polybrene–poly(vinyl sulfonate)-coated capillary and (b) a Polybrene–dextran sulfate–Polybrene-coated capillary. Conditions: background electrolyte, 1 mol L^{-1} formic acid (pH 2.0); sample injection, 3.5 kPa for 60 s.
Reprinted with permission of Wiley-VCH from Ref. 92.

for cationic metabolites depends on the purpose of the study. The PB–PVS CE-MS method is suitable for high-throughput analysis, whereas the PB–DS–PB approach provides a more information-rich metabolic profile with a longer analysis time.

A differential metabolomics strategy with CE-MS was used to assess the efficacy of nutritional intervention to attenuate oxidative stress induced by strenuous exercise.[93] A healthy volunteer was recruited to perform a submaximal prolonged ergometer cycling trial until volitional exhaustion with frequent blood collection over a 6 h time interval, which included pre-, during and post-exercise periods while at rest. A follow-up study was performed on the same subject after high-dose oral intake of *N*-acetyl-L-cysteine before performing the same exercise protocol under standardized conditions. Time-dependent changes in global metabolism of filtered red blood cell lysates by CE-MS were measured and revealed significant attenuation of cellular oxidation associated with high-dose oral *N*-acetyl-L-cysteine intake. In this study, untargeted metabolite profiling allowed the identification and quantification of several putative early- and late-stage biomarkers that reflected oxidative stress inhibition because of nutritional intervention, including oxidized glutathione, reduced glutathione, 3-methylhistidine, L-carnitine, *O*-acetyl-L-carnitine and creatine.

6.5 Conclusion

The use of CE-MS for metabolomics has increased in recent years. Currently, an ideal analytical platform does not exist and it is unlikely that there will be one available for comprehensive metabolomics in the near future. Because CE is particularly suited for the separation of highly polar and charged metabolites, the metabolic information that it provides is complementary to other analytical platforms such as LC-MS.

Most CE-MS analyses are performed using an ESI interface, which is suitable for the ionization of polar compounds. Recently, APCI and atmospheric pressure photoionization have been employed as MS interfaces. Both of these methods can generate ions from non-polar compounds and can be used with a wider range of MS buffer components and additives. Therefore, these interfaces can be used to increase the applicability of CE-MS.

The sheath-flow interface is commonly used for CE-MS because it can be easily and reproducibly constructed. However, the use of a sheathless interface offers numerous advantages and this approach can provide additional information in the field of metabolomics.

Microchip CE systems are promising because of their speed, high resolution and excellent sensitivity. However, because of difficulties with coupling with MS, this system is widely used only for targeted metabolomic studies and not for non-targeted analysis, which require the very high resolution of CE.

References

1. Q. Z. Wang, C. Y. Wu, T. Chen, X. Chen and X. M. Zhao, *Appl. Microbiol. Biotechnol.*, 2006, **70**, 151.
2. C. Barbas, E. P. Moraes and A. Villasenor, *J. Pharm. Biomed. Anal.*, 2011, **55**, 823.
3. W. B. Dunn, N. J. Bailey and H. E. Johnson, *Analyst*, 2005, **130**, 606.
4. K. Dettmer, P. A. Aronov and B. D. Hammock, *Mass Spectrom. Rev.*, 2007, **26**, 51.

5. H. K. Kim, Y. H. Choi and R. Verpoorte, *Trends Biotechnol.*, 2011, **29**, 267.
6. M. R. Monton and T. Soga, *J. Chromatogr. A*, 2007, **1168**, 237.
7. R. Ramautar, O. A. Mayboroda, G. W. Somsen and G. J. de Jong, *Electrophoresis*, 2011, **32**, 52.
8. R. Ramautar, G. W. Somsen and G. J. de Jong, *Electrophoresis*, 2009, **30**, 276.
9. K. Isoo and S. Terabe, *Anal. Chem.*, 2003, **75**, 6789.
10. S. R. Bean, G. L. Lookhart and J. A. Bietz, *J. Agric. Food Chem.*, 2000, **48**, 318.
11. D. W. Armstrong, G. Schulte, J. M. Schneiderheinze and D. J. Westenberg, *Anal. Chem.*, 1999, **71**, 5465.
12. T. Hasunuma, K. Harada, S. Miyazawa, A. Kondo, E. Fukusaki and C. Miyake, *J. Exp. Bot.*, 2010, **61**, 1041.
13. N. Ishii, K. Nakahigashi, T. Baba, M. Robert, T. Soga, A. Kanai, T. Hirasawa, M. Naba, K. Hirai, A. Hoque, P. Y. Ho, Y. Kakazu, K. Sugawara, S. Igarashi, S. Harada, T. Masuda, N. Sugiyama, T. Togashi, M. Hasegawa, Y. Takai, K. Yugi, K. Arakawa, N. Iwata, Y. Toya, Y. Nakayama, T. Nishioka, K. Shimizu, H. Mori and M. Tomita, *Science*, 2007, **316**, 593.
14. E. Nevedomskaya, R. Ramautar, R. Derks, I. Westbroek, G. Zondag, I. Pluim, A. M. Deelder and O. A. Mayboroda, *J. Proteome Res.*, 2010, **9**, 4869.
15. R. Ramautar, O. A. Mayboroda, A. M. Deelder, G. W. Somsen and G. J. de Jong, *J. Chromatogr. B*, 2008, **871**, 370.
16. S. Sato, T. Soga, T. Nishioka and M. Tomita, *Plant J.*, 2004, **40**, 151.
17. T. Soga, Y. Ohashi, Y. Ueno, H. Naraoka, M. Tomita and T. Nishioka, *J. Proteome Res.*, 2003, **2**, 488.
18. C. Barbas, M. Vallejo, A. Garcia, D. Barlow and M. Hanna-Brown, *J. Pharm. Biomed. Anal.*, 2008, **47**, 388.
19. J. Hou, J. Zheng and S. A. Shamsi, *J. Chromatogr. A*, 2007, **1159**, 208.
20. Y. Ishihama, H. Katayama and N. Asakawa, *Anal. Biochem.*, 2000, **287**, 45.
21. H. Ozaki, N. Itou, S. Terabe, Y. Takada, M. Sakairi and H. Koizumi, *J. Chromatogr. A*, 1995, **716**, 69.
22. K. Koezuka, H. Ozaki, N. Matsubara and S. Terabe, *J. Chromatogr. B*, 1997, **689**, 3.
23. W. Bragg and S. A. Shamsi, *J. Chromatogr. A*, 2011, **1218**, 8691.
24. G. Xie, M. Su, P. Li, X. Gu, C. Yan, Y. Qiu, H. Li and W. Jia, *Electrophoresis*, 2007, **28**, 4459.
25. J. Zheng, W. Bragg, J. Hou, N. Lin, S. Chandrasekaran and S. A. Shamsi, *J. Chromatogr. A*, 2009, **1216**, 857.
26. E. Barcelo-Barrachina, E. Moyano and M. T. Galceran, *Electrophoresis*, 2004, **25**, 1927.
27. N. Johannesson, L. Olsson, D. Backstrom, M. Wetterhall, R. Danielsson and J. Bergquist, *Electrophoresis*, 2007, **28**, 1435.
28. M. L. Riekkola, *Electrophoresis*, 2002, **23**, 3865.

29. I. Bjornsdottir, J. Tjornelund and S. H. Hansen, *Electrophoresis*, 1998, **19**, 2179.
30. K. Vuorensola, H. Siren, R. Kostiainen and T. Kotiaho, *J. Chromatogr. A*, 2002, **979**, 179.
31. C. Huhn, R. Ramautar, M. Wuhrer and G. W. Somsen, *Anal. Bioanal. Chem.*, 2010, **396**, 297.
32. R. Ramautar, E. Nevedomskaya, O. A. Mayboroda, A. M. Deelder, I. D. Wilson, H. G. Gika, G. A. Theodoridis, G. W. Somsen and G. J. de Jong, *Mol. BioSyst.*, 2011, **7**, 194.
33. R. Ramautar, A. A. van der Plas, E. Nevedomskaya, R. J. Derks, G. W. Somsen, G. J. de Jong, J. J. van Hilten, A. M. Deelder and O. A. Mayboroda, *J. Proteome Res.*, 2009, **8**, 5559.
34. J. Ohnesorge, C. Sanger-van de Griend and H. Watzig, *Electrophoresis*, 2005, **26**, 2360.
35. A. Staub, J. Schappler, S. Rudaz and J. L. Veuthey, *Electrophoresis*, 2009, **30**, 1610.
36. G. Bringmann, I. Kajahn, C. Neususs, M. Pelzing, S. Laug, M. Unger and U. Holzgrabe, *Electrophoresis*, 2005, **26**, 1513.
37. J. Han, R. M. Danell, J. R. Patel, D. R. Gumerov, C. O. Scarlett, J. P. Speir, C. E. Parker, I. Rusyn, S. Zeisel and C. H. Borchers, *Metabolomics*, 2008, **4**, 128.
38. E. E. Baidoo, P. I. Benke, C. Neususs, M. Pelzing, G. Kruppa, J. A. Leary and J. D. Keasling, *Anal. Chem.*, 2008, **80**, 3112.
39. J. Ding and P. Vouros, *Anal. Chem.*, 1999, **71**, 378A.
40. J. L. Edwards and R. T. Kennedy, *Anal. Chem.*, 2005, **77**, 2201.
41. V. Sanz-Nebot, F. Benavente, E. Balaguer and J. Barbosa, *Electrophoresis*, 2003, **24**, 883.
42. K. Vuorensola, J. Kokkonen, H. Siren and R. A. Ketola, *Electrophoresis*, 2001, **22**, 4347.
43. T. Soga and D. N. Heiger, *Anal. Chem.*, 2000, **72**, 1236.
44. T. Soga, K. Igarashi, C. Ito, K. Mizobuchi, H. P. Zimmermann and M. Tomita, *Anal. Chem.*, 2009, **81**, 6165.
45. Y. R. Chen and G. R. Her, *Rapid Commun. Mass Spectrom.*, 2003, **17**, 437.
46. Z. Kele, G. Ferenc, E. Klement, G. K. Toth and T. Janaky, *Rapid Commun. Mass Spectrom.*, 2005, **19**, 881.
47. M. S. Kriger, K. D. Cook and R. S. Ramsey, *Anal. Chem.*, 1995, **67**, 385.
48. V. Sanz-Nebot, E. Balaguer, F. Benavente and J. Barbosa, *Electrophoresis*, 2005, **26**, 1457.
49. A. D. Zamfir, N. Dinca, E. Sisu and J. Peter-Katalinic, *J. Sep. Sci.*, 2006, **29**, 414.
50. H. J. Issaq, G. M. Janini, K. C. Chan and T. D. Veenstra, *J. Chromatogr. A*, 2004, **1053**, 37.
51. P. Cao and M. Moini, *J. Am. Soc. Mass Spectrom*, 1997, **8**, 561.
52. Z. Chen, B. Boggess and H. C. Chang, *J. Mass Spectrom.*, 2007, **42**, 244.

53. L. H. Shi, Y. X. Jin, D. C. Moon, S. K. Kim and S. R. Park, *Electrophoresis*, 2009, **30**, 1661.
54. M. Moini, *Anal. Chem.*, 2007, **79**, 4241.
55. J. M. Busnel, B. Schoenmaker, R. Ramautar, A. Carrasco-Pancorbo, C. Ratnayake, J. S. Feitelson, J. D. Chapman, A. M. Deelder and O. A. Mayboroda, *Anal. Chem.*, 2010, **82**, 9476.
56. K. Faserl, B. Sarg, L. Kremser and H. Lindner, *Anal. Chem.*, 2011, **83**, 7297.
57. R. Haselberg, C. K. Ratnayake, G. J. de Jong and G. W. Somsen, *J. Chromatogr. A*, 2010, **1217**, 7605.
58. M. Kawai, Y. Iwamuro, R. Iio-Ishimaru, S. Chinaka, N. Takayama and K. Hayakawa, *Anal. Sci.*, 2011, **27**, 857.
59. R. B. Cole, *Electrospray Ionization Mass Spectrometry: Fundamentals, Instrumentation and Applications*, Wiley, New York, 1997.
60. S. L. Nilsson, C. Andersson, P. J. Sjoberg, D. Bylund, P. Petersson, M. Jornten-Karlsson and K. E. Markides, *Rapid Commun. Mass Spectrom.*, 2003, **17**, 2267.
61. P. Hommerson, A. M. Khan, G. J. de Jong and G. W. Somsen, *Mass Spectrom. Rev.*, 2011, **30**, 1096.
62. S. J. Bos, S. M. van Leeuwen and U. Karst, *Anal. Bioanal. Chem.*, 2006, **384**, 85.
63. P. Hommerson, A. M. Khan, G. J. DeJong and G. W. Somsen, *J. Am. Soc. Mass Spectrom*, 2009, **20**, 1311.
64. K. A. Hanold, S. M. Fischer, P. H. Cormia, C. E. Miller and J. A. Syage, *Anal. Chem.*, 2004, **76**, 2842.
65. Y. Tanaka, K. Otsuka and S. Terabe, *J. Pharm. Biomed. Anal.*, 2003, **30**, 1889.
66. Y. Takada, M. Sakairi and H. Koizumi, *Anal. Chem.*, 1995, **67**, 1474.
67. J. Schappler, D. Guillarme, J. Prat, J. L. Veuthey and S. Rudaz, *Electrophoresis*, 2007, **28**, 3078.
68. J. Hernandez-Borges, C. Neususs, A. Cifuentes and M. Pelzing, *Electrophoresis*, 2004, **25**, 2257.
69. Y. Deng, H. Zhang and J. Henion, *Anal. Chem.*, 2001, **73**, 1432.
70. W. C. Sung, H. Makamba and S. H. Chen, *Electrophoresis*, 2005, **26**, 1783.
71. F. Foret and P. Kusy, *Electrophoresis*, 2006, **27**, 4877.
72. P. Hoffmann, U. Hausig, P. Schulze and D. Belder, *Angew. Chem. Int. Ed.*, 2007, **46**, 4913.
73. W. C. Sung, S. Y. Huang, P. C. Liao, G. B. Lee, C. W. Li and S. H. Chen, *Electrophoresis*, 2003, **24**, 3648.
74. J. Kameoka, H. G. Craighead, H. Zhang and J. Henion, *Anal. Chem.*, 2001, **73**, 1935.
75. F. A. Li, J. L. Huang and G. R. Her, *Electrophoresis*, 2008, **29**, 4938.
76. M. Sugimoto, A. Hirayama, T. Ishikawa, M. Robert, R. Baran, K. Uehara, K. Kawai, T. Soga and M. Tomita, *Metabolomics*, 2010, **6**, 27.
77. M. Katajamaa and M. Oresic, *BMC Bioinf.*, 2005, **6**, 179.

78. A. Nordstrom, G. O'Maille, C. Qin and G. Siuzdak, *Anal. Chem.*, 2006, **78**, 3289.
79. C. A. Smith, E. J. Want, G. O'Maille, R. Abagyan and G. Siuzdak, *Anal. Chem.*, 2006, **78**, 779.
80. R. C. De Vos, S. Moco, A. Lommen, J. J. Keurentjes, R. J. Bino and R. D. Hall, *Nat. Protocols*, 2007, **2**, 778.
81. R. Baran, H. Kochi, N. Saito, M. Suematsu, T. Soga, T. Nishioka, M. Robert and M. Tomita, *BMC Bioinf.*, 2006, **7**, 530.
82. R. Baran, M. Robert, M. Suematsu, T. Soga and M. Tomita, *BMC Bioinf*, 2007, **8**, 72.
83. M. Sugimoto, D. T. Wong, A. Hirayama, T. Soga and M. Tomita, *Metabolomics*, 2010, **6**, 78.
84. M. Sugimoto, S. Kikuchi, M. Arita, T. Soga, T. Nishioka and M. Tomita, *Anal. Chem.*, 2005, **77**, 78.
85. M. Sugimoto, A. Hirayama, M. Robert, S. Abe, T. Soga and M. Tomita, *Electrophoresis*, 2010, **31**, 2311.
86. R. Lee, A. S. Ptolemy, L. Niewczas and P. Britz-McKibbin, *Anal. Chem.*, 2007, **79**, 403.
87. A. Hirayama, K. Kami, M. Sugimoto, M. Sugawara, N. Toki, H. Onozuka, T. Kinoshita, N. Saito, A. Ochiai, M. Tomita, H. Esumi and T. Soga, *Cancer Res.*, 2009, **69**, 4918.
88. K. Harada, Y. Ohyama, T. Tabushi, A. Kobayashi and E. Fukusaki, *J. Biosci. Bioeng.*, 2008, **105**, 249.
89. C. Desiderio, D. V. Rossetti, I. Messana, B. Giardina and M. Castagnola, *Electrophoresis*, 2010, **31**, 1894.
90. U. Ceglarek, A. Leichtle, M. Brugel, L. Kortz, R. Brauer, K. Bresler, J. Thiery and G. M. Fiedler, *Mol. Cell. Endocrinol.*, 2009, **301**, 266.
91. T. Soga, R. Baran, M. Suematsu, Y. Ueno, S. Ikeda, T. Sakurakawa, Y. Kakazu, T. Ishikawa, M. Robert, T. Nishioka and M. Tomita, *J. Biol. Chem.*, 2006, **281**, 16768.
92. R. Ramautar, J. S. Torano, G. W. Somsen and G. J. de Jong, *Electrophoresis*, 2010, **31**, 2319.
93. R. Lee, D. West, S. M. Phillips and P. Britz-McKibbin, *Anal. Chem.*, 2010, **82**, 2959.

CHAPTER 7

Microchip Technology in Metabolomics

TIINA M. SIKANEN

University of Helsinki, Faculty of Pharmacy, Division of Pharmaceutical Chemistry, Viikinkaari 5E, 00014 Helsinki, Finland
Email: tiina.sikanen@helsinki.fi

7.1 Device Miniaturization Through Microfabrication

Miniaturization of analytical instrumentation is often suggested to be one of the most promising approaches to rise to the challenge of high-throughput screening. In this chapter, miniaturization is considered from the viewpoint of microfabrication technology, which allows for new perspectives in device design in terms of process integration and parallelization. Miniaturization of analytical instruments *via* microfabrication dates back to the progressive development of silicon microfabrication processes in the early 1980s. This progress eventually led to an explosive growth of semiconductor technology so that all key electrical components (*e.g.*, transistors, diodes, integrated circuits, solar cells) could be fabricated at such low cost that computers, mobile phones and other electrical supplies became available to average users. From the late 1990s, the same microfabrication processes have been increasingly applied to the production of miniaturized devices for chemical analysis, often referred to as microfluidic total analysis systems (μTAS) or lab(oratory)-on-a-chip technology.[1–3] In addition to chemical analysis, microfabricated electromechanical systems are widely developed for various biomedical or biological applications and this approach is commonly abbreviated to BioMEMS.[4–6]

RSC Chromatography Monographs No. 18
Chromatographic Methods in Metabolomics
Edited by Tuulia Hyötyläinen and Susanne Wiedmer

Published by the Royal Society of Chemistry, www.rsc.org

Although these acronyms largely refer to the same technology, one should bear in mind that BioMEMS is by default concerned with the engineering of mechanical parts for biomedical applications, whereas lab-on-a-chip is more focused on miniaturization and integration of chemical laboratory (primarily separation) processes and experiments into a single microfluidic chip.

Microfluidics by definition is a field of study that encompasses the physics of fluid behaviour on small scales and the engineering aspects of the design and fabrication of devices for controlling the flow of small amounts of fluids. Through microfabrication, all critical unit operations of a (bio)chemical analysis (*e.g.*, sample preparation, reaction, separation and detection of the reaction products) can be combined on a single microfluidic chip through multiple interconnected microchannels and reservoirs.[3] This results in negligible dead volumes (*i.e.*, no time lag) at the intersections between the different unit operations and thus increased sample throughput. At best, sub-millisecond separations with nearly baseline resolution have been reported,[7] suggesting a theoretical throughput of up to thousands of samples per hour. In practice, for all but the most extreme applications, the total cycle time is of the order of 1 min per analysis with a more realistic throughput being tens of samples per hour. In addition to the improved speed of a single analysis, the possibility of parallelization of multiple concurrent analyses is highly advantageous for high-throughput screening.

To date, a variety of microfluidic separation systems have been developed, inspired by the introduction of the first chip-based gas chromatography (GC)[8] and capillary electrophoresis (CE)[9,10] systems in the late 1970s and early 1990s, respectively. In general, it is the improved speed of analysis, specifically, that justifies the use of miniaturized separation systems instead of more traditional solutions. The possibility of sampling very small volumes within the microfluidic channels is also a great advantage when analysing, for instance, cellular contents. Even if sample volume is not the limiting factor, savings in terms of the consumption of expensive (derivatization) reagents or the production of toxic waste that come together with miniaturization are advantageous. Despite the proven potential of microchip-based separation systems in high-throughput analysis, the technology has hardly found use in the field of small-molecule analysis or metabolomics *per se*.[11] Instead, microchip technology has mainly been applied in research fields such as proteomics,[3,12,13] where the target analytes, *i.e.*, peptides and proteins, are easily derivatized with appropriate chromophores to facilitate sensitive detection by laser-induced fluorescence (LIF) microscopy. Microchip capillary electrophoresis (MCE) in combination with LIF detection, in particular, has found its place in rapid screening of proteomic samples, although shortage of automation has significantly slowed commercialization of the technology. Many of the endogenous metabolites, in turn, are not inherently fluorescent or do not have functional groups that allow fluorescence coupling reactions, yet they need to be identified at trace-level concentrations. This poses the major limitation for application of microchip technology to metabolomics, as MCE-LIF, the 'gold standard' for miniaturized separation systems, is not directly feasible. Of the alternative detection techniques, electrospray ionization mass spectrometry (ESI-MS) provides by far the best sensitivity and selectivity in chip-based separation systems, as in traditional systems, but suffers from relatively complex realization of the electrospray emitter as an integral part of the separation chip.[14,15] Thus, microfabricated ESI emitters have found the most widespread use as direct infusion interfaces and are only rarely integrated with separation systems.[16] Direct UV (absorbance) detection based on linear imaging of the entire microchannel[17–19] is another option for the detection of microchip-based separations (primarily MCE), but only if high sensitivity is not an issue. This renders electrochemical (EC) detection the most likely method of choice for metabolite monitoring as both the EC electrodes and the control instrumentation required are easily implemented by microfabrication means.[20] Although the sensitivity and selectivity of EC are not as good as those of ESI-MS, these are not compromised because of miniaturization either, and typically the detection sensitivity achieved is similar to or better than that of the traditional EC systems.

In this chapter, past applications of microchip technology to metabolomics are reviewed from the viewpoint of the detection sensitivity obtained and with improved throughput as a default value. In addition, the vast potential of lab-on-chip technology in terms of process integration is considered with the help of selected applications where not only separation of metabolites (in body fluids) but also enzymatic reactions are performed on the same microfluidic chip. On the microscale, certain differences emerge as a result of downscaling, *e.g.*, the flow is laminar instead of turbulent, the diffusion times are essentially short and capillary forces often dominate over body forces.[21] These features add unique challenges (*e.g.*, mixing, scaling of pressure-driven flow), but also many opportunities (*e.g.*, fast diffusion, capillary flow) that are not feasible with traditional systems. For instance, the possibility of exploiting microfluidics in cell culturing with *in situ* analysis[22,23] or building up microfluidic tissue-based organomimetic systems[24] certainly appear to be fascinating future approaches for targeted metabolomics. In addition to electric field-driven and pressure-driven separation systems, there are a few marginal, but enabling, technologies, such as digital microfluidics based on electrical actuation of nanolitre droplets on a planar array of electrodes,[25,26] the application of which to metabolomics is also briefly touched upon in this chapter. Most of all, this chapter provides the reader with an overview of the main technological advantages and disadvantages of lab-on-a-chip technology from the viewpoint of metabolomics.

7.2 Microchip-based Separation Systems

7.2.1 The Choice of the Microfabrication Material

The first miniaturized analytical systems[8,27] were fabricated on silicon by etching techniques already established in semiconductor technology. Despite these well-established methods, silicon suffers from two main disadvantages that limit its use as a structural material in microfabricated separation systems, namely the optical opacity to visible light and the electrical conductivity that precludes electric field-driven applications, such as MCE. The advantages of glass and quartz over silicon are therefore unambiguous, including not only optical transparency and electrical insulation, but also the well-known surface chemistry resembling that of fused-silica capillaries so that basically all traditional surface modification techniques are easily applied to altering the surface properties of glass.[28] However, glass microfabrication (primarily etching) has been less explored and is generally more difficult than processing of silicon. For instance, implementation of a sharp, monolithic ESI emitter tip on the outlet of a glass microchannel has long remained a major challenge and researchers have ended up gluing or screwing external nanospray needles to the microchannel outlet to allow for off-chip ESI-MS detection.[14,15] Only fairly recently have a few groups independently developed different microfabrication approaches for the realization of sharp on-chip emitter tips made of glass.[29–31] In terms of optical detection, however, quartz-based microdevices are superior to any other material as they allow for both absorbance and fluorescence detection in the low-UV range.[32]

Currently, polymer microfabrication has been adopted beside glass/quartz and silicon processing allowing rapid and simple approaches to low-cost chip fabrication.[33,34] The commonly used bulk polymers (*i.e.*, plastics) provide a wide variety of materials properties,[35] and are often classified into three main categories according to their mechanical properties:

- hard and mechanically strong thermoplastics, *e.g.*, poly(methyl methacrylate) (PMMA), polycarbonate (PC), poly(ethylene terephthalate) (PET), cycoolefin copolymer (COC), polystyrene (PS) and Parylene C;
- hard, but brittle thermoset plastics, *e.g.*, epoxy polymer SU-8, polyimide (PI); and
- soft and flexible elastomers, *e.g.*, polydimethylsiloxane (PDMS).

Similarly, the polymer microfabrication methods can be divided into two main categories:[28]

- replication from a master (hot embossing/imprinting, injection moulding, casting); and
- direct machining methods [laser ablation/micromachining, micromilling, chemical vapour deposition (CVD), (UV) lithography].

Many thermoplastics can be patterned by both replication (embossing and injection moulding) and direct machining methods (laser ablation, micromilling), whereas thermosets are almost solely patterned by UV lithography and elastomers by casting.[33] The direct machining methods are generally slower or more complex (with respect to the required instrumentation), but often allow more advanced microstructures than replication methods. The differences between different microfabrication materials and methods play a particular role if one wants to add structural complexity to a polymer-based separation system. For instance, bonding of multiple layers or implementation of a sharp, needle-like emitter tip as an integral (monolithic) part of the microchannel are also challenges for polymer microfabrication. In general, lithographically defined polymers (*e.g.*, epoxy polymer SU-8) provide the best structural accuracy and feature resolution, yet their limited film thickness results in poor mechanical strength. On the other hand, silicon substrates patterned with lithographically defined (SU-8) microstructures, as negative masters for PDMS casting, also provide fairly high accuracy and resolution in PDMS devices. For a more comprehensive view of the different microfabrication materials and methods, the reader is referred to relevant literature.[36,37]

At present, fabrication of a microchannel is possible in practice by any available method from any available polymer. There are also plenty of surface modification techniques that allow fine tuning of the surface properties of the common polymers towards any desired physical, chemical or biological function.[28,38] In addition, polymers are electrically insulating, which favours their use in electric field-driven applications. However, one should bear in mind that most polymers, apart from PDMS, PI and the recently introduced set of commercial inorganic–organic hybrid polymers (Ormocers),[39] are not transparent to UV light, which prevents their use in absorbance detection. Autofluorescence originating from the polymer itself is another limiting factor with respect to the sensitivity of fluorescence detection even in the visible range.[40,41] Most polymers, some (*e.g.*, PDMS, PMMA) more than others, are also prone to swelling and degradation upon exposure to organic solvents,[42,43] which may hinder their use in LC or ESI-MS applications where the use of organic additives is often required.

In addition to inherent materials properties, there are also a few material related features of microfabricated structures that have an effect on the operation of microfluidic separation devices, such as channel cross-section shape and surface roughness. The former mainly affects the packing of stationary phase material (for LC or solid-phase extraction), whereas the latter is more of an issue with optical and electric field-driven systems. The typical microchannel cross-section shapes achieved through direct micromachining of polymers are square and rectangular (Figure 7.1a), whereas the polymer replication methods also reproduce an inverse copy of the master geometry, typically vertical-walled structures.[36,37] Also, glass (and silicon) microchannels have vertical walls if plasma (anisotropic) etching is used. In contrast, the cross-section shape resulting from isotropic etching of glass (or silicon) is semicircular (Figure 7.1b). In all cases, the microchannel cross-section typically features at least two sharp corners and these appear particularly problematic when packing the stationary phase inside the microchannel.[44,45] It is difficult to fill the vertical-walled periphery of the microchannel evenly, hence large void volumes often appear in the corners. The total absence of cylindrical-shaped channels is mainly due to the challenges of fabricating them on microchips.

In electric field-driven systems, the effect of the microchannel cross-section shape on the stability of the electroosmotic flow (EOF) or the separation

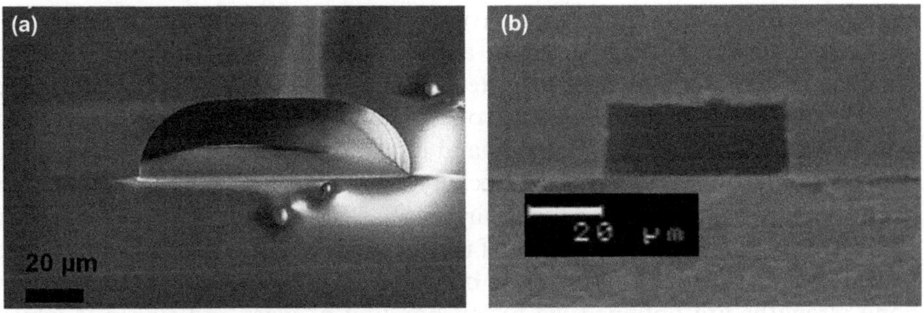

Figure 7.1 Typical cross-section shapes of (a) glass-based and (b) polymer-based (SU-8) microchannels patterned by isotropic etching and UV lithography, respectively.
(a) Reprinted with permission from Ref. 31. Copyright 2012 American Chemical Society. (b) Courtesy of T. Sikanen.

efficiency is probably less pronounced, but the aspect ratio (height-to-width) is a critical factor when controlling the fluidic resistance.[46] As the fluidic resistance scales to the power of three to the microchannel height (h^3), wide and shallow microchannels are superior to narrow and deep microchannels in the elimination of pressure-induced flow. In addition, the surface roughness may play a role with respect to the stability of the EOF. On the other hand, the smoother the surface, the smaller the optical loss (scattering of light) at the interface. In general, polymer processing results in relatively smooth surfaces compared with etching of glass (or silicon), hence the use of polymers is especially advantageous when integrating passive microoptical elements (*e.g.*, planar waveguides, microlenses and mirrors) with microfluidic separation systems.[47,48] Although this technology has not yet found its full potential in chemical analysis, its benefits to optical detection of small molcecules are inevitable and include an increase in the optical pathlength and focusing of both incoming (excitation) and outgoing (emission) light beam in addition to reduction of the optical losses.[49] The poor adhesion of thin-film metals on some polymers may also pose restrictions on the use of EC detection as the electrodes cannot be patterned directly on these materials. Often, the problem is circumvented by patterning the thin-film electrodes on a glass substrate, which is then attached to the polymer-based separation system.[50] Another approach is to use microwires, aligned through an on-chip electrode alignment channel, as working electrodes for EC detection.[51]

The bonding of microchannels to prepare enclosed systems is yet another issue to be accounted for.[52] Heat-curable PDMS, in particular, has found widespread use in microfluidic applications thanks to its convenient bonding based on reversible self-adhesion to almost any material.[53] However, the bond strength of two self-adhered PDMS substrates is of the order of 100 kPa and suffices for low-pressure applications, such as MCE, only. Through chemical treatment (*e.g.*, oxygen plasma) of the surfaces to be bonded, irreversible

PDMS sealing with bond strength averaging 300 kPa[53] can also be achieved. For other polymers and for glass and silicon, the microchannel bonding remains more of an issue and separate adhesive bonding techniques need to be developed.[52] Additionally, all walls of the same material are often preferred because of the better analysis quality and repeatability associated with uniform surface properties, these being especially important for electric field-driven applications that rely on a uniform surface charge. Typically, polymer bonding techniques[52] permit the fabrication of microchannels with all (four) walls of the same material with reasonably good bond strengths. However, by far the best bond strengths are obtained *via* fusion bonding of glass and silicon,[54] which renders them the most feasible fabrication material for microchip LC and other applications where high pressure tolerance is required.

Finally, another important aspect of the microchip format is the packing density, which is very similar the different microfabrication materials and methods, but affects the overall cost efficiency of the microchip technology. In order to achieve the most economic approach for the fabrication of the separation chips, it is advantageous to include as many as possible individual separation systems on a 4 in (silicon) wafer, which is the standard starting point (substrate) for nearly all microfabrication processes regardless of the structural material of the final microchip. This inevitably leads to a situation where the length of the separation channel is limited by the diameter of the wafer unless the channel is folded into a serpentine shape. Although addition of microchannel turns improves the packing density and is extremely straightforward in terms of fabrication, it has a less favourable effect on the separation efficiency (Figure 7.2).[55] In general, folding of the separation channel introduces an additional geometric contribution to analyte dispersion through lateral variations in migration distance (and electric field strength). Therefore, compensation structures, such as tapering curves, need to be added to meandering microchannels in order to preserve the separation efficiency of both electric field-driven and pressure-driven separation systems.[56,57]

With the above-mentioned limitations in mind, the choice of the appropriate microfabrication material and method is typically made on the basis of certain governing factors, such as minimum feature size, overall cost and the speed and complexity of the fabrication process. In addition, transparency to visible light and inherent biocompatibility and/or preknown surface chemistry significantly ease the practical work, for instance, filling of the microchannel and prevention/promotion of the unwanted/wanted interactions on the microchannel walls. In general, polymers are useful materials for rapid and simple chip fabrication at significantly lower cost than is achievable through glass or silicon processing. However, when extreme performance in terms of optical detection or pressure tolerance is desired, glass (quartz) is often the material of choice. Needless to say, silicon only rarely proves useful for the fabrication of microchip-based separation devices because its opaqueness and electrical conductivity prevent both optical and EC or ESI-MS detection, respectively.

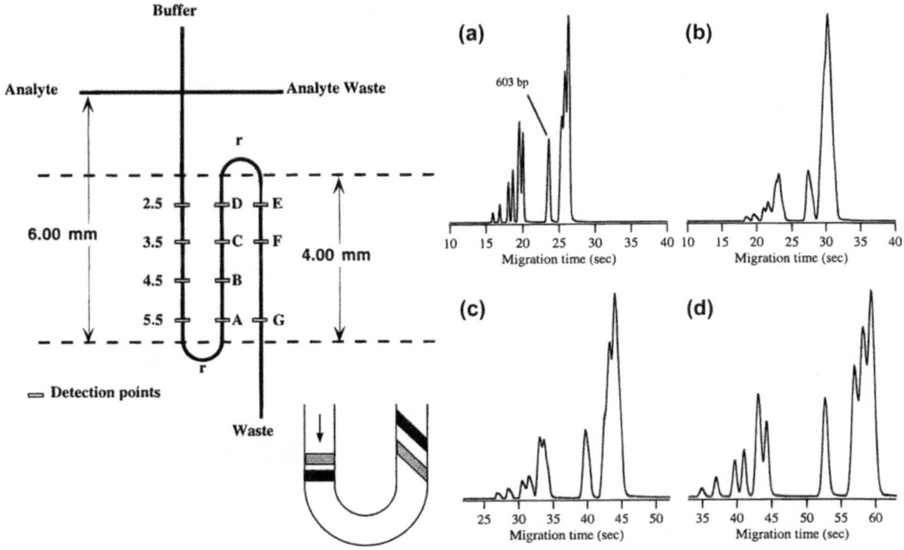

Figure 7.2 A meandering microchannel incorporating two turns with insert of the microchannel turn illustrating the excess dispersion introduced by the turn as the molecules situated along the inside of the turn move more quickly through the turn than those along the outside. The numbered detection points indicate the distance from the injection cross in millimetres, and the letters represent the locations of detection points. Electropherograms measured prior to the first turn (a), after first turn (b), prior to second turn (c) and after the second turn (d) clearly show the dispersion effect in meandering microchannels.
Reprinted with permission from Ref. 55. Copyright 1998 American Chemical Society.

7.2.2 Notes on the Detection Sensitivity

In general, the two most important features of a detection method are selectivity and sensitivity. Whereas the assay selectivity is mainly determined by the detection method used, the sensitivity of a microfluidic separation system also greatly depends on the fabrication material. Both the materials properties and the possibility of simultaneous fabrication of the detection unit as an integral part of the microfluidic chip are of great importance. As discussed in Section 7.2.1, the following material-related limitations should be taken into account:

- *Optical detection:* quartz is practically the only available material for absorbance detection in the deep-UV range accompanied by glass and PDMS in the near-UV range; most polymers are only applicable to absorbance/fluorescence detection in the visible range.
- *EC:* adhesion of the thin-film metals required for electrodes is poor on certain polymers.

- *ESI-MS:* implementation of a sharp (monolithic) ESI emitter tip to the outlet of a separation channel is possible with only a few microfabrication techniques and materials (*e.g.*, PI, SU-8, glass); the compatibility of some polymers (*e.g.*, PDMS, PMMA) with organic solvents is limited.

When developing miniaturized separation systems, one should also bear in mind that even if the microfluidic separation techniques are well established, the detection is still strongly dependent on the conventional, macroscale instrumentation, as can be seen from Figure 7.3. Apart from microfabricated EC

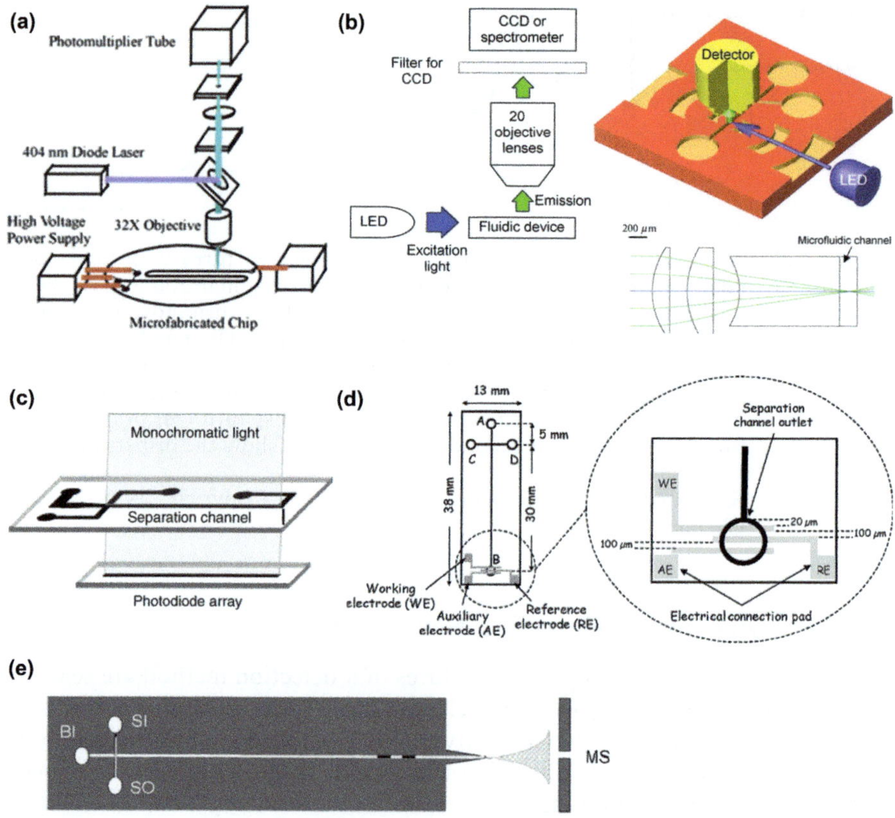

Figure 7.3 The different detection setups used for monitoring of microfluidic separation. (a) Fluorescence detection by epifluorescence microscopy. Reproduced from Ref. 67. Copyright 1999 Royal Society of Chemistry. (b) Fluorescence detection with help of planar, integrated optics (microlenses). Reprinted with permission from Ref. 68. Copyright 2004 Elsevier. (c) UV absorbance detection with a linear photodiode array. Reprinted with permission from Ref. 62. Copyright 2006 Elsevier. (d) EC detection by end-channel amperometric detector. Reprinted with permission from Ref. 50. Copyright 2009 Elsevier. (e) ESI chip in front of a mass spectrometer. Reprinted from Ref. 30 with permission of John Wiley & Sons, Inc. ©2007.

electrodes, ion sources or the sparsely used integrated microoptical elements, the active parts of detectors are external to the microchip. Hence the detection sensitivity is often compromised by the decrease in the detection volume that comes together with miniaturization. For instance, there is an inherent difference in the concentration sensitivity between microchip CE and LC techniques. This is because of their fundamentally different approach for sample introduction. In microchip CE, the sample is commonly introduced through an on-chip injection channel which intersects the main separation channel so that the injection volume is equal to the volume of the intersection.[58] In this manner, the sample zone becomes very narrow – which is the primary source of the inherent high efficiency of MCE, as will be discussed in Section 7.2.3 – but also the volume of the injected sample plug is ultimately small, typically only a few tens to hundreds of picolitres. In theory, the small volume should not be a problem as the sensitivity of the detection method used scales with sample concentration. In practice, with the short optical pathlengths in microfluidic systems, absorbance detection offers only limited sensitivity unless the pathlength is increased through the use of, for instance, microfabricated multireflection cells.[59,60] Another approach for on-chip absorbance detection is to use linear photodiode arrays which have been successfully applied for the simultaneous monitoring of the entire separation channel at a concentration level of 10^{-5} M.[19,61–63] Similarly, the sensitivity in fluorescence detection is dependent on the optical pathlength and does not directly benefit from miniaturization. However, the high excitation power and ease of focusing make LIF feasible for highly sensitive detection, as the background noise decreases along with reduced detection volume.[21] In microchip LC, in turn, inherent sample preconcentration results from the retention of the sample components on the stationary phase. Thus, the sample can be loaded into the separation column in much larger volumes (typically nanolitres at the minimum) and this allows for inherently better sensitivity in microchip LC than in MCE. Often, sample loading is performed with the help of an external injector (*e.g.*, LC autosampler), but LC microchips with on-chip injectors have also been reported.[64–66] The volume of an on-chip injector loop is typically tens of nanolitres,[64,65] which is less than those of external injectors, but still about 100 times larger than the volume of the common on-chip injector geometries used in MCE.

The narrow sample zone (peak width), eminently characteristic of MCE, also has its downside if the time constant of the detector becomes the limiting factor for the detection sensitivity. In many MS analysers, for example, the number of temporal data points is not sufficient for reliable sampling of the very narrow peaks (often <1 s) as they pass by the MS detector. Often, for reliable sampling of such narrow peaks, the dwell time has to be decreased to a minimum, which inevitably reduces the detection sensitivity. Thus, targeted monitoring of a single metabolite (or a few) in a tandem MS (MS/MS) configuration with inherently short dwell times appears to be the most reasonable approach for the combination of MCE and ESI-MS.[69,70] Alternatively, where monitoring of a broad mass range (multiple metabolites) is desired, one should use a mass spectrometer with inherent spectral resolution at a high data acquisition rate, such as a time-of-flight (TOF) instrument.[71] In EC and optical detection, the temporal resolution is usually not an issue as signal recording can be performed using high-frequency data loggers. Also, in the case of MS detection, one should bear in mind that the ionization process *per se* usually benefits from miniaturization through improved efficiency of the downscaled process (*e.g.*, ESI *versus* nanospray).[72,73] Hence the best sensitivity is usually obtained when coupling microchip-based separation systems to MS detection. In addition to microfluidic separation systems, microfabricated ion sources can be used for direct infusion experiments and some of them also in combination with traditional separation systems,[14,15] when the time constant of the MS detector becomes less of an issue. For instance, the NanoMate 100 system from Advion Biosciences based on a microfabricated array of silicon emitters allow detection limits commonly in the low or even sub-nanomolar range for drugs and their metabolites and are therefore widely used to replace traditional nanospray needles.[16]

In addition to optical, EC and MS detection, other detection modes have also been used in the context of microfluidic separation systems, although only rarely applied to metabolomics. For instance, a microfluidic blood-sampling device made of PDMS has been coupled to radiological detection in order to allow the determination of the kinetics of fluorodeoxyglucose (^{18}F) in the mouse brain and myocardial tissues *via* quantitative small-animal positron emission tomography (PET).[74] On rare occasions, the reaction products of an on-chip enzymatic reaction have been monitored by off-chip nuclear magnetic resonance (NMR)[75] or Fourier transform infrared (FT-IR) spectroscopy,[76] or a MEMS-based thermal biosensor.[77] The feasibility of using microcoil-based probes for NMR spectroscopy[78,79] and of microchips fully or partly made of calcium fluoride for IR spectroscopy[80,81] has also been shown, but not applied to metabolite detection. Direct image analysis based on fluorescence, bioluminescence or chemiluminescence detection is another relevant technique for targeted metabolite monitoring, particularly in the context of cell-based assays that are briefly discussed in Section 7.4.

7.2.3 Microchip Capillary Electrophoresis

The initial excitement about microfabricated separation systems in the early 1990s goes back to Monnig and Jorgenson, who first explained that resolution in CE is independent of column length and band spreading is proportional to the square root of separation time and diffusion coefficient.[82] In practice, one should use the highest possible voltage before the mobile phase overheats and the shortest possible column in order to reach maximum efficiency. Thus, the limiting factor with traditional CE capillaries is the injection volume (*i.e.*, the length of the injected sample plug), which compels one to use relatively long columns to deal with the large injection volume. Here, microfabrication provided the long-awaited solution in the form of on-chip injection systems with crossed microchannels. Thanks to the intersecting injection and separation channels, the sample can be introduced across the separation channel (not lengthwise) so that the initial length of the sample zone is diminished to only a few tens of micrometres (Figure 7.4a).[58] This eventually allows for the use of shorter columns (typically a few centimetres in length) in chip-based CE separations, which results in significant improvements in analysis times without sacrificing the high resolution. Moreover, the short column length allows for the application of significantly higher electric field strengths with the same instrumentation. Another important aspect, which in fact permits the use of elevated electric field strengths, is the more efficient elimination of excessive (Joule) heat on the planar chip-based systems thanks to their larger thermal mass than that of traditional CE capillaries.[83] Over the years, even sub-millisecond separations with baseline resolution have been demonstrated following careful device design in order to allow the application of electric field

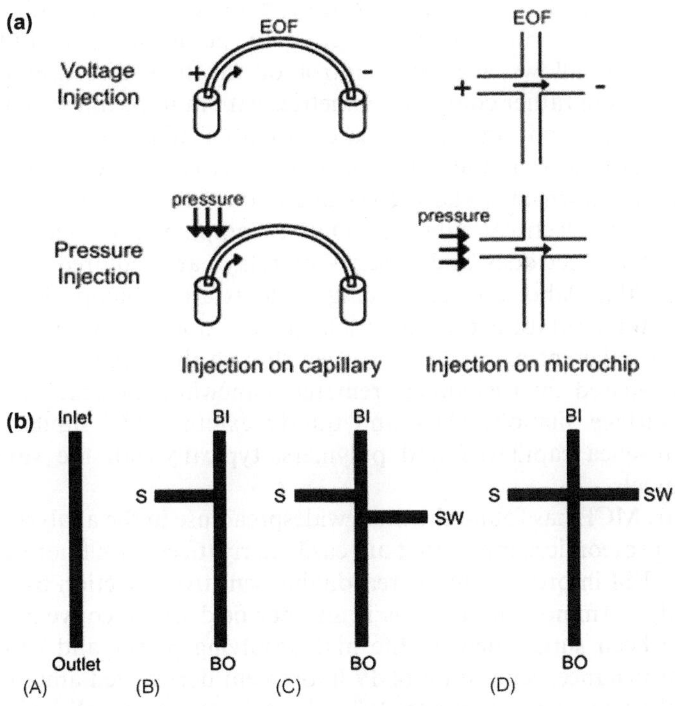

Figure 7.4 (a) Schematic illustration of the differences between sample injection on capillaries (lengthwise) and on microchips (across the separation channel). (b) Common separation channel designs for MCE: (A) a conventional CE design features a single microchannel with a single inlet and outlet, (B) a single T injector, (C) a double-T injector with an offset between the sample inlet and sample waste arms and (D) a simple cross injector. S = sample inlet; SW = sample waste; BI = buffer inlet; BO = buffer outlet. Reprinted with permission from Ref. 58, Copyright 2012 Elsevier.

strengths as high as 53 kV cm^{-1} to the separation column (8.6 kV applied to the microchip).[7] In practice, a more reasonable cycle time for an MCE analysis lies within 1 min per sample with typical applied electric field strengths not exceeding 1 kV cm^{-1}.

The common injector geometries used in MCE are depicted in Figure 7.4b and include simple cross, T and double-T cross intersections, the latter two usually being preferred as they allow for injection of a fully developed sample plug and thus elimination of electrophoretic discrimination.[58] By adding a small deviation between the injection side arms (*i.e.*, simple cross *versus* double-T), one can easily increase the injection volume for better concentration sensitivity. In addition to precise sampling, electric field-driven separation systems that rely on EOF are attractive from both microfabrication and actuation viewpoints. Namely, EOF does not produce a back-pressure as high as that of pressure-driven flow and this allows certain flexibility in terms of microchannel bonding. Even self-adhesive PDMS with typical bond strengths of the order of 100 kPa only can be used to prepare sealed microchannels for MCE, in addition to the adhesively bonded microfabrication materials that withstand much higher pressures. From the fluid actuation point of view, EOF appears attractive as it allows accurate control of the flow direction and precise sampling, even in rather complex geometries with multiple interconnections, by simply switching on an appropriate potential difference. Calculation of electrical resistances in the different side arms can be done based on the microchannel dimensions and the conductivity of the background electrolyte, in analogy with electrical circuits.[84] Only a computer-controlled high-voltage power supply, electrodes and a series of relays are required for generation of flow so that laborious interfacing with external pumps is avoided. Of the common microfabrication materials, most if not all have also been shown to maintain inherent cathodic EOF even though the origin of surface charge in polymer-based microchannels remains somewhat unclear.[35,85] Thanks to the free surface silanols, glass and quartz exhibit EOF similar to that in traditional silica capillaries and polymers, typically half the value of glass microchannels.

Thus far, MCE has found the most widespread use in the analysis of peptides and other proteomic samples that are easily derivatized with fluorescent dyes or ionized by ESI in order to allow reasonably sensitive detection by LIF or MS, respectively.[3] Amino acid analysis is another field where conventional techniques have been superseded by the high resolving power and throughput of MCE. For instance, separation of 19 fluorescent derivatized amino acids, with theoretical plate numbers up to 10^6 m^{-1}, has been accomplished by on-chip micellar electrokinetic chromatography (MEKC)-LIF in less than 3 min.[86] In addition to proteomics, the MCE techniques have been applied to clinical and forensic analysis,[87] and to some extent in metalomics *per se*. For instance, native fluorescence detection following MCE separation of inherently fluorescent riboflavins harvested off-chip from *Escherichia coli* cells showed LIF detection limits in the sub-micromolar range (10^{-8}–10^{-7} M).[88] In another study, MCE separation of 12 fluorescent derivatized inflammatory biomarkers (peptides) in microdissected human skin biopsies was obtained in about 2 min with LIF detection limits in the nanomolar range (10^{-9} M).[89] For reference, detection limits down to sub-picomolar level (10^{-13} M), corresponding to detection of only a few hundred molecules, have been reported for MCE-LIF analysis of the bare fluorescein dye.[90] Instead, fluorescence detection induced by a light-emitting diode has provided of detection limits in the low-micromolar range (10^{-6} M) for fluorescent derivatized plant glucosinolates separated by MCE.[91]

In addition to fluorescence, label-free detection of the metabolites of microbial cultured samples,[63] bisphenol A and other phenolic compounds in waters[18] and the nitrite and nitrate concentrations in human serum[19,62] and plasma[61] has been accomplished at micromolar (10^{-5} M) detection levels using linear photodiode arrays for UV absorbance detection. By coupling on-line isotachophoretic preconcentration (32-fold) with the MCE-UV analysis, two orders of magnitude better detection sensitivities (*i.e.*, at the 10^{-7} M level) have been reported for flavonoids.[92]

Label-free EC detection following MCE separation has also proven particularly useful approach for the quantification of dopamine, catechol and other neurotransmitters with typical detection limits in the range 10^{-6}–10^{-7} M (Figure 7.5a).[87] However, the lowest detection limits are generally reported in the context of MS detection. For instance, detection limits in the low nanomolar range (10^{-9} M) have been achieved for cytochrome P450 substrates and metabolites using tandem MS detection.[69,70] MCE-ESI-MS analysis of neuropeptides (Figure 7.5b)[29] and carnitines in human urine[93] have also been accomplished using concentrations in the low micromolar range, although in these studies determination of the actual detection limits was not performed.

To obtain an overall idea of the detection limits feasible for each type of detection mode, selected examples of metabolite monitoring by MCE (and pressure-driven techniques) are listed in Table 7.1. A more comprehensive view on the use of MCE in metabolomics can be found elsewhere.[3,11] Regardless of the detection method or the chip material used, high separation efficiency is a default value in MCE provided that attention is paid to eliminating hydrostatic pressure anomalies[21] and unwanted surface interactions (non-specific adsorption) within the system.[28,38] In addition to metabolite monitoring of body fluids and of off-chip manipulated tissue samples and enzymatic assays, the MCE technology has also been applied to the separation of the cellular metabolites following on-chip lysis of individual cells. This approach is described in more detail in Section 7.4.

7.2.4 Microchip Liquid Chromatography

Thus far, the number of microfabricated pressure-driven separation systems remains limited[110,111] and most effort has gone into the development of microchips for MCE and other electric field-driven applications.[112] As briefly discussed in Section 7.2.2, a possible reason for this is the problems related to

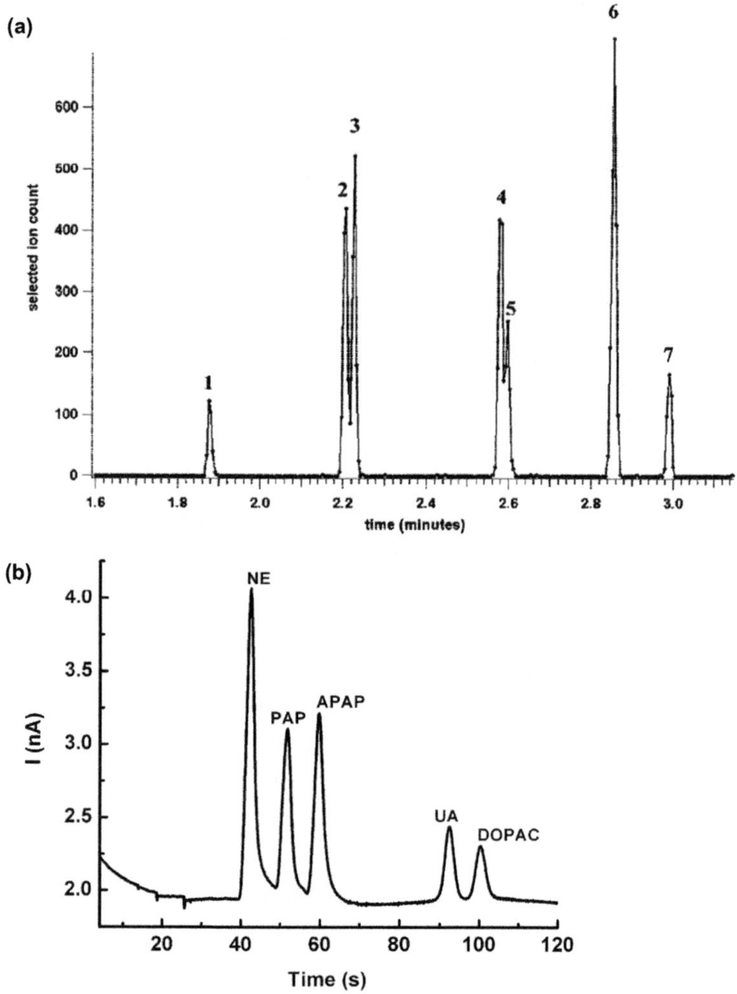

Figure 7.5 (a) MCE separation of norepinephrine (NE), acetaminophen (APAP), 4-aminophenol (PAP), uric acid (UA) and 3,4-dihydroxyphenylacetic acid (DOPAC) with EC detection on a gold electrode array. Reproduced from Ref. 94. Copyright 2009 Royal Society of Chemistry. (b) MCE separation of fluorescein (1), methionine–enkephalin (2), leucine–enkephalin (3), angiotensin II (4), neurotensin (5), bradykinin (6) and thymopentin (7) with on-chip ESI-MS detection. The separation channel length was as long as 20.5 cm in order to allow for better peak characterization by MS in addition to improved separation efficiency. Reprinted with permission from Ref. 29. Copyright 2008 American Chemical Society.

the packing of the stationary phase inside a microchannel in a homogeneous way.[113] In principle, there are four different ways to introduce the stationary phase into and to perform LC in a microfabricated (sealed) channel, namely (1) coating only the channel walls for open-channel LC, (2) use of microbeads

Table 7.1 Selected applications of the MCE and microchip LC or SPE techniques in metabolomics, categorized by the detection method used and the detection limits obtained, where the sensitivity of an assay has been reported.

Limit of detection (mol L^{-1})	*UV absorbance (linear imaging systems)*	*Fluorescence microscopy*	*Electrochemical*	*Mass spectrometry*
10^{-4}	MEKC-UV		MCE-EC: Creatinine, creatine and uric acid in human urine[95]	
10^{-5}	Bisphenol A and other phenols[18] MCE-UV: Nitrite and nitrate in human plasma and serum[19,61,62] Antimicrobial metabolites[63]		MCE-EC: Glucose and lactate in human serum[96] MEKC-EC	
10^{-6}		MCE-FL (derivatized): Plant glucosinolates[91] MCE-LIF (derivatized): L-Amino acids[97]	Natural flavonoids[98] MCE-EC: Glucose,[99] aminophenols,[94] carbohydrates[100] Antibiotics[51] Caffeine and theophylline in rat serum and urine[101]	
10^{-7}	ITP-MCE-UV: Flavonoids[92]	MCE-LIF (native fluorescence): Riboflavins in *E. coli* cells[88]	MCE-EC: Morphine and codeine in human urine[102] Neurotransmitters in mouse brain homogenates[103]/standards[50,104] Mercaptopurine[105] HPLC-EC: Catechins[64]	
10^{-8}			MCE-EC (enzymatic oxidation): Neurotransmitters[106]	SPE-ESI-MS: Anticancer agents in cell incubation mixture[107] Imipramine[108]
10^{-9}		MCE-LIF (derivatized): Inflammatory biomarkers (peptides)[89]		MCE-ESI-MS/MS: Drug metabolites in HLM Incubation mixture[69]
10^{-10}				HPLC-chip-ESI-MS/MS:
10^{-11}				Steroids (oxime derivatized) in mouse plasma[109]

(particles) for packing of channels, (3) heat- or UV-initiated polymerization of a porous polymer monolith *in situ* and (4) use of microfabricated pillar arrays as solid-phase supports (Figure 7.6). Of these different techniques, the first (coating) only rarely provides a high enough surface area for efficient separation based on chromatographic interactions.[114] The other two approaches, in turn, readily suffer from problems regarding the packing density, specifically at microchannel corners, with therefore impaired separation efficiency.[44,45] In general, reversed-phase (C_{18}) microparticles have shown the best performance as packing materials for microchip LC.[66,115] However, for fixing the beads inside a microchannel, microfabricated frits need to be patterned to the channel outlet to act as mechanical barriers for holding the beads.[116] Another approach is to use magnetic core beads.[117] As an alternative to packing of beads, *in situ* polymerization of a porous monolith inside a microfluidic channel by heat or

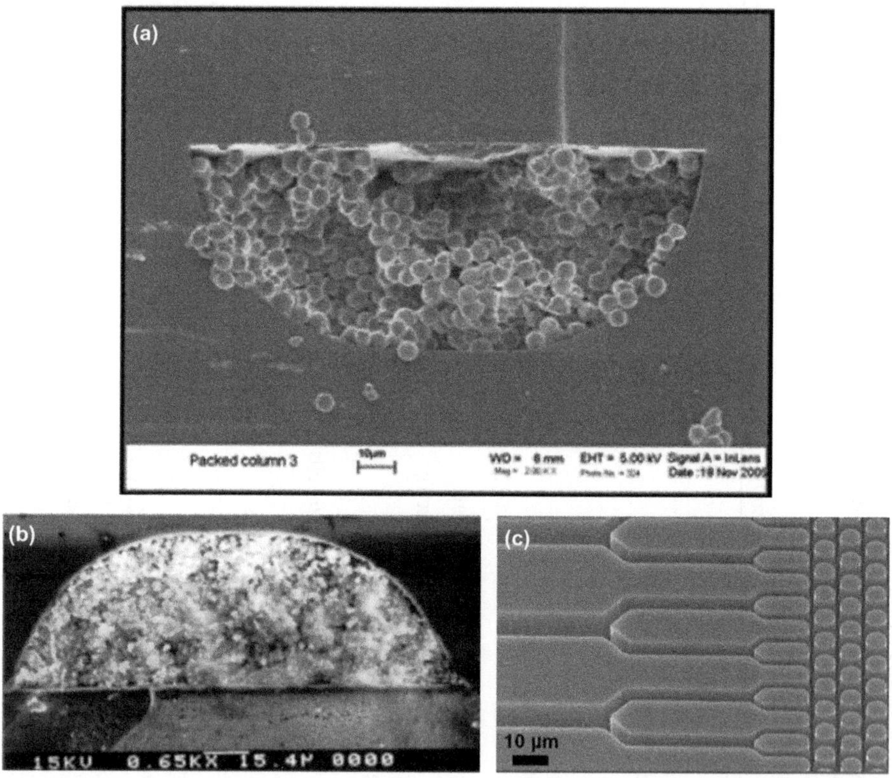

Figure 7.6 SEM images of a cross-section through a microfluidic LC channel (a) packed with 5 µm particles and (b) packed by *in situ* photopolymerization of methacrylate monomers. (a) Reprinted with permission from Ref. 66. Copyright 2006 American Chemical Society. (b) Reprinted from Ref. 124 with permission of John Wiley & Sons, Inc. ©2003. (c) SEM image of an interconnection between an inlet channel and a silicon micropillar array used for microchip LC. Reproduced from Ref. 125. Copyright 2012 Royal Society of Chemistry.

UV irradiation offers several attractive features. These include frit-free construction with no risk of leakage of the solid phase material and easy tuning of the surface chemistry by simply changing the monomer composition.[118] UV-initiated photopolymerization, in particular, allows the formation of monoliths within precisely defined regions of a microfluidic device by simply masking the unexposed areas. A maskless approach is to use laser-induced photopolymerization.[119] In addition to LC, porous polymer monoliths are widely used in chip-based capillary electrochromatography.[120,121] The porosity of the polymer monoliths is also easily altered by simply changing the porogen content to permit, for instance, lower back-pressures than those in packed microchannels for actuation by electrokinetic means and thus removing the need for coupling to external pumps. In addition to monoliths, the microfabricated pillar arrays provide lower back-pressures than those of packed microchannels. However, attention needs to be paid to the fact that sometimes the surface area on pillar arrays need to be increased *via* porosification of the pillar walls,[122] although separation efficiencies as high as 10^6 theoretical plates have also been reported for non-porous pillar arrays.[123]

The lack of integrated on-chip micropumps, capable of producing high enough pumping pressures, is another limiting factor with respect to the development of on-chip LC systems.[110] On the microscale, it is simply much easier to work with electrokinetic flow as electrodes and a power supply usually suffices for fluid actuation. Instead, the use of pressure-driven flow is much more challenging from an engineering viewpoint because connecting to external high-pressure pumps suffers from rather complex couplings, such as gluing of pressure-tight capillary connectors on to the microchannel inlets and outlets. Even if microfabricated solutions for capillary couplings also exist,[126] their pressure tolerance is often limited. Worth attention here is also the fact that, as the channel dimensions are reduced along with miniaturization, ever increasing pressures need to be applied to the pressure-driven systems, whereas less and less voltage is required to run the electric field-driven systems.[46] The use of pressure-driven flow is further aggravated by the need for smaller beads (packing) or longer channels to compensate for miniaturization of the separation column. In other words, one must increase the surface area for chromatographic interaction in order to achieve a reasonable separation performance. This eventually increases the pressures required for running a microchip-based LC system and rendering integration of microfabricated on-chip pumps rather inefficient. At best, pumping pressures up to about 1000 kPa at typical flow rates in the region of 100 nL min^{-1} have been reached by using, for instance, electrokinetic[66] or electrochemical (Figure 7.7a)[127] control.

An alternative approach to pumping is to use centrifugal force for actuation of chromatographic systems on spinning disks.[128] In addition to pumping, special attention needs to be paid to mixing of fluids in the laminar flow regime, which is governed by the relatively slow diffusion. Although miniaturization generally decreases the diffusion distances and thus favours rapid mixing, speciality microstructures (*e.g.*, split-and-combine, Figure 7.7b) are often needed to speed up mixing of the eluents (the mobile phase) if pumping is

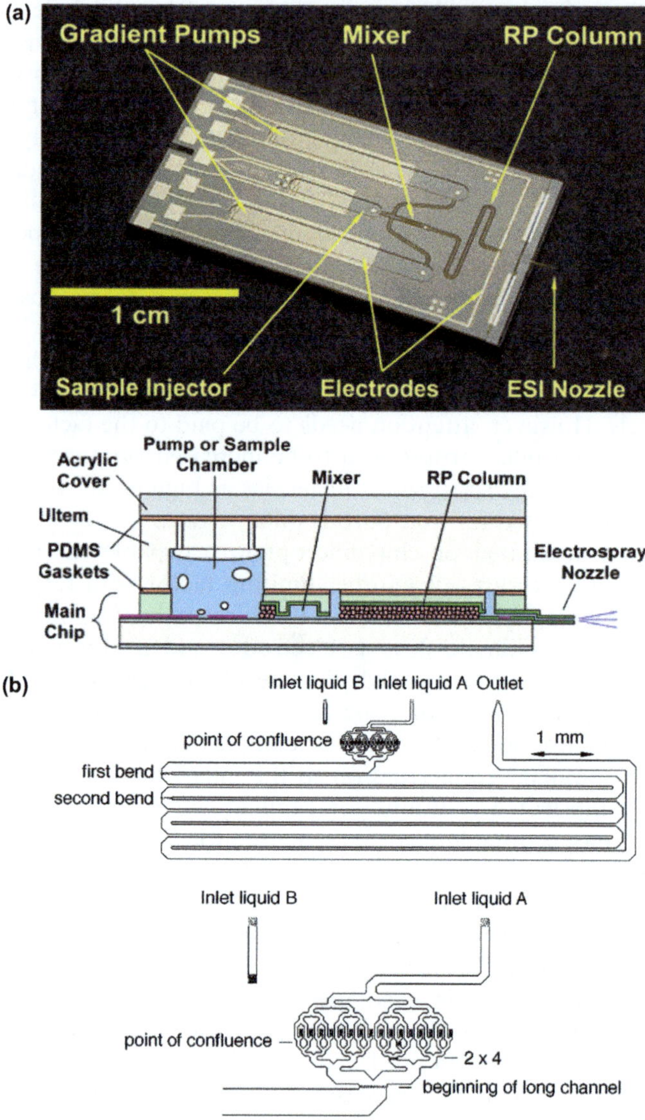

Figure 7.7 (a) An LC chip incorporating all functions on-chip: pumping based on electrolysis of water (galvanostatic control), mixing structures, sample injection loop, separation column packed with C_{18} microbeads and an ESI emitter. Reprinted with permission from Ref. 127. Copyright 2005 American Chemical Society. (b) A micromixer design based on splitting of two main flow streams into smaller fluid streams to allow rapid diffusion and subsequently combining the mixed fluid streams into one. Reproduced from Ref. 130. Copyright 1999 Royal Society of Chemistry.

performed on-chip. Hence the question remains of whether pumps and valves should be incorporated on-chip or whether coupling to external pumps, mixers and automated sampling systems is the more reasonable approach after all. All possible combinations of LC microchips are covered by the previous literature, *i.e.*, on-chip pumping and mixing combined with on-chip injection,[66,127] off-chip pumping and mixing combined with on-chip injection[65] and off-chip pumping and mixing combined with off-chip injection.[115,116,129] A fully integrated LC device featuring all functions (pumping, mixing, injection, separation and ESI) on a single chip has also been reported, although it was only applied to the separation of proteomic peptide digests (Figure 7.7a).[127]

Despite the complexity associated with microchip-based LC, pressure-driven systems often perform in a more repeatable manner than electric field-driven systems in terms of process automation. As an indication of this, the rare commercial solutions are actually based on chromatographic separation chips.[115,131] For instance, the HPLC-chip/MS system from Agilent Technologies, featuring an LC column packed with microbeads integrated with an on-chip ESI emitter, has been used for the separation of steroids,[109] anthocyanins,[132] *N*-glycans[133] and oligosaccharides[134] (Figure 7.8). At best, detection limits as low as a few tens of picomoles per litre (10^{-10}–10^{-11} M) have been obtained for oxime-derivatized steroids.[109] Although microchip-LC-ESI-MS generally provides by far the best detection sensitivity in small-molecule analysis (Table 7.1), most applications of the Agilent HPLC-chip have been focused on screening of the proteomic biomarkers, mainly because the stationary phase was optimized for separation of peptides in the first place.

In addition to ESI, other ionization techniques, such as atmospheric pressure chemical ionization (APCI)[135] and photoionization (APPI),[136] have also been the target for miniaturization and are also feasible for monolithic coupling to an on-chip LC column.[116] The feasibility of a microfabricated heated nebulizer (*i.e.*, the APCI and APPI ion source) for metabolomics has also been demonstrated *via* analysis of anabolic steroids in human urine.[137] Detection limits in the low nanomolar range were obtained for anabolic steroids using the miniaturized APPI/MS, although in this case the separation of steroids was performed off-chip with a commercial capillary LC instrument.

Another commercial solution for on-chip LC, from Eksigent, features eight parallel microfluidic channel and has been used for the chiral separation of benzoin and (*R*,*S*)-dihydrobenzoin enantiomers prior to UV absorbance detection by individual diode arrays at each channel.[131] Non-commercial HPLC chips, featuring glass microchannels with monolithic silica[65] or PS/PDMS microchannels packed with microbeads,[64] have also been used for the separation of catechins prior to EC detection. At best, detection limits in the sub-micromolar range (10^{-7} M) have been reported for (+)-catechin, epicatechin and epigallocatechin gallate.[64] Non-porous pillar arrays have also been used for LC separation of steroids with off-chip UV detection[129] or of pharmaceuticals with off-chip ESI-MS.[125] Overall, similarly to MCE, most microchip LC systems have been targeted at monitoring of proteomic biomarkers with only rare applications to metabolite monitoring.

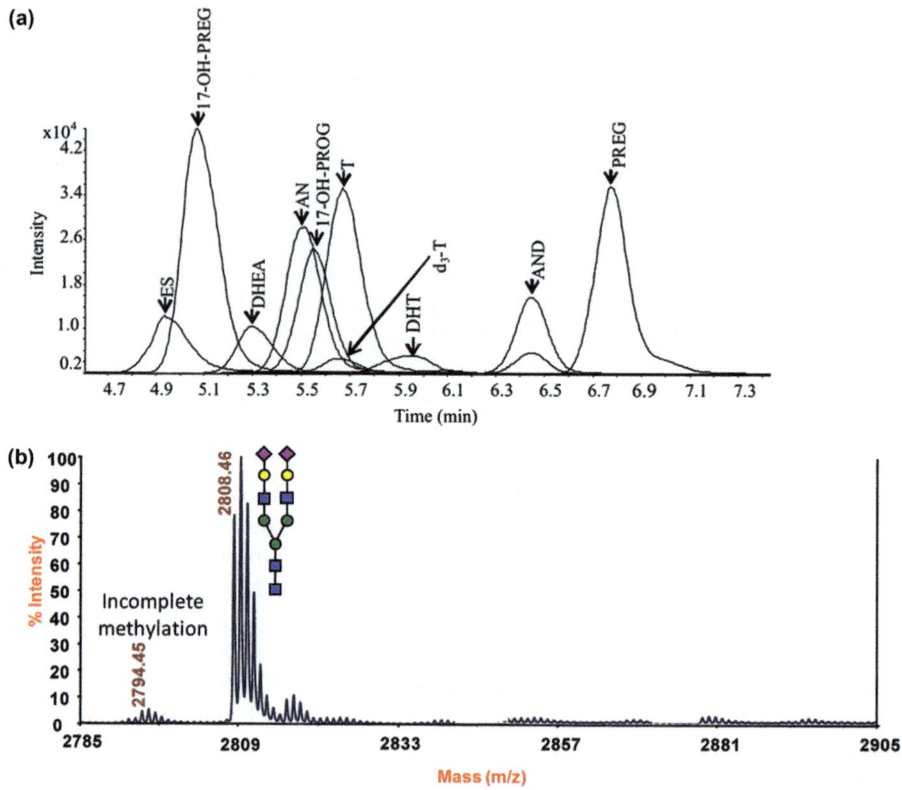

Figure 7.8 On-chip LC separations of (a) steroids and (b) *N*-glycans achieved using the Agilent HPLC-chip/MS setup. (a) Reprinted with permission from Ref. 109. Copyright 2012 Elsevier. (b) Reprinted with permission from Ref. 133. Copyright 2010 American Chemical Society.

7.2.5 On-chip Sample Preparation

As discussed in Section 7.2.2, the common drawback of miniaturization, particularly with MCE systems, is the relatively poor concentration sensitivity originating from the small injection volume. Hence there is a great need for coupling sample preconcentration as an integral part of the chip-based separation systems.[138] In metabolomics, sample precleaning is also required in many places as the salts and proteins in body fluids and tissue homogenates may impair the separation performance or detection sensitivity. Of the different on-chip preconcentration techniques, solid-phase extraction (SPE) provides by far the highest enrichment factors, at best up to 80–500-fold,[139] with inherent, simultaneous sample desalting. In principle, the solid support structures for SPE can be realized by the same techniques as with LC microchips.[113] In addition to on-chip preconcentration prior to MCE or microchip LC separation, chip-based SPE has been extensively used, as an integral part of various microreactors, for the purification and preconcentration of the reaction products (Table 7.1). For instance, anticancer agents[107] and vitamin E metabolites[140] in cell incubation matrices have been monitored using direct infusion to on-chip SPE and eventually to off-chip ESI-MS. In another study, on-chip SPE was integrated with an enzymatic microreactor incorporating hydrogel-encapsulated human liver microsomes (HLMs) in order to purify the products of enzyme incubation.[141] An array of monolithic columns in a polymeric substrate was also used for on-chip SPE of imipramine prior to off-chip ESI-MS detection.[108] Thus far, most SPE chips have been actuated by pressure-driven flow and only a few approaches have attempted to integrate on-chip SPE with electric field-driven separation systems.[119,142,143] In these studies, the solid phase was implemented as an integral part of a standard MCE chip *via in situ* photopolymerization of porous monoliths at the injection cross by a laser or UV lamp (Figure 7.9a). At best, enrichment factors exceeding 10^5-fold have been reported for fluorescence-derivatized glycopeptides and oligosaccharides.[119] However, the extraction time required for such high enrichment factors (~5 min) increased the overall cycle time of the subsequent MCE-LIF analysis. In further work, in order to maintain the high throughput of MCE, the extraction time was reduced to as low as 10–25 s, although at the cost of enrichment factors obtained that were 15–23-fold for pharmaceuticals detected by on-chip ESI-MS.[143]

In addition to SPE, electrokinetic techniques such as field amplification stacking (FAS)[144,145] and isotachophoresis (ITP)[92,146] have been used for on-chip sample preconcentration and precleaning, mainly in combination with MCE. Common to all electrokinetic techniques is that one must fill the microchannel with two or more different electrolyte solutions to create appropriate electric field gradients for sample stacking. However, techniques such as FAS and ITP are only applicable to charged analytes, although some sweeping techniques[147] also allow for stacking of neutral molecules. At best, enrichment factors of up to 2200–2900-fold have been reported for oligosacchararides by large volume sample stacking (Figure 7.9b),[148] whereas electrokinetic techniques generally result in an order of magnitude lower enrichment factors (typically 10–100-fold) than those of SPE.[138] For instance, 32-fold concentration has been obtained for flavonoids by ITP (Table 7.1).[92]

The application of microfluidic devices to two-phase (organic–aqueous) liquid–liquid extraction (LLE) has also attracted some interest, although most of the work has been focused just on the development of off-line sample preparation. For instance, LLE performed on a glass microchip uniting parallel fluid streams of two immiscible solvents has been used for the extraction of amphetamine-type stimulants in urine, followed by off-chip GC analysis (Figure 7.9c).[149] On-chip LLE with on-line fluorescence detection has also been performed by feeding aqueous droplets into a continuous organic phase,[150] or trapping an organic droplet in a cavity next to the microchannel filled with aqueous phase.[151] In the latter approach, preconcentration factors of up to 2000-fold were reported for a fluorescent rhodamine B derivative. Droplet-based LLE has also been performed *via* electrowetting on a dielectric digital microfluidic (DMF) platform, which enables not only the combination of two

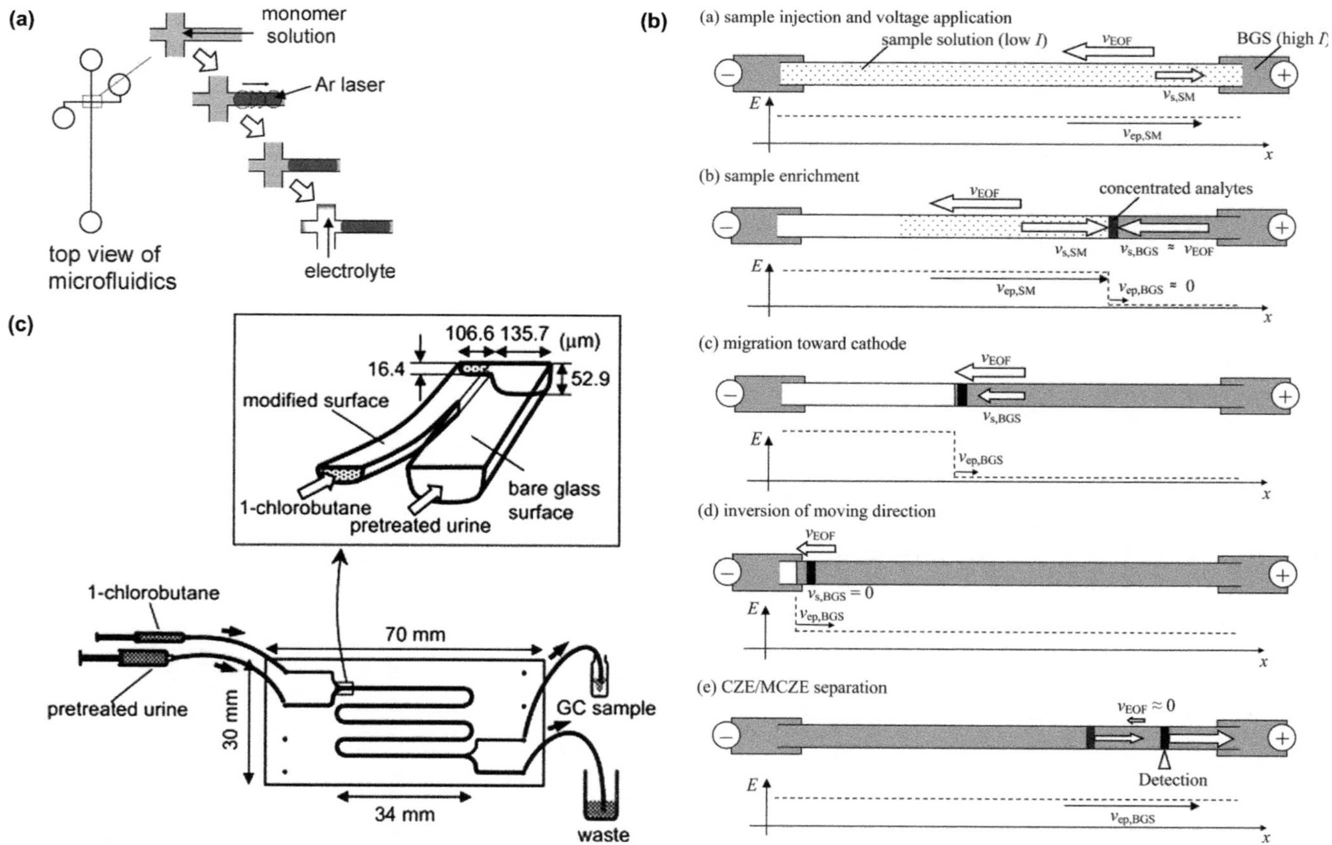

Figure 7.9 (a) An MCE device with *in situ* photopolymerized porous polymer monolith at the injection cross for on-line SPE. Reprinted with permission from Ref. 119. Copyright 2008 American Chemical Society. (b) Schematic diagram of the operating principle of large-volume sample stacking. Reprinted with permission from Ref. 148. Copyright 2010 American Chemical Society. (c) A microfluidic chip used for on-chip liquid–liquid extraction of amphetamine-type stimulants from urine to 1-chlorobutane. Reprinted with permission from Ref. 149. Copyright 2006 Elsevier.

immiscible droplets of very low volume (approximately nanolitre level), but also phase separation with subsequent image- or UV-based concentration measurement once the extraction process is completed.[152] Even three-phase (aqueous–organic–aqueous) LLE with image-based fluorescence detection has been accomplished with open-to-air designer droplet geometries built up by hydrophobic/hydrophilic surface-chemistry patterning.[153] The more common approach for three-phase LLE is, however, to implement mechanical barriers, such as porous polymer membranes, to support the middle (organic) phase.[69,154,155] By assembling the membrane on top of a microchannel inlet, on-line extraction prior to MCE-LIF or MCE-ESI-MS of (spiked) fluorescein derivatives[155] or analgesic metabolites in human urine,[69] respectively, has been accomplished with micromolar detection levels. Electrically assisted three-phase LLE over a polypropylene membrane bonded to the interface of two adjacent PMMA microchannels was also coupled to off-chip ESI-MS detection for monitoring of the real-time metabolism of amitriptyline in rat liver microsomes.[154] In addition, membrane-based approaches have been used to allow for on-chip ESI-MS detection of pharmaceuticals in phosphate-buffered saline following on-chip desalting.[156] In this case the sample was pumped through a poly(vinylidene fluoride) membrane assembled on top of a microchannel inlet.

7.3 Microfluidic Enzyme Assays in Metabolomics

Microfluidic separation systems *per se* undoubtedly hold great promise for high-throughput analysis in the field of metabolomics, but the full potential of the microchip technology is hardly realized without considering the possibility of implementing a microfluidic enzyme reactor or cell-based assay as an integral part of the separation chip. On the microscale, however, special attention needs to be paid to mixing of the reagents so that diffusion will not become the limiting factor with respect to the reaction turnover rate. As already discussed in Section 7.2.3, mixing in the laminar flow regime is slow as it occurs through diffusion only, and this needs to be accounted for in the device design. In addition to efficient mixing of reagents, precise temperature control is often needed in order to preserve enzyme activity. An easy approach for thermostating a microfluidic reactor is to use thermofoil-type heaters attached to the bottom of the chip and to monitor the temperature using a surface thermometer.[157] However, in the reported microfluidic enzyme assays, temperature control is only rarely accounted for and most of the assays have been performed at room temperature. Yet microfabricated platforms featuring multiple interconnecting microchannels undoubtedly provide a more convenient approach for performing multistep and parallel reactions than the conventional off-line techniques.

7.3.1 Microfluidic Enzyme Reactors

In its simplest form, a microfluidic reactor comprises of two channels that are united in a Y or T configuration allowing two different reagent streams (*i.e.*, the

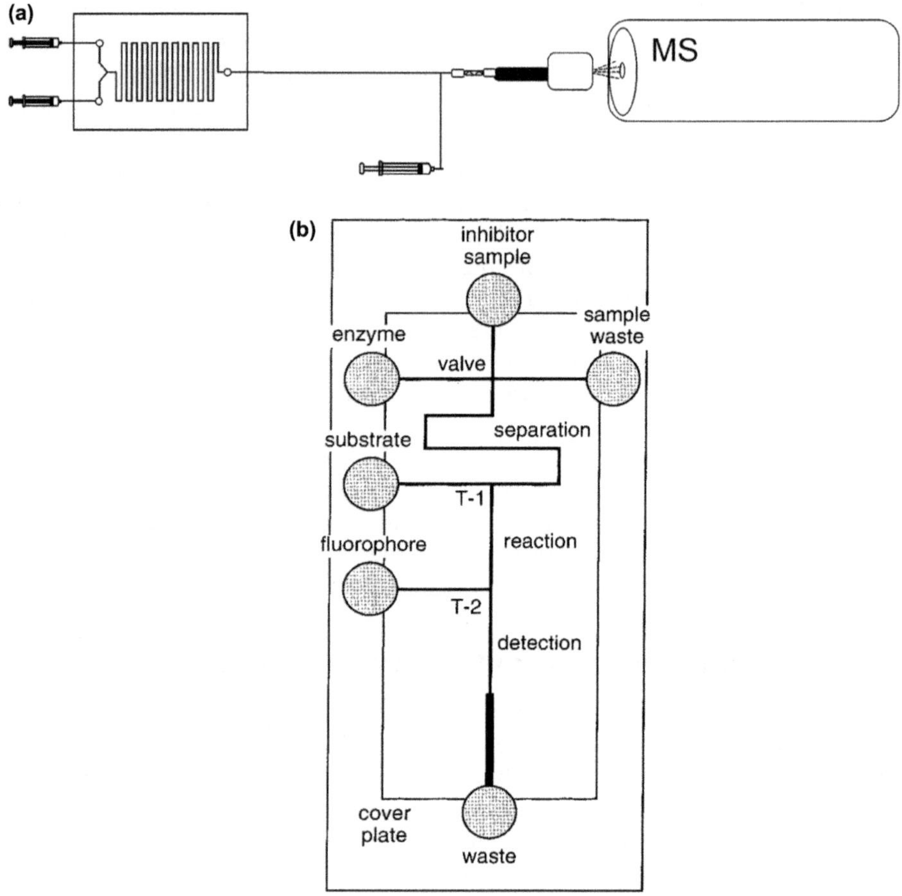

Figure 7.10 (a) A pressure-driven microfluidic Y reactor coupled to off-line ESI-MS for monitoring of the products of an enzymatic reaction. Reprinted with permission from Ref. 157. Copyright 2003 American Chemical Society. (b) An electric field-driven microfluidic chip featuring multiple interconnecting channels to allow multistep enzyme reaction (*i.e.*, sequential addition of inhibitors and substrates) and post-reaction derivatization of the reaction products. Reprinted with permission from Ref. 165. Copyright 1999 American Chemical Society.

enzyme and the substrate) to mix and react with one another (Figure 7.10a).[157,158] The two most common approaches in metabolomics have been to use the enzyme microreactors for the determination of the kinetic parameters of the cytochrome P450-mediated metabolism of drugs[141,157,159] or of the oxidation of glucose.[158,160] However, often the performance parameters of the on-chip enzyme assays have also been validated based on enzymatic conversion of fluorescent model compounds.[161–163] Although most work has been focused on the development of experimental setups, numerical techniques have also been applied to the simulation of, for instance, laccase-catalysed L-dopa oxidation in an oxygen-saturated water solution using different residence times in a Y-shaped microreactor.[164]

In addition to Y (and T) reactors, microchannels with multiple interconnections have been developed with a view to performing multistep on-chip enzyme assays.[161–163,165] One of the first designs reported made use of a mixing T to introduce a continuous flow of a non-fluorescent substrate, resorufin β-D-galactopyranoside, into the background electrolyte (BGE) prior to addition of a continuous flow of the enzyme, β-galactosidase, through the injection channel.[162] The enzyme kinetic parameters were determined by quantifying the fluorescent metabolite, resorufin, by LIF at elevated substrate concentrations obtained by changing the mixing ratio of the BGE and the substrate *via* voltages applied at each reservoir. In the second-generation device (electrokinetically actuated), an enzyme inhibitor was first injected into a continuous stream of acetylcholinesterase-doped BGE and allowed to react with the enzyme prior to addition of the substrate, acetylthiocholine, *via* a second injection cross (Figure 7.10b).[165] Finally, a third injection cross was used to stop the reaction with the addition of acetonitrile containing a post-reaction reagent to form a fluorescent derivative of the metabolite, thiocholine. A microchannel configuration somewhat similar to the first-generation device, but actuated by pressure-driven flow instead of EOF, was also developed by another group for monitoring of the removal of a phosphate group from 6,8-difluoro-4-methylumbelliferyl phosphate by human T-cell phosphatase.[163]

Another approach for performing in-solution enzyme assays made use of multiphase fluid flow by transporting the reagents inside aqueous droplets (plugs) surrounded by an immiscible carrier fluid (Figure 7.11a).[161] Thanks to the very small volume (sub-microlitre) of the droplets fed into the carrier fluid, mixing of reagents by diffusion was fast, allowing the determination of single-turnover kinetics of ribonuclease A with better than millisecond time resolution *via* fluorescence microscopy. Another droplet-based approach used for performing on-chip enzyme assays relies on digital microfluidics (DMF) on an array of electrodes. The DMF approach has been exploited for, for instance, the quantification of succinylacetone, a marker for hepatorenal tyrosinaemia, in dried blood spots from newborns by on-chip ESI-MS (Figure 7.11b)[166] or monitoring of enzymatic glucose oxidation in various biofluids by colorimetric detection.[167,168] Determination of the (pre-steady-state) enzyme kinetic parameters has also been accomplished using the DMF approach.[169,170]

In addition to in-solution assays, the effective microchannel part can be coated[158] or packed with, for instance, microbeads[160] or hydrogel[141,159,160,171] to bear immobilized enzymes. For instance, biotinylated phospholipid bilayer coatings[158] or packing of aminopropyltriethoxysilane-modified glass microbeads have been used to link streptavidin-conjugated alkaline phosphatase or glucose oxidase (GOx), respectively, in PDMS microchannels. Poly(ethylene glycol) (PEG) hydrogels[141,159] and tetramethoxysilane sol–gels[171] have also been used for the encapsulation of HLMs. Some work has also reported not only the enzymatic reaction, but also preconcentration/clean-up of the reaction products

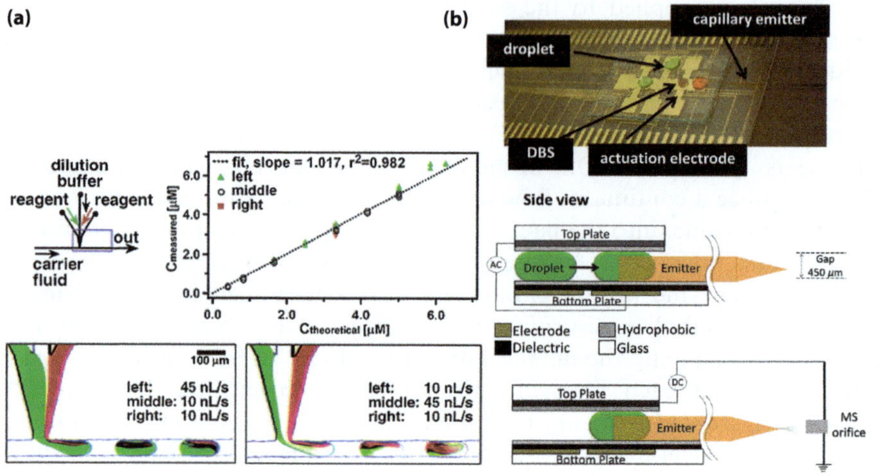

Figure 7.11 (a) A droplet-based approach for studying enzyme kinetics in aqueous droplets fed into immiscible carrier fluid. Reprinted with permission from Ref. 161. Copyright 2003 American Chemical Society. (b) A digital microfluidic platform used for monitoring metabolic biomarkers in dried blood spots from newborns by ESI-MS. Reprinted with permission from Ref. 166. Copyright 2012 American Chemical Society.

within the subsequent on-chip SPE column[157] or post-reaction fluorescence derivatization[158,160] on the same microchip prior to detection (Figure 7.12a). Integration of an on-line cytotoxicity assay, exploiting on-chip cultured human hepatocellular carcinoma cells (HepG2), with prior on-chip incubation of acetaminophen with encapsulated HLMs, has also been reported for more advanced systems.[141,171] In one design, the HLM incubation solution was analysed by on-line MCE-UV parallel to the cytotoxicity assay,[171] whereas the other approach included an on-line, on-chip SPE on microbeads prior to off-chip ESI-MS detection of the metabolites (Figure 7.12b).[141] Even if most designs are in theory amenable to integration with a downstream separation channel, microreactors bearing entrapped enzymes or microsomes are only rarely coupled to on-chip separation systems.[171] Instead, the enzyme incubation solution is directly infused to detection by fluorescence microscopy,[158,160,162,163,165] off-chip ESI-MS[141,157,159] or image-based analysis,[161] all allowing the monitoring of micromolar metabolite concentrations. The other approach for performing on-chip enzyme reactions, *i.e.*, to utilize the microfluidic separation channel for (mainly electrokinetic) flow injection analysis, is presented in the following section.

7.3.2 Microfluidic Enzyme Assays with MCE Separation

In addition to bare MCE, the standard chip design with intersecting injection and separation channels has also been used for performing on-chip enzyme assays with on-line MCE separation of the reaction products. Probably the

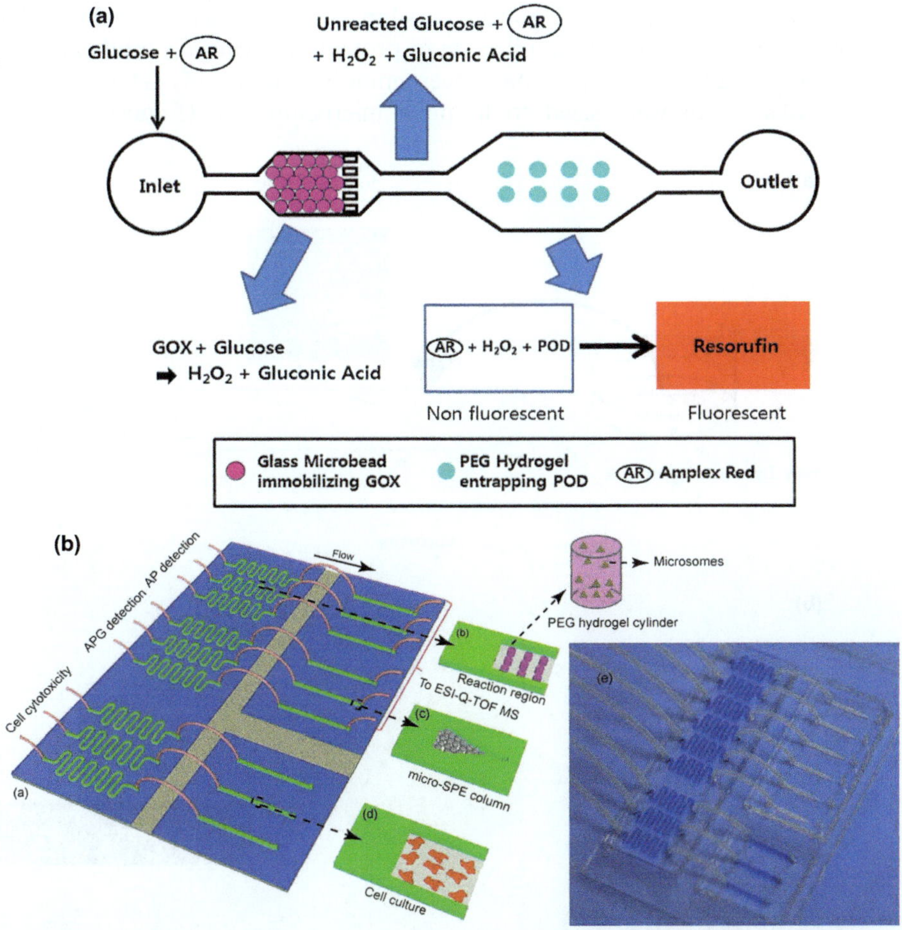

Figure 7.12 (a) Schematic illustration of a microfluidic chip used for performing sequential bienzymatic reaction of glucose, *i.e.*, glucose oxidation within the reaction chamber and detection *via* conversion of non-fluorescent Amplex Red into fluorescent resorufin within the reaction chamber. Reprinted with permission from Ref. 160. Copyright 2009 Elsevier. (b) Schematic diagram of a microfluidic chip capable of performing parallel HepG2 cell cytotoxicity assay and enzyme incubation in HLMs integrated with on-chip SPE to desalt and concentrate the metabolites prior to ESI-MS detection. Reproduced from Ref. 141. Copyright 2012 Royal Society of Chemistry.

simplest approach is to mix all reaction components in one microchannel well (inlet) and continuously follow the progress of reaction by injecting aliquots of the mixture into the separation channel to separate the substrate and the metabolite(s).[172,173] This approach has been taken to follow, for instance, the protein kinase A-mediated phosphorylation of a fluorescent labelled heptapeptide, kemptide,[172] and the hydrolysis of fluorescein mono-β-D-glucuronide with β-glucuronidase.[173]

Another approach for integrating on-chip enzyme assays with MCE has emerged through the inherent adhesion of proteins and cells on solid supports. For instance, adhesive HepG2 have been immobilized on poly-L-lysine-coated glass disks which were sized to fit on a microchip well (Figure 7.13a).[174]

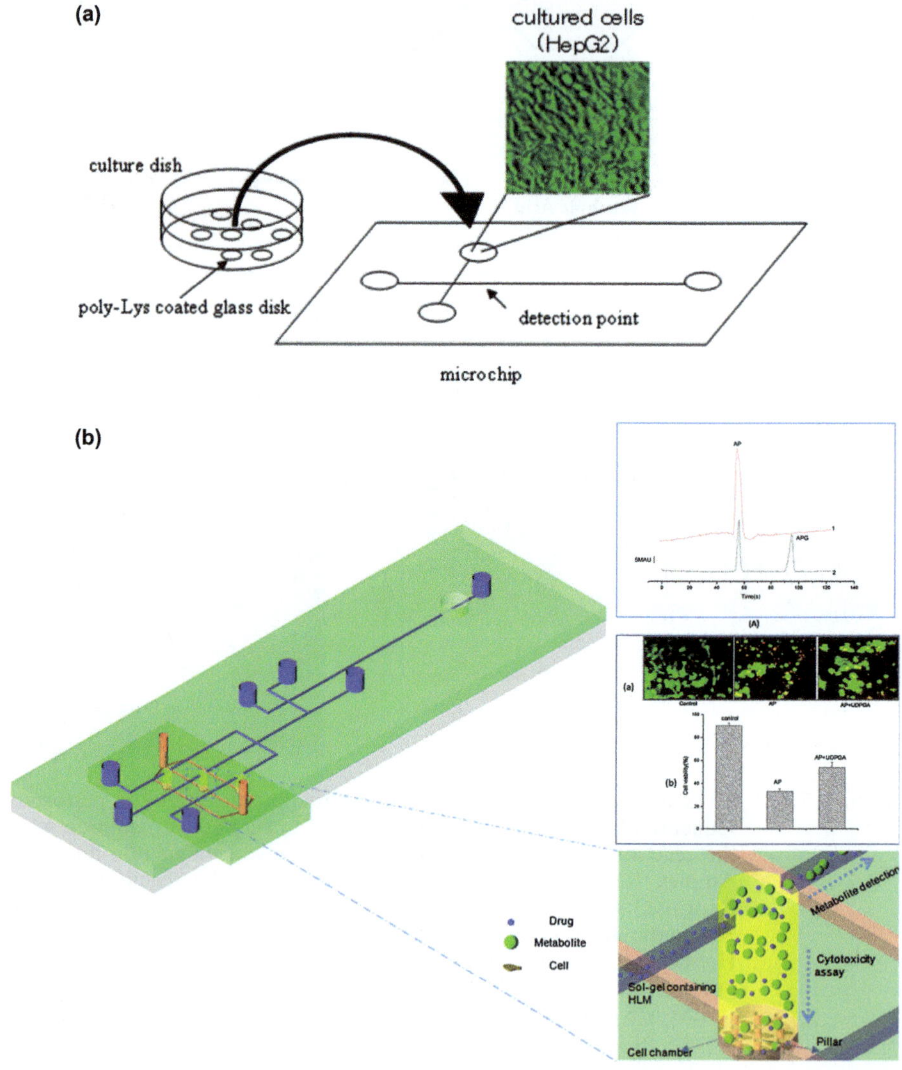

Figure 7.13 (a) Schematic diagram of an MCE chip used for on-line monitoring of CYP450-mediated metabolism reactions on HepG2 cells adhering to a microchannel well. Reprinted with permission from Ref. 174. Copyright 2009 American Institute of Physics. (b) Schematic diagram of a microfluidic chip capable of performing parallel (A) MCE-UV separation of acetaminophen (AP) and its glucuronide (APG) and (B) HepG2 cell cytotoxicity assay following on-chip AP incubation with HLMs. Reproduced from Ref. 171. Copyright 2009 Royal Society of Chemistry.

This permitted on-chip injection of representative samples of the cell incubation matrix and, thus, determination of the kinetics of CYP1A1-mediated ethoxyresorufin *O*-deethylation by on-line MCE-LIF. The strong non-specific adsorption of proteins on PET was also utilized to immobilize glucose oxidase (GOx) effectively on the separation channel walls.[175] In this case, only the substrate, glucose, was injected into the separation channel and hydrogen peroxide, liberated during the enzyme-mediated oxidation, was quantified by MCE-EC. Electrophoretically mediated microanalysis (EMMA) has also been performed on-chip by injecting the enzyme, leucine aminopeptidase, into a continuous stream of the substrate, L-leucine-β-naphthylamide, in BGE.[176] In this case, the differential electrophoretic mobilities of the enzyme and substrate allowed for efficient mixing in an open microchannel, after which a zero potential (stopped-flow) incubation was performed to liberate the fluorescent metabolite, β-naphthylamide. Also worthy of attention is the microreactor design exploiting sol–gel-encapsulated HLMs that was briefly described in the previous section and incorporates parallel cytotoxicity assay and an MCE unit downstream of the enzyme (HLM) reactor (Figure 7.13b).[171]

7.3.3 Parallel Microreactors

In addition to individual microreactors performing one enzyme assay at a time, parallel microfluidic processors have also been developed to study enzyme kinetics. In one approach, a microplate reader-compatible microchip was developed to facilitate dosing of the reagents and detection of the metabolites.[177] The device comprised four assay modules each of which had six individual microreactors to perform enzyme (alkaline phosphatase) reactions at six different substrate (*p*-nitrophenyl phosphate) concentrations. In order to deliver different substrate concentration to each reactor, a microfluidic split-and-mix network actuated by pressure-driven flow was integrated with the reactor array so that appropriate concentrations could be diluted using only two different substrate solutions of high and low concentration. In this manner, altogether 24 parallel enzyme reactions (six reactions per module) could be monitored simultaneously by a microplate reader. Another type of substrate dilution system was established on a pneumatically controlled PDMS chip featuring metering valves for dosing of 11 parallel microreactors (Figure 7.14a).[178] This approach allowed the determination of the enzyme kinetic parameters and also the effect of varying inhibitor concentrations of β-galactosidase mediated metabolism of resorufin-β-D-galactopyranoside with a single experiment.

Yet another dilution system based on diffusion was established by combining two streams of liquid (substrate solution and pure buffer) at a Y-junction and allowing them to diffuse into each other as they flowed downstream side-by-side (Figure 7.14b).[158] After a fixed distance (2 cm), the main channel was split into 12 smaller channels that emanated from the main flow so that enzymes immobilized on the walls of the small channels were incubated at 12 different substrate concentrations ranging over a factor of 33 (from the first to the last). In addition to pressurized systems, centrifugal forces have also been

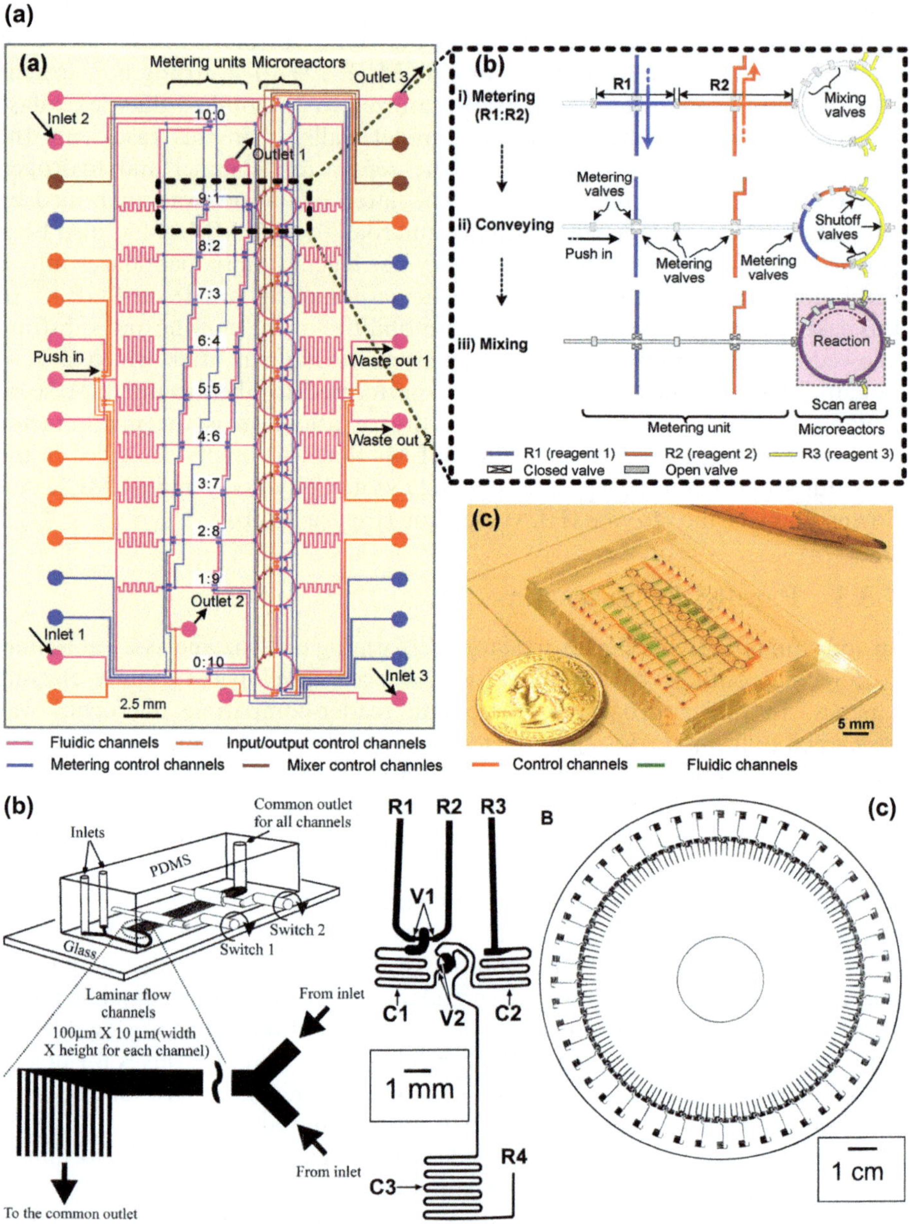

Figure 7.14 (a) An integrated microfluidic system for parallel processing of 11 simultaneous enzymatic reactions. Reprinted with permission from Ref. 178. Copyright 2009 American Chemical Society. (b) A dilution microchip used for a one-shot measurement of phosphatase activity in 12 parallel channels at 12 different substrate concentrations delivered by a Y-shaped mixing channel. Reprinted with permission from Ref. 158. Copyright 2002 American Chemical Society. (c) A microfluidic, disk-shaped chip featuring 48 parallel enzyme reactors actuated by centrifugal forces. Reprinted with permission from Ref. 179. Copyright 1999 American Chemical Society.

exploited for actuation of a disk-shaped chip featuring 48 parallel enzyme reactors (Figure 7.14c).[179] In this case, each reactor had individual fluid inlets near the centre of the disk from where the reagents were pumped outwards as the disk was rotated on the shaft of a simple motor at 60–3000 rpm (flow rates varying between 5 nL s^{-1} and >0.1 mL s^{-1}) prior to detection by a centrifugal colorimetric instrument. The gating of flow was achieved by the use of passive valves based on capillary forces so that enzyme reactions could be performed in a multistep fashion (*i.e.*, mixing of the enzyme and the inhibitor prior to addition of the substrate).

7.3.4 Biomimetic Microreactors

In addition to assays exploiting biological material, microfluidic approaches have also taken to predict the (phase I) metabolic pathways of drugs *via* electrochemical conversion[180] or photocatalysis.[181] In the former design, only the electrochemical cell featuring palladium and platinum electrodes for oxidation of amodiaquine in a reaction volume of only 9.6 nL was integrated on chip with on-line coupling to an external LC-MS instrument (Figure 7.15a).[180] In the latter approach, a micropillar array made of silicon and coated with an anatase-phase TiO_2 nanolayer was monolithically integrated with an on-chip emitter to permit on-line ESI-MS analysis of the oxidation products of drugs produced *via* UV photocatalysis on TiO_2 (Figure 7.15b).[181] In both studies, the results were in good agreement with those obtained by traditional electrochemical assays. Yet another approach for electrochemically assisted conversion of drugs was assembled on a PMMA chip incorporating a reaction chamber with two electrode strips, one of which contained the human cytochrome P450 3A4 covalently bound to gold *via* a thiol-based self-assembled monolayer.[182] The chip was used for electrochemical determination of the K_M values of four known CYP3A4 substrates and the results were in good agreement with those obtained on human liver microsomes.

7.4 Microfluidic Organomimetic Systems in Metabolomics

Cell- and tissue-based assays along with clinical diagnosis systems are rapidly becoming the largest field of application for microfluidic systems.[22,184,185] Currently, microengineering techniques are largely used for the fabrication of controlled microenvironments for short- or long-term culture of various types of living cells and for maintenance of their tissue-specific functions. In order to ensure adequate nutrition during cell incubation, gas-permeable PDMS has proved to be a particularly useful substrate material. A common approach for cell trapping inside a microfluidic chamber is to coat the walls with poly-L-lysine or, alternatively, to use hydrogels for cell encapsulation. Sometimes inherent adhesion of cells may suffice for trapping the cells. In addition to cellular assays, the feasibility of microfluidics for small-scale intact tissue-based

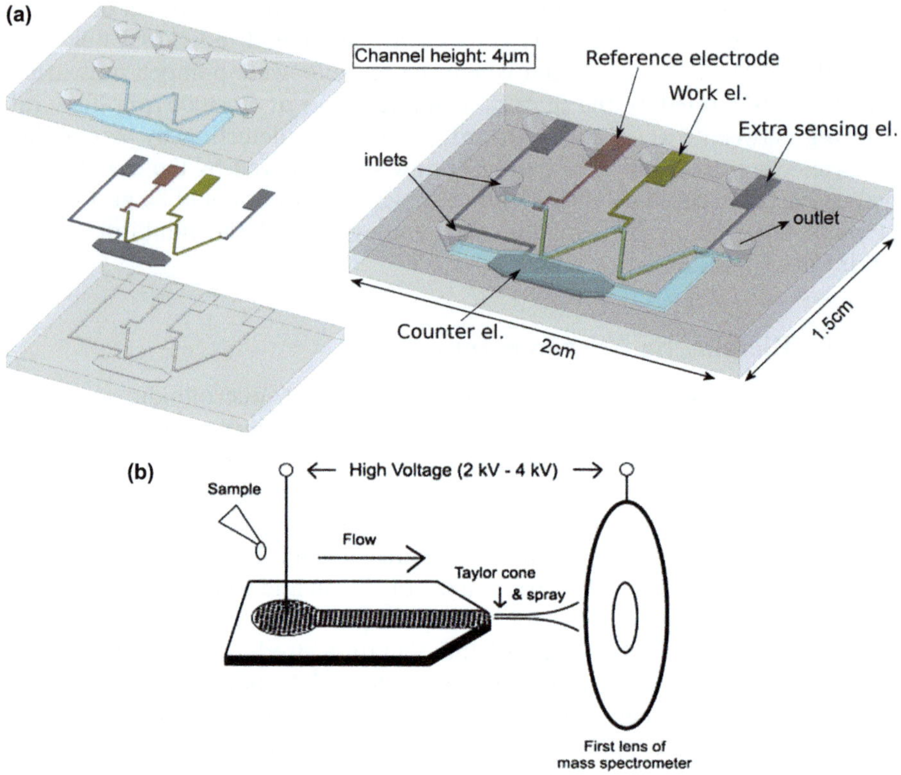

Figure 7.15 (a) Schematic diagram of the glass chip used for electrochemical oxidation of drugs: bottom layer with recess etched wells for the electrodes and top wafer with inlets and fluidic channels. Reprinted with permission of the Royal Society of Chemistry from Ref. 180. (b) Schematic diagram of the silicon micropillar array-ESI chip coated with TiO_2 nanolayer and used for on-chip drug oxidation *via* photocatalysis (on TiO_2).[181] Reproduced from Ref. 183. Copyright 2007 Royal Society of Chemistry.

drug metabolism studies has been demonstrated with scarce human tissue, such as precision-cut liver slices assembled inside a PDMS chip for continuous perfusion with drugs (Figure 7.16a).[186] By connecting microchambers supporting different organ slices in series for sequential perfusion, even inter-organ interactions can be measured *in vitro*,[187] which is generally very difficult, if not impossible, to study in conventional systems. Provided that the viability and leakage of liver-specific enzymes are carefully addressed, it is likely that this approach also provides better *in vitro–in vivo* correlation than that of the traditional *in vitro* enzyme assays using cells or subcellular fractions.

In most cases, on-chip cell metabolism has been monitored by targeted detection of a single metabolite using, for instance, bioluminescence-,[188] chemiluminescence-[189,190] or fluorescence-based[191–198] image analysis or impedance[194,199,200] and other electrochemical techniques.[201–203] From the technical point of view, most cell- and tissue-based assays should be readily feasible for integration with a microfluidic separation system, but in reality, separation of the metabolites produced often relies on conventional off-chip instrumentation, such as HPLC-UV detection[204] or HPLC-MS.[186,187,205,206] MS imaging matrix-assisted laser desorption/ionization (MALDI) of peptide release from neuronal cells within microfluidic devices has also been reported[207] and, on rare occasions, NMR has been used for off-chip identification of the on-chip produced metabolites.[75] In the following, the organomimetic systems incorporating cells and tissues are only briefly touched upon with help of selected examples where the microchip has an effective role in separation or other post-reaction manipulation of the metabolites produced. For a more comprehensive view on microfluidic organomimetic systems, the reader is referred to the relevant literature.[11,22,184,185,208]

Similarly to microreactors incorporating immolized enzymes, cell incubation assays have also been coupled to on-line, off-chip ESI-MS in order to identify cellular metabolites.[107,140] As the high selectivity of MS enables several compounds to be monitored at the same time, on-line separation of the metabolites is not always necessary. Instead, in order to avoid suppression of the ESI signal, the cell incubation matrix needs to be eliminated, hence on-line SPE is often implemented on-chip prior to ESI-MS detection (Figure 7.16b).[107,140] Such setups have been used for studying the metabolism of vitamin E in human lung epithelial A549 cells[140] and of the anticancer agent genistein in human breast adenocarcinoma MCF-7 cells.[107,140]

In addition to cell suspensions, monitoring of the biochemical processes of single cells has also been accomplished using microfluidic devices coupled to ESI-MS. For instance, automated real-time analysis of individual erythrocytes following cell lysis at the injection cross of a glass-based MCE-ESI chip has been reported (Figure 7.16c).[209] The cell lysis was performed by a rapid decrease in the osmotic pressure (dilution of the buffer) and application of an increased electric field at the injection cross during electrokinetic cell migration from the sample inlet towards the microchannel intersection. Subsequently, on-line MCE separation of the cell contents was performed prior to on-chip ESI-MS detection. In another study, MCE-ESI-MS analysis of cellular content was performed on off-chip produced human cell lysates.[210] Although MS allows the direct identification of the cellular metabolites, the original idea of the on-line analysis of on-chip lysed cells was presented using LIF detection of cytosolic fluorescent dyes, Oregon Green and carboxyfluorescein, from Jurkat T cells.[211] Later, somewhat similar approaches, but with electrically or electromechanically induced cell lysis, were made in order to monitor the lysates by both fluorescence and electrochemical detection.[212] At best, throughputs of up to 12–13 cells per minute have been reported.[209,211]

Another type of approach for manipulation of cells is to use digital microfluidics on a planar array of electrodes (Figure 7.16d).[213] This approach allows the dosing of nanolitre droplets of cell incubation reagents and actuation of cell suspensions (droplets) with very high concentrations ($>10^6\,mL^{-1}$) without affecting the cell viability or proliferation. Although in this work the analysis of the cellular contents was performed using off-line MALDI-MS, the technique is

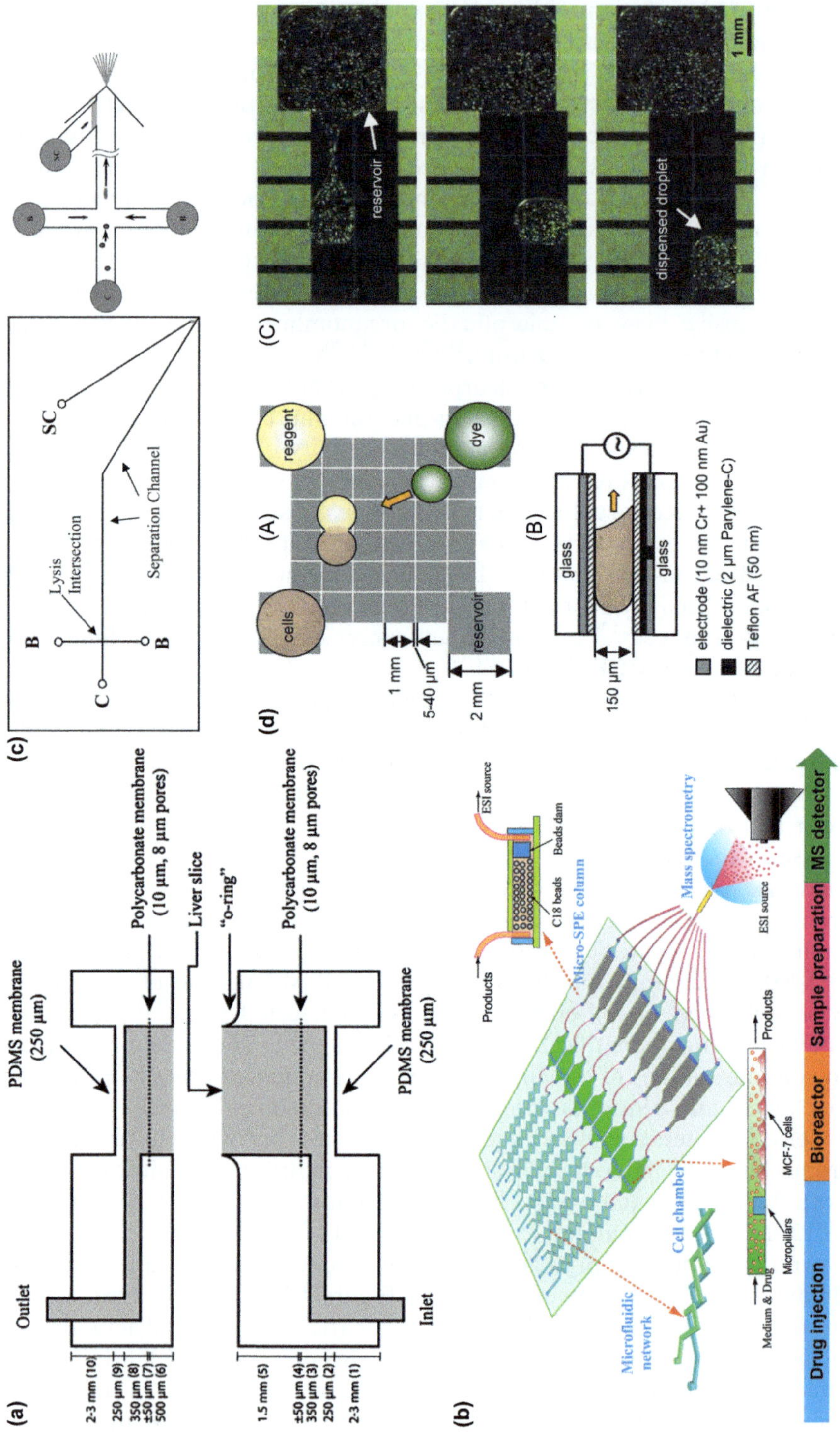
(a)
Outlet
Inlet
PDMS membrane (250 µm)
Polycarbonate membrane (10 µm, 8 µm pores)
Liver slice
"o-ring"
2-3 mm (10)
250 µm (9)
350 µm (8)
±50 µm (7)
500 µm (6)
1.5 mm (5)
±50 µm (4)
350 µm (3)
250 µm (2)
2-3 mm (1)
(b)
Microfluidic network
Cell chamber
Micro-SPE column
Mass spectrometry
Products
ESI source
C18 beads
Beads dam
Medium & Drug
Micropillars
MCF-7 cells
Drug injection
Bioreactor
Sample preparation
MS detector
(c)
B
C
SC
Lysis Intersection
Separation Channel
(d)
(A)
cells
reagent
dye
reservoir
1 mm
5-40 µm
2 mm
(B)
glass
150 µm
electrode (10 nm Cr+ 100 nm Au)
dielectric (2 µm Parylene-C)
Teflon AF (50 nm)
(C)
reservoir
dispensed droplet
1 mm

readily feasible for integration with on-line sample preparation by LLE[152] and on-chip ESI-MS.[166] In addition to digital microfluidics, encapsulation of cells in aqueous droplets has been accomplished *via* multiphase fluid flow (*i.e.*, aqueous droplets fed into an immiscible carrier flow).[214] This approach has been used for monitoring the effects of antibiotics on the bacterial cell metabolism by downstream fluorescence microscopy.

7.5 Future Perspectives

In the light of the recent progress in the field of miniaturized total (bio)chemical analysis systems (μTAS), it is clear that microfluidics and microchip-based technologies show high feasibility for metabolomics, particularly *via* implementation of enzyme assays as an integral part of a separation chip. However, after two decades of intensive research, the number of true μTAS platforms that combine all (bio)analytical key functions on a single chip remains small. For one part, this is because apart from soft lithography, fabrication of μTAS devices often requires complex and expensive cleanroom processing, which slows their development and utilization for routine laboratory analyses. Hence the rising low-cost technologies, such as 3D printing, are likely to promote the development of more user-specific devices in standard laboratory conditions. Another, probably less recognized, drawback is the fact that researchers often seize to build up very exact, and therefore too complex, miniature models of the traditional biological and analytical systems instead of going beyond the customary solutions and creating exclusively new approaches to the task at hand. In this respect, techniques such as digital microfluidics are only rare success stories, whereas MCE, the most widely used technology, suffers from poor usability because of the lack of automated sample handling. Even if one must not look down on the high separation efficiency of MCE, it is worthy of attention that the truly enabling techniques should be quickly accepted. Thus, not only the benefits, but also the pitfalls, that come with miniaturization should be

Figure 7.16 (a) Schematic diagrams (cross-sectional view) of a PDMS biochip with embedded polycarbonate membranes to support precision-cut liver slices. Reprinted from Ref. 186 with permission of John Wiley & Sons, Inc. ©2010. (b) Schematic diagram of a chip-ESI-MS system incorporating a microfluidic network for culture medium and drug injections, cell culture chambers, on-chip micro-SPE columns for sample desalting and purifying and a capillary coupling to ESI-Q-TOF-MS. Reprinted with permission from Ref. 107. Copyright 2012 American Chemical Society. (c) Schematic diagrams of the cell lysis on MCE-ESI-MS microchip. The arrows indicate the direction and relative magnitude of EOF. C = cells, B = buffer, SC = EO pump side channel. Reprinted with permission from Ref. 209. Copyright 2010 American Chemical Society. (d) Schematic diagrams of a DMF device used for cell-based assays (A, B) with a sequence of pictures from a movie where a 150 nL droplet containing ∼260 cells is dispensed (C). Cells were labelled with calcein AM, which fluoresces green. Reproduced from Ref. 213. Copyright 2008 Royal Society of Chemistry.

clearly established. In addition to automated operation of the chips, there is a lot to be done in the fields of detection development and sample preparation in order to improve the sensitivity of microchip-based analyses. Finally, the importance of materials science *per se* should also be accounted for, as it impacts on all aspects of microchip technology, from microfabrication *via* fluid actuation and surface interactions to detection sensitivity.

References

1. K. Huikko, R. Kostiainen and T. Kotiaho, *Eur. J. Pharm. Sci.*, 2003, **20**, 149.
2. K. Ohno, K. Tachikawa and A. Manz, *Electrophoresis*, 2008, **29**, 4443.
3. G. T. Roman and R. T. Kennedy, *J. Chromatogr. A*, 2007, **1168**, 170.
4. E. E. Nuxoll and R. A. Siegel, *IEEE Eng. Med. Biol. Mag.*, 2009, **28**, 31.
5. T. James, M. S. Mannoor and D. V. Ivanov, *Sensors*, 2008, **8**, 6077.
6. A. C. R. Grayson, R. S. Shawgo, A. M. Johnson, N. T. Flynn, Y. W. Li, M. J. Cima and R. Langer, *Proc IEEE*, 2004, **92**, 6.
7. S. C. Jacobson, C. T. Culbertson, J. E. Daler and J. M. Ramsey, *Anal. Chem.*, 1998, **70**, 3476.
8. S. C. Terry and J. B. Angell, *Des. Biomed. Appl. Solid State Chem. Sens. Workshop*, 1978, 207.
9. A. Manz, D. J. Harrison, E. M. J. Verpoorte, J. C. Fettinger, A. Paulus, H. Ludi and H. M. Widmer, *J. Chromatogr.*, 1992, **593**, 253.
10. D. J. Harrison, A. Manz, Z. H. Fan, H. Ludi and H. M. Widmer, *Anal. Chem.*, 1992, **64**, 1926.
11. J. R. Kraly, R. E. Holcomb, Q. Guan and C. S. Henry, *Anal. Chim. Acta*, 2009, **653**, 23.
12. J. Lee, S. A. Soper and K. K. Murray, *J. Mass Spectrom.*, 2009, **44**, 579.
13. S. L. S. Freire and A. R. Wheeler, *Lab Chip*, 2006, **6**, 1415.
14. T. Sikanen, S. Franssila, T. J. Kauppila, R. Kostiainen, T. Kotiaho and R. A. Ketola, *Mass Spectrom. Rev.*, 2010, **29**, 351.
15. S. Koster and E. Verpoorte, *Lab Chip*, 2007, **7**, 1394.
16. C. E. C. A. Hop, *Curr. Drug Metab.*, 2006, **7**, 557.
17. M. Vlckova, F. Kalman and M. A. Schwarz, *J. Chromatogr. A*, 2008, **1181**, 145.
18. S. Wakida, K. Fujimoto, H. Nagai, T. Miyado, Y. Shibutani and S. Takeda, *J. Chromatogr. A*, 2006, **1109**, 179.
19. T. Miyado, Y. Tanaka, H. Nagai, S. Takeda, K. Saito, K. Fukushi, Y. Yoshida, S. Wakida and E. Niki, *J. Chromatogr. A*, 2004, **1051**, 185.
20. R. M. Wightman, *Science*, 1988, **240**, 415.
21. D. Janasek, J. Franzke and A. Manz, *Nature*, 2006, **442**, 374.
22. K. Sato, K. Mawatari and T. Kitamori, *Lab Chip*, 2008, **8**, 1992.
23. A. Ros, W. Hellmich, K. Leffhalm, A. Sischka and D. Anselmetti, *Mol. Cell. Proteomics*, 2005, **4**, S336.
24. D. Huh, G. A. Hamilton and D. E. Ingber, *Trends Cell Biol.*, 2011, **21**, 745.

25. R. B. Fair, *Microfluidics Nanofluidics*, 2007, **3**, 245.
26. M. J. Jebrail, M. S. Bartsch and K. D. Patel, *Lab Chip*, 2012, **12**, 2452.
27. A. Manz, Y. Miyahara, J. Miura, Y. Watanabe, H. Miyagi and K. Sato, *Sens. Actuators B. Chem.*, 1990, **1**, 249.
28. D. Belder and M. Ludwig, *Electrophoresis*, 2003, **24**, 3595.
29. J. S. Mellors, V. Gorbounov, R. S. Ramsey and J. M. Ramsey, *Anal. Chem.*, 2008, **80**, 6881.
30. P. Hoffmann, U. Hausig, P. Schulze and D. Belder, *Angew. Chem. Int. Ed.*, 2007, **46**, 4913.
31. L. Sainiemi, T. Sikanen and R. Kostiainen, *Anal. Chem.*, 2012, **84**, 8973.
32. P. Schulze and D. Belder, *Anal. Bioanal. Chem.*, 2009, **393**, 515.
33. H. Becker and C. Gartner, *Anal. Bioanal. Chem.*, 2008, **390**, 89.
34. H. Becker and L. E. Locascio, *Talanta*, 2002, **56**, 267.
35. H. Shadpour, H. Musyimi, J. Chen and S. A. Soper, *J. Chromatogr. A*, 2006, **1111**, 238.
36. S. Franssila, *Introduction to Microfabrication*, Wiley, Chichester, 2004.
37. M. Madou, *Fundamentals of Microfabrication*, CRC Press, Boca Raton, FL, 1997.
38. H. Makamba, J. H. Kim, K. Lim, N. Park and J. H. Hahn, *Electrophoresis*, 2003, **24**, 3607.
39. T. Sikanen, S. Aura, L. Heikkila, T. Kotiaho, S. Franssila and R. Kostiainen, *Anal. Chem.*, 2010, **82**, 3874.
40. T. Sikanen, L. Heikkila, S. Tuomikoski, R. A. Ketola, R. Kostiainen, S. Franssila and T. Kotiaho, *Anal. Chem.*, 2007, **79**, 6255.
41. A. Piruska, I. Nikcevic, S. H. Lee, C. Ahn, W. R. Heineman, P. A. Limbach and C. J. Seliskar, *Lab Chip*, 2005, **5**, 1348.
42. K. Huikko, P. Ostman, K. Grigoras, S. Tuomikoski, V. M. Tiainen, A. Soininen, K. Puolanne, A. Manz, S. Franssila, R. Kostiainen and T. Kotiaho, *Lab Chip*, 2003, **3**, 67.
43. Z. Meng, S. Qi, S. A. Soper and P. A. Limbach, *Anal. Chem.*, 2001, **73**, 1286.
44. S. Jung, A. Hoeltzel, S. Ehlert, J. Mora, K. Kraiczek, M. Dittmann, G. P. Rozing and U. Tallarek, *Anal. Chem.*, 2009, **81**, 10193.
45. S. Jung, S. Ehlert, J. Mora, K. Kraiczek, M. Dittmann, G. P. Rozing and U. Tallarek, *J. Chromatogr. A*, 2009, **1216**, 264.
46. S. C. Jacobson and C. T. Culbertson, in *Separation Methods in Microanalytical Systems*, J. P. Kutter and Y. Fintschenko, CRC Press, Boca Raton, FL, 2006, pp. 19–54.
47. K. B. Mogensen, H. Klank and J. P. Kutter, *Electrophoresis*, 2004, **25**, 3498.
48. K. B. Mogensen, J. El-Ali, A. Wolff and J. P. Kutter, *Appl. Opt.*, 2003, **42**, 4072.
49. E. Verpoorte, *Lab Chip*, 2003, **3**, 42N.
50. M. Castano-Alvarez, M. T. Fernandez-Abedul, A. Costa-Garcia, M. Agirregabiria, L. J. Fernandez, J. M. Ruano-Lopez and B. Barredo-Presa, *Talanta*, 2009, **80**, 24.

51. C. D. Garcia and C. S. Henry, *Anal. Chem.*, 2003, **75**, 4778.
52. F. Niklaus, G. Stemme, J. Q. Lu and R. J. Gutmann, *J. Appl. Phys.*, 2006, **99**, 031101.
53. M. A. Eddings, M. A. Johnson and B. K. Gale, *J. Micromech. Microeng.*, 2008, **18**, 067001.
54. Z. Xiao, G. Wu, Z. Li, C. Zhang, Y. Hao and Y. Wang, *Sens. Actuators A Phys.*, 1999, **72**, 46.
55. C. T. Culbertson, S. C. Jacobson and J. M. Ramsey, *Anal. Chem.*, 1998, **70**, 3781.
56. C. H. Tsai, C. H. Tai, L. M. Fu and F. B. Wu, *J. Micromech. Microeng.*, 2005, **15**, 377.
57. Y. Sato, G. Irisawa, M. Ishizuka, K. Hishida and M. Maeda, *Meas. Sci. Technol.*, 2003, **14**, 114.
58. J. M. Karlinsey, *Anal. Chim. Acta*, 2012, **725**, 1.
59. E. Verpoorte, A. Manz, H. Ludi, A. E. Bruno, F. Maystre, B. Krattiger, H. M. Widmer, B. H. Vanderschoot and N. F. Derooij, *Sens. Actuators B Chem.*, 1992, **6**, 66.
60. H. Salimi-Moosavi, Y. T. Jiang, L. Lester, G. McKinnon and D. J. Harrison, *Electrophoresis*, 2000, **21**, 1291.
61. T. Miyado, S. Wakida, H. Aizawa, Y. Shibutani, T. Kanie, M. Katayama, K. Nose and A. Shimouchi, *J. Chromatogr. A*, 2008, **1206**, 41.
62. T. Miyado, Y. Tanaka, H. Nagai, S. Takeda, K. Saito, K. Fukushi, Y. Yoshida, S. Wakida and E. Niki, *J. Chromatogr. A*, 2006, **1109**, 174.
63. E. Guihen and J. D. Glennon, *J. Chromatogr. A*, 2005, **1071**, 223.
64. A. Ishida, M. Natsume and T. Kamidate, *J. Chromatogr. A*, 2008, **1213**, 209.
65. A. Ishida, T. Yoshikawa, M. Natsume and T. Kamidate, *J. Chromatogr. A*, 2006, **1132**, 90.
66. I. M. Lazar, P. Trisiripisal and H. A. Sarvaiya, *Anal. Chem.*, 2006, **78**, 5513.
67. L. D. Hutt, D. P. Glavin, J. L. Bada and R. A. Mathies, *Anal. Chem.*, 1999, **71**, 4000.
68. J. Seo and L. P. Lee, *Sens. Actuators B Chem.*, 2004, **99**, 615.
69. N. Nordman, T. Sikanen, M. Moilanen, S. Aura, T. Kotiaho, S. Franssila and R. Kostiainen, *J. Chromatogr. A*, 2011, **1218**, 739.
70. T. Sikanen, S. Tuomikoski, R. A. Ketola, R. Kostiainen, S. Franssila and T. Kotiaho, *Anal. Chem.*, 2007, **79**, 9135.
71. S. Fritzsche, P. Hoffmann and D. Belder, *Lab Chip*, 2010, **10**, 1227.
72. M. Wilm and M. Mann, *Anal. Chem.*, 1996, **68**, 1.
73. M. S. Wilm and M. Mann, *Int. J. Mass Spectrom. Ion Processes*, 1994, **136**, 167.
74. H. Wu, G. Sui, C. Lee, M. L. Prins, W. Ladno, H. Lin, A. S. Yu, M. E. Phelps and S. Huang, *J. Nucl. Med.*, 2007, **48**, 837.
75. D. A. Ouattara, J. Prot, A. Bunescu, M. Dumas, B. Elena-Herrmann, E. Leclerc and C. Brochot, *Mol. Biosyst.*, 2012, **8**, 1908.

76. R. A. Shaw, C. Rigatto, M. Reslerova, S. L. Ying, A. Man, B. Schattka, C. F. Battrell, J. Matthewson and C. Mansfield, *Analyst*, 2009, **134**, 1224.
77. L. Wang, D. M. Sipe, Y. Xu and Q. Lin, *J. Microelectromech. Syst.*, 2008, **17**, 318.
78. B. Sorli, J. F. Chateaux, M. Pitaval, H. Chahboune, B. Favre, A. Briguet and P. Morin, *Meas. Sci. Technol.*, 2004, **15**, 877.
79. C. Massin, F. Vincent, A. Homsy, K. Ehrmann, G. Boero, P. A. Besse, A. Daridon, E. Verpoorte, N. F. de Rooij and R. S. Popovic, *J. Magn. Reson.*, 2003, **164**, 242.
80. S. Kulka, N. Kaun, J. R. Baena, J. Frank, P. Svasek, D. Moss, M. J. Vellekoop and B. Lendl, *Anal. Bioanal. Chem.*, 2004, **378**, 1735.
81. T. Pan, R. T. Kelly, M. C. Asplund and A. T. Woolley, *J. Chromatogr. A*, 2004, **1027**, 231.
82. C. A. Monnig and J. W. Jorgenson, *Anal. Chem.*, 1991, **63**, 802.
83. N. J. Petersen, R. P. H. Nikolajsen, K. B. Mogensen and J. P. Kutter, *Electrophoresis*, 2004, **25**, 253.
84. K. Seiler, Z. H. H. Fan, K. Fluri and D. J. Harrison, *Anal. Chem.*, 1994, **66**, 3485.
85. J. K. Beattie, *Lab Chip*, 2006, **6**, 1409.
86. C. T. Culbertson, S. C. Jacobson and J. M. Ramsey, *Anal. Chem.*, 2000, **72**, 5814.
87. E. Verpoorte, *Electrophoresis*, 2002, **23**, 677.
88. B. F. Liu, H. Hisamoto and S. Terabe, *J. Chromatogr. A*, 2003, **1021**, 201.
89. T. M. Phillips and E. F. Wellner, *Electrophoresis*, 2007, **28**, 3041.
90. G. Ocvirk, T. Tang and D. J. Harrison, *Analyst*, 1998, **123**, 1429.
91. M. Fouad, M. Jabasini, N. Kaji, K. Terasaka, M. Tokeshi, H. Mizukami and Y. Baba, *Electrophoresis*, 2008, **29**, 2280.
92. B. Ma, X. Zhou, G. Wang, H. Huang, Z. Dai, J. Qin and B. Lin, *Electrophoresis*, 2006, **27**, 4904.
93. Y. Deng, J. Henion, J. Li, P. Thibault, C. Wang and D. J. Harrison, *Anal. Chem.*, 2001, **73**, 639.
94. R. E. Holcomb, J. R. Kraly and C. S. Henry, *Analyst*, 2009, **134**, 486.
95. C. D. Garcia and C. S. Henry, *Analyst*, 2004, **129**, 579.
96. J. Wang, M. P. Chatrathi and G. E. Collins, *Anal. Chim. Acta*, 2007, **585**, 11.
97. J. Qiao, L. Qi, H. Yan, Y. Li and X. Mu, *Electrophoresis*, 2013, **34**, 409.
98. R. W. Hompesch, C. D. Garcia, D. J. Weiss, J. M. Vivanco and C. S. Henry, *Analyst*, 2005, **130**, 694.
99. C. D. Garcia and C. S. Henry, *Anal. Chim. Acta*, 2004, **508**, 1.
100. N. E. Hebert, W. G. Kuhr and S. A. Brazill, *Electrophoresis*, 2002, **23**, 3750.
101. Q. L. Zhang, H. Z. Lian, W. H. Wang and H. Y. Chen, *J. Chromatogr. A*, 2005, **1098**, 172.
102. Q. Zhang, J. Xu, X. Li, H. Lian and H. Chen, *J. Pharm. Biomed. Anal.*, 2007, **43**, 237.

103. M. Vlckova and M. A. Schwarz, *J. Chromatogr. A*, 2007, **1142**, 214.
104. H. Yu, F. He, Y. Lu, Y. Hu, H. Zhong and X. Xia, *Talanta*, 2008, **75**, 43.
105. C. M. Chen, Y. H. Ho, S. M. Wu, G. L. Chang and C. H. Lin, *J. Nanosci. Nanotechnol.*, 2009, **9**, 718.
106. M. A. Schwarz, *Electrophoresis*, 2004, **25**, 1916.
107. Q. Chen, J. Wu, Y. Zhang and J. Lin, *Anal. Chem.*, 2012, **84**, 1695.
108. A. M. Tan, S. Benetton and J. D. Henion, *Anal. Chem.*, 2003, **75**, 5504.
109. L. Ahonen, P. Keski-Rahkonen, T. Saarelainen, J. Paviala, R. A. Ketola, S. Auriola, M. Poutanen and R. Kostiainen, *Anal. Chim. Acta*, 2012, **721**, 115.
110. J. P. Kutter, *J. Chromatogr. A*, 2012, **1221**, 72.
111. A. de Mello, *Lab Chip*, 2002, **2**, 48N.
112. D. P. Wu, J. H. Qin and B. C. Lin, *J. Chromatogr. A*, 2008, **1184**, 542.
113. D. Peterson, *Lab Chip*, 2005, **5**, 132.
114. J. Kutter, S. Jacobson and J. Ramsey, *J. Microcolumn Sep.*, 2000, **12**, 93.
115. H. Yin, K. Killeen, R. Brennen, D. Sobek, M. Werlich and T. van de Goor, *Anal. Chem.*, 2005, **77**, 527.
116. M. Haapala, V. Saarela, J. Pol, K. Kolari, T. Kotiaho, S. Franssila and R. Kostiainen, *Anal. Chim. Acta*, 2010, **662**, 163.
117. Y. H. Tennico and V. T. Remcho, *Electrophoresis*, 2010, **31**, 2548.
118. F. Svec, *J. Sep. Sci.*, 2004, **27**, 1419.
119. S. Yamamoto, S. Hirakawa and S. Suzuki, *Anal. Chem.*, 2008, **80**, 8224.
120. T. B. Stachowiak, F. Svec and J. M. J. Frechet, *J. Chromatogr. A*, 2004, **1044**, 97.
121. F. Svec, *J. Sep. Sci.*, 2005, **28**, 729.
122. W. De Malsche, S. De Bruyne, J. O. De Beeck, S. Eeltink, F. Detobel, H. Gardeniers and G. Desmet, *J. Sep. Sci.*, 2012, **35**, 2010.
123. W. De Malsche, J. O. De Beeck, S. De Bruyne, H. Gardeniers and G. Desmet, *Anal. Chem.*, 2012, **84**, 1214.
124. T. B. Stachowiak, T. Rohr, E. F. Hilder, D. S. Peterson, M. Q. Yi, F. Svec and J. M. J. Frechet, *Electrophoresis*, 2003, **24**, 3689.
125. L. Sainiemi, T. Nissila, R. Kostiainen, S. Franssila and R. A. Ketola, *Lab Chip*, 2012, **12**, 325.
126. V. Saarela, S. Franssila, S. Tuomikoski, S. Marttila, P. Ostman, T. Sikanen, T. Kotiaho and R. Kostiainen, *Sens. Actuators B Chem.*, 2006, **114**, 552.
127. J. Xie, Y. Miao, J. Shih, Y. C. Tai and T. D. Lee, *Anal. Chem.*, 2005, **77**, 6947.
128. M. Madou, J. Zoval, G. Y. Jia, H. Kido, J. Kim and N. Kim, *Annu. Rev. Biomed. Eng.*, 2006, **8**, 601.
129. W. De Malsche, S. De Bruyne, J. O. De Beek, P. Sandra, H. Gardeniers, G. Desmet and F. Lynen, *J. Chromatogr. A*, 2012, **1230**, 41.
130. F. G. Bessoth, A. J. deMello and A. Manz, *Anal. Commun.*, 1999, **36**, 213.
131. P. Sajonz, W. Schafer, X. Gong, S. Shultz, T. Rosner and C. J. Welch, *J. Chromatogr. A*, 2007, **1145**, 149.

132. R. Flamini, M. De Rosso, A. Smaniotto, A. Panighel, A. D. Vedova, R. Seraglia and P. Traldi, *Rapid Commun. Mass Spectrom.*, 2009, **23**, 2891.
133. W. R. Alley Jr, M. Madera, Y. Mechref and M. V. Novotny, *Anal. Chem.*, 2010, **82**, 5095.
134. N. Tao, E. J. DePeters, J. B. German, R. Grimm and C. B. Lebrilla, *J. Dairy Sci.*, 2009, **92**, 2991.
135. P. Ostman, S. J. Marttila, T. Kotiaho, S. Franssila and R. Kostiainen, *Anal. Chem.*, 2004, **76**, 6659.
136. T. J. Kauppila, P. Ostman, S. Marttila, R. A. Ketola, T. Kotiaho, S. Franssila and R. Kostiainen, *Anal. Chem.*, 2004, **76**, 6797.
137. L. L. Ahonen, M. Haapala, V. Saarela, S. Franssila, T. Kotiaho and R. Kostiainen, *Rapid Commun. Mass Spectrom.*, 2010, **24**, 958.
138. J. Lichtenberg, N. F. de Rooij and E. Verpoorte, *Talanta*, 2002, **56**, 233.
139. R. D. Oleschuk, L. L. Shultz-Lockyear, Y. Ning and D. J. Harrison, *Anal. Chem.*, 2000, **72**, 585.
140. D. Gao, H. Wei, G. Guo and J. Lin, *Anal. Chem.*, 2010, **82**, 5679.
141. S. Mao, D. Gao, W. Liu, H. Wei and J. Lin, *Lab Chip*, 2012, **12**, 219.
142. C. Cakal, J. P. Ferrance, J. P. Landers and P. Caglar, *Anal. Chim. Acta*, 2011, **690**, 94.
143. N. Nordman, B. Barrios-Lopez, S. Laurén, P. Suvanto, T. Kotiaho, S. Franssila, R. Kostiainen and T. Sikanen, *Proc. MicroTAS*, Freiburg, Germany, October 27–31, 2013.
144. S. Jacobson and J. Ramsey, *Electrophoresis*, 1995, **16**, 481.
145. J. J. Li, C. Wang, J. F. Kelly, D. J. Harrison and P. Thibault, *Electrophoresis*, 2000, **21**, 198.
146. S. Fanali, C. Desiderio, E. Olvecka, D. Kaniansky, M. Vojtek and A. Ferancova, *HRC J. High Resolut. Chromatogr.*, 2000, **23**, 531.
147. J. P. Quirino, S. Terabe and P. Bocek, *Anal. Chem.*, 2000, **72**, 1934.
148. T. Kawai, K. Sueyoshi, F. Kitagawa and K. Otsuka, *Anal. Chem.*, 2010, **82**, 6504.
149. H. Miyaguchi, M. Tokeshi, Y. Kikutani, A. Hibara, H. Inoue and T. Kitamori, *J. Chromatogr. A*, 2006, **1129**, 105.
150. P. Mary, V. Studer and P. Tabeling, *Anal. Chem.*, 2008, **80**, 2680.
151. H. Chen, Q. Fang, X. F. Yin and Z. L. Fang, *Lab Chip*, 2005, **5**, 719.
152. P. A. L. Wijethunga, Y. S. Nanayakkara, P. Kunchala, D. W. Armstrong and H. Moon, *Anal. Chem.*, 2011, **83**, 1658.
153. V. Jokinen, R. Kostiainen and T. Sikanen, *Adv. Mater.*, 2012, **24**, 6240.
154. N. J. Petersen, S. T. Foss, H. Jensen, S. H. Hansen, C. Skonberg, D. Snakenborg, J. P. Kutter and S. Pedersen-Bjergaard, *Anal. Chem.*, 2011, **83**, 44.
155. T. Sikanen, S. Pedersen-Bjergaard, H. Jensen, R. Kostiainen, K. E. Rasmussen and T. Kotiaho, *Anal. Chim. Acta*, 2010, **658**, 133.
156. N. Lion, V. Gobry, H. Jensen, J. S. Rossier and H. Girault, *Electrophoresis*, 2002, **23**, 3583.

157. S. Benetton, J. Kameoka, A. Tan, T. Wachs, H. Craighead and J. D. Henion, *Anal. Chem.*, 2003, **75**, 6430.
158. H. B. Mao, T. L. Yang and P. S. Cremer, *Anal. Chem.*, 2002, **74**, 379.
159. J. C. Zguris, L. J. Itle, D. Hayes and M. V. Pishko, *Biomed. Microdevices*, 2005, **7**, 117.
160. D. N. Kim, Y. Lee and W. Koh, *Sens. Actuators B Chem.*, 2009, **137**, 305.
161. H. Song and R. F. Ismagilov, *J. Am. Chem. Soc.*, 2003, **125**, 14613.
162. A. G. Hadd, D. E. Raymond, J. W. Halliwell, S. C. Jacobson and J. M. Ramsey, *Anal. Chem.*, 1997, **69**, 3407.
163. M. Kerby and R. L. Chien, *Electrophoresis*, 2001, **22**, 3916.
164. M. Tisma, B. Zelic, D. Vasic-Racki, P. Znidarsic-Plazl and I. Plazl, *Chem. Eng. J.*, 2009, **149**, 383.
165. A. G. Hadd, S. C. Jacobson and J. M. Ramsey, *Anal. Chem.*, 1999, **71**, 5206.
166. S. C. C. Shih, H. Yang, M. J. Jebrail, R. Fobel, N. McIntosh, O. Y. Al-Dirbashi, P. Chakraborty and A. R. Wheeler, *Anal. Chem.*, 2012, **84**, 3731.
167. V. Srinivasan, V. K. Pamula and R. B. Fair, *Anal. Chim. Acta*, 2004, **507**, 145.
168. V. Srinivasan, V. K. Pamula and R. B. Fair, *Lab Chip*, 2004, **4**, 310.
169. E. M. Miller and A. R. Wheeler, *Anal. Chem.*, 2008, **80**, 1614.
170. K. P. Nichols and H. J. G. E. Gardeniers, *Anal. Chem.*, 2007, **79**, 8699.
171. B. Ma, G. Zhang, J. Qin and B. Lin, *Lab Chip*, 2009, **9**, 232.
172. C. B. Cohen, E. Chin-Dixon, S. Jeong and T. T. Nikiforov, *Anal. Biochem.*, 1999, **273**, 89.
173. D. E. Starkey, A. Han, J. J. Bao, C. H. Ahn, K. R. Wehmeyer, M. C. Prenger, H. B. Halsall and W. R. Heineman, *J. Chromatogr. B*, 2001, **762**, 33.
174. M. Kato, T. Yamamoto, M. Sekimoto, M. Degawa and T. Toyo'oka, *J. Appl. Phys.*, 2009, **105**, 102019.
175. A. Liu, T. Zhou, F. He, J. Xu, Y. Lu, H. Chen and X. Xia, *Lab Chip*, 2006, **6**, 811.
176. S. A. Zugel, B. J. Burke, F. E. Regnier and F. E. Lytle, *Anal. Chem.*, 2000, **72**, 5731.
177. J. H. Kang and J. K. Park, *Sens. Actuators B Chem.*, 2005, **107**, 980.
178. S. Jambovane, E. C. Duin, S. K. Kirn and J. W. Hong, *Anal. Chem.*, 2009, **81**, 3239.
179. D. C. Duffy, H. L. Gillis, J. Lin, N. F. Sheppard and G. J. Kellogg, *Anal. Chem.*, 1999, **71**, 4669.
180. M. Odijk, A. Baumann, W. Lohmann, F. T. G. van den Brink, W. Olthuis, U. Karst and A. van den Berg, *Lab Chip*, 2009, **9**, 1687.
181. T. Nissila, L. Sainiemi, M. Karikko, M. Kemell, M. Ritala, S. Franssila, R. Kostiainen and R. A. Ketola, *Lab Chip*, 2011, **11**, 1470.
182. A. Fantuzzi, E. Capria, L. H. Mak, V. R. Dodhia, S. J. Sadeghi, S. Collins, G. Somers, E. Hug and G. Gilardi, *Anal. Chem.*, 2010, **82**, 10222.

183. T. Nissila, L. Sainiemi, T. Sikanen, T. Kotiaho, S. Franssila, R. Kostiainen and R. A. Ketola, *Rapid Commun. Mass Spectrom.*, 2007, **21**, 3677.
184. P. M. van Midwoud, E. Verpoorte and G. M. M. Groothuis, *Integr. Biol.*, 2011, **3**, 509.
185. J. Park, K. Kim, J. Lee, H. Kim and D. Huh, *Biomed. Eng. Lett.*, 2012, **2**, 88.
186. P. M. van Midwoud, G. M. M. Groothuis, M. T. Merema and E. Verpoorte, *Biotechnol. Bioeng.*, 2010, **105**, 184.
187. P. M. van Midwoud, M. T. Merema, E. Verpoorte and G. M. M. Groothuis, *Lab Chip*, 2010, **10**, 2778.
188. S. K. Yoo, J. H. Lee, S. Yun, M. B. Gu and J. H. Lee, *Biosens. Bioelectron.*, 2007, **22**, 1586.
189. R. Davidsson, B. Johansson, V. Passoth, M. Bengtsson, T. Laurell and J. Emneus, *Lab Chip*, 2004, **4**, 488.
190. D. O. Fesenko, T. V. Nasedkina, D. V. Prokopenko and A. D. Mirzabekov, *Biosens. Bioelectron.*, 2005, **20**, 1860.
191. S. Faley, K. Seale, J. Hughey, D. K. Schaffer, S. VanCornpernolle, B. McKinney, F. Baudenbacher, D. Unutmaz and J. P. Wikswo, *Lab Chip*, 2008, **8**, 1700.
192. J. V. Rocheleau, G. M. Walker, W. S. Head, O. P. McGuinness and D. W. Piston, *Proc. Natl. Acad. Sci. U. S. A.*, 2004, **101**, 12899.
193. P. Hersen, M. N. McClean, L. Mahadevan and S. Ramanathan, *Proc. Natl. Acad. Sci. U. S. A.*, 2008, **105**, 7165.
194. N. Klauke, G. Smith and J. M. Cooper, *Lab Chip*, 2007, **7**, 731.
195. H. Zhu, G. Stybayeva, M. Macal, E. Ramanculov, M. D. George, S. Dandekar and A. Revzin, *Lab Chip*, 2008, **8**, 2197.
196. T. D. Oblak, J. A. Meyer and D. M. Spence, *Analyst*, 2009, **134**, 188.
197. A. M. Clark, K. M. Sousa, C. Jennings, O. A. MacDougald and R. T. Kennedy, *Anal. Chem.*, 2009, **81**, 2350.
198. Y. S. Heo, L. M. Cabrera, C. L. Bormann, G. D. Smith and S. Takayama, *Lab Chip*, 2012, **12**, 2240.
199. R. Gomez, R. Bashir and A. K. Bhunia, *Sens. Actuators B Chem.*, 2002, **86**, 198.
200. R. Gomez-Sjoberg, D. T. Morisette and R. Bashir, *J. Microelectromech. Syst.*, 2005, **14**, 829.
201. W. Cheng, N. Klauke, H. Sedgwick, G. L. Smith and J. M. Cooper, *Lab Chip*, 2006, **6**, 1424.
202. W. Satoh, S. Takahashi, F. Sassa, J. Fukuda and H. Suzuki, *Lab Chip*, 2009, **9**, 35.
203. N. Hu, C. Wu, D. Ha, T. Wang, Q. Liu and P. Wang, *Biosens. Bioelectron.*, 2013, **40**, 167.
204. A. Carraro, W. Hsu, K. M. Kulig, W. S. Cheung, M. L. Miller, E. J. Weinberg, E. F. Swart, M. Kaazempur-Mofrad, J. T. Borenstein, J. P. Vacanti and C. Neville, *Biomed. Microdevices*, 2008, **10**, 795.

205. P. Chao, T. Maguire, E. Novik, K. C. Cheng and M. L. Yarmush, *Biochem. Pharmacol.*, 2009, **78**, 625.
206. E. Novik, T. J. Maguire, P. Chao, K. C. Cheng and M. L. Yarmush, *Biochem. Pharmacol.*, 2010, **79**, 1036.
207. K. Jo, M. L. Heien, L. B. Thompson, M. Zhong, R. G. Nuzzo and J. V. Sweedler, *Lab Chip*, 2007, **7**, 1454.
208. R. Baudoin, A. Corlu, L. Griscom, C. Legallais and E. Leclerc, *Toxicol. In Vitro*, 2007, **21**, 535.
209. J. S. Mellors, K. Jorabchi, L. M. Smith and J. M. Ramsey, *Anal. Chem.*, 2010, **82**, 967.
210. N. Nordman, T. Sikanen, S. Aura, S. Tuomikoski, K. Vuorensola, T. Kotiaho, S. Franssila and R. Kostiainen, *Electrophoresis*, 2010, **31**, 3745.
211. M. A. McClain, C. T. Culbertson, S. C. Jacobson, N. L. Allbritton, C. E. Sims and J. M. Ramsey, *Anal. Chem.*, 2003, **75**, 5646.
212. T. Chao and A. Ros, *J. R. Soc. Interface*, 2008, **5**, S139.
213. I. Barbulovic-Nad, H. Yang, P. S. Park and A. R. Wheeler, *Lab Chip*, 2008, **8**, 519.
214. J. Q. Boedicker, L. Li, T. R. Kline and R. F. Ismagilov, *Lab Chip*, 2008, **8**, 1265.

CHAPTER 8

Data Handling

TUULIA HYÖTYLÄINEN AND MATEJ OREŠIČ*

VTT Technical Research Centre of Finland, Tietotie 2, P.O. Box 1000, 02044 VTT, Espoo, Finland
*Email: matej.oresic@vtt.fi

8.1 Introduction

Over recent years, metabolomics has rapidly emerged as a key technology for life science and biomedical research. In addition to the developments of analytical technologies for metabolomics, another key contributing factor to this rise of this field is advances in metabolomics data processing and bioinformatics.[1]

Given the chemical diversity of metabolites and their wide dynamic range of concentrations, it is not feasible to cover the complete metabolome of a specific biological system under investigation with a single analytical technique. Application of multiple complementary analytical techniques is therefore needed for a comprehensive coverage of the metabolome. Additionally, owing to the large amounts of information produced by the analytical methods applied to metabolomics, data processing and analysis are a major challenge of any metabolomics pipeline. Software and computational tools therefore play an essential role in metabolomics data processing, analysis and interpretation.

The most commonly applied analytical methodologies in metabolomics for molecular separation and identification are based on mass spectrometry (MS), often combined with chromatographic methods such as liquid chromatography (LC) and gas chromatography (GC). Nuclear magnetic resonance (NMR) spectroscopy, which is not chemically selective but generally offers lower

RSC Chromatography Monographs No. 18
Chromatographic Methods in Metabolomics
Edited by Tuulia Hyötyläinen and Susanne Wiedmer

Published by the Royal Society of Chemistry, www.rsc.org

sensitivity than MS-based approaches, is also widely used for fingerprinting.[2] In terms of metabolite coverage, two kinds of strategies are used in metabolomics, namely targeted selective analysis and more comprehensive non-targeted profiling methods.[3] In the targeted analysis, only preselected metabolites are analysed with an optimized analytical protocol. However, although this approach allows very sensitive and robust determination of the selected metabolites, it gives relatively limited information on the global metabolic profile. The non-targeted approaches aim to cover within a single analysis many metabolites representing multiple functional classes. The analytical methods for non-targeted metabolomics are typically semiquantitative and it is not possible to optimize the method for all compounds. Advanced data analysis tools are required for non-targeted profiling, since the amount of data is large and typically includes a large number of unidentified metabolites.

Data preprocessing is one of the critical steps in metabolomics, with its main goal being reduction of complexity and extraction of the most important features from the raw data, *i.e.*, to identify and quantify the relevant biological signals corresponding to metabolites present in the sample. The first steps of data processing include signal processing, data normalization, transformation and assessment followed by application of statistical methods for comparison of groups and the construction of predictive models.[4,5] Several software packages have been developed for data preprocessing.[6]

For most analytical instruments, commercial software tools exist for the automated initial processing of metabolomics data, especially for the platforms based on chromatographic separation coupled with mass spectrometry.[6–10] Samples can be processed automatically, including peak detection, integration and alignment. Data treatment in targeted analyses is more straightforward than in non-targeted analyses and is typically performed using the instrument vendor's software. Typically, metabolite-specific mass-to-charge ratio (m/z) ions and specific retention time windows are given for each metabolite in order to enable the software first to detect the targeted peaks in all data files and then to assign them a peak area. This method is very accurate and can be quantitative if using calibration curves and corresponding stable isotope-labelled internal standards for each target metabolite.

8.2 Data Acquisition and Handling

Most commonly, the structure of metabolomics data is three-dimensional (Figure 8.1). For example, the data acquired from a chromatographic method coupled with mass spectrometry include values for retention time, peak intensity (area or height) and peak m/z. In GC-MS-based approaches, the spectra comprise several m/z fragments from a single metabolite (Figure 8.1b) whereas in LC-MS there is less fragmentation and isotopic peaks can be detected (Figure 8.1a). In some methods, such as comprehensive two-dimensional GC combined with MS (GC×GC-MS), there is an additional dimension, as there are two retention times instead of one, making the data analysis even more

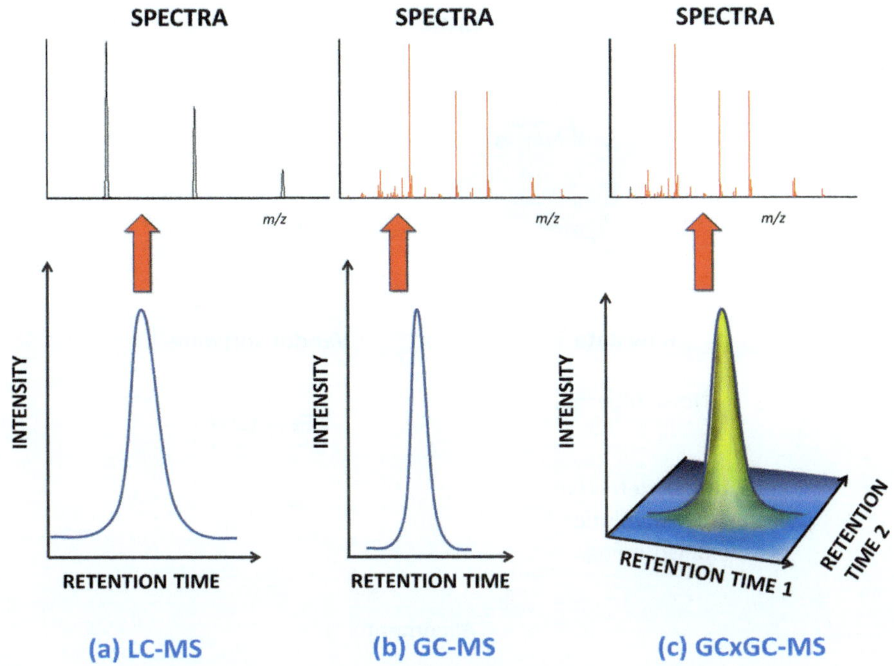

Figure 8.1 Structures of different types of metabolomics data. (a) LC-MS, (b) GC-MS, (c) GCxGC-MS.

challenging (Figure 8.1c). In the initial data processing, the following steps are typically required (Figure 8.2):

- noise removal
- peak detection
- peak deconvolution
- peak alignment
- peak normalization and quantification
- peak labelling/identification.

The first step is needed in order to filter out the noise, *i.e.*, to separate the analytical signal, the background signal and the noise. The last two result from the instrumental interference and random variations in the analytical system, *e.g.*, the detector. In peak detection, the correct form of the signal is identified and the intensity of the peak is calculated. In LC-MS analysis, the isotopic elemental composition must also be taken into consideration and the so-called deisotoping algorithms are applied to group the isotopic peaks with the corresponding monoisotopic peak. In GC-MS analyses, peak deconvolution algorithms need to be applied. Deconvolution uses mathematical multivariate curve resolution procedures to obtain accurate mass spectra of specific compounds in the presence of coeluting peaks.

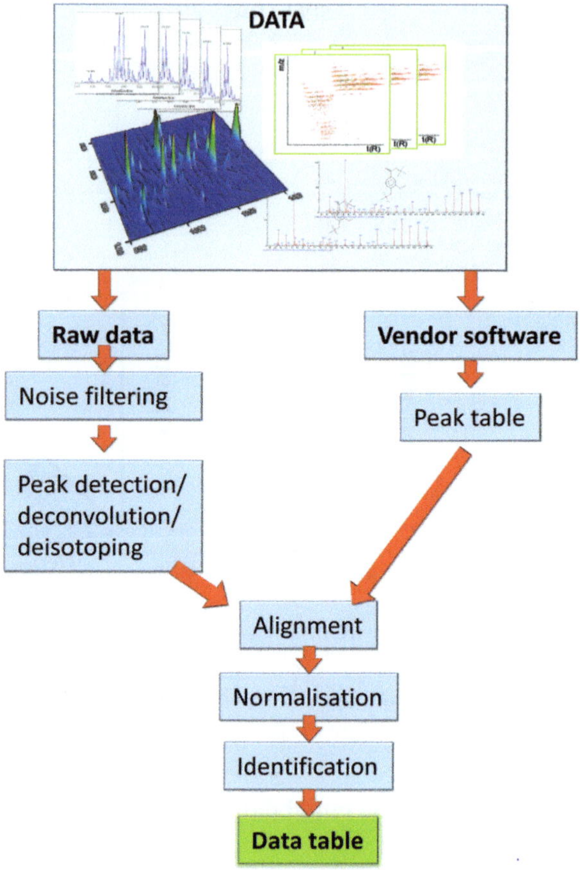

Figure 8.2 Metabolomics data processing steps.

Alignment algorithms are used for identifying the peaks which are associated with a specific metabolite across all samples analysed. After the alignment, the next step is normalization of the data. Peak normalization is used for the correction of methodological and instrumental variations in order to permit quantitative comparisons of peak values (*i.e.*, metabolite concentrations) across the different samples analysed.

There are two ways of handling the acquired analytical raw data: the raw data can either be converted to a specific open data format such as NetCDF (network common data format) or mzData[11] and exported for external processing, or the data are processed *in situ* using the vendor software of the instrument manufacturer.

8.2.1 Peak Detection and Deconvolution

The aim of peak detection and deconvolution is to identify and quantify the signals corresponding to the metabolites present in the biological sample.

An effective peak detection method identifies the true signals while avoiding the false positives. Moreover, peak detection and deconvolution reduce the complexity of the data and make the subsequent data analysis feasible.[4] Deconvolution is needed because the signals of different metabolites often overlap owing to the complexity of the biological matrix and coelution in the separation system. The basic idea of deconvolution is that the raw signal is a superposition of single-metabolite contributions from the metabolite responses in that signal. Deconvolution is a mathematical operation that exploits differences in mass spectral information to separate overlapping peaks, thereby enhancing the analytical resolution.

The instrument vendor software is usually well suited for peak detection and deconvolution. Although these software packages generally work well also for alignment and normalization, they may lack the flexibility needed for optimizing and automating the processing of large sample sets.[12] Also, different software solutions have been shown to have large differences both in the numbers of components identified and in the qualitative results.[12] Therefore, in metabolomics studies it is common that the raw data are transferred directly into specialized software for the data processing.

Several programs have functions for performing peak detection and deconvolution. Here we list a few popular software package:

- MZmine 2[8] is open-source software that divides the peak detection into three parts. The first step, Exact Mass Detection, detects the exact mass of each ion in the mass spectrum, creating a list of unique masses. The second step, Chromatogram Builder, connects data points of successive scans to form continuous chromatograms. In the third step, the peaks inside each built chromatogram are deconvoluted using one of the multiple available algorithms.
- XCMS[7] has two algorithms for peak detection: Matched Filter and centWave. Matched Filter bins the data into slices and determines the signal in each slice by taking the maximum intensity at each retention time. Then it filters each slice using a second-derivative Gaussian as the model shape. The algorithm centWave finds the regions containing potentially interesting masses and uses a continuous wavelet transform to detect the peaks with different widths. The intensity of the peak is obtained by the maximum value of the centroid peak in the estimated peak boundaries.
- apLCMS[13] first groups features based on m/z and applies a kernel density-based iterative splitting to separate them. The same kernel density-based iterative splitting is also applied in the retention time dimension.
- MAVEN[10] has an automatic peak detection algorithm that produces an extracted ion chromatogram using a binary search in each (m/z, retention time) slice. Peak boundaries are determined as points where the sign of the derivative changes or the intensity falls below the baseline.
- MsInspect[14] has several peak detection algorithms to find features in a selected scan or to detect co-eluting peaks in features.

8.2.2 Peak Alignment

In LC-MS, GC-MS and other techniques [*e.g.*, GC×GC-time-of-flight (TOF)MS, capillary electrophoresis (CE)-MS, MS, Fourier transform (FT)-MS, NMR spectroscopy, matrix-assisted laser desorption/ionization (MALDI)], the peaks from multiple analyses have to be aligned before they can be compared across the samples. As there is always a difference in the data, *e.g.*, in the retention times due to analytical drift, alignment is required for the accurate correction in order to be able to refer to the same metabolite or component.

Most of the available metabolomics software packages are designed for the data treatment of LC-MS and GC-MS data. They provide capabilities to align the chromatograms to account for instrumental variation of retention time. Internal standards and/or endogenous metabolites can be used across the analysed samples to align the peaks, based on their retentions and spectra of the peaks. However, this type of approach does not work well for the alignment of very large data sets. In GC-MS, it is also possible to use retention index (RI) calculation, which can correctly align chromatograms even over long periods of time because the retention indices are very reproducible. In LC-MS, however, the utilization of retention indices is not as straightforward, because the retention times are very dependent on the analytical conditions and instrumentation, unlike in GC-based approaches. In two-dimensional separations, such as GC×GC-MS, the alignment is particularly difficult because of the inherently higher relative variability of the retention times in the very short second-dimension time window.[15]

Most alignment algorithms work in a pair-wise manner by aligning either only pairs of samples or multiple samples against a selected reference sample or a template.[4] Either alignment methods can use raw data, or alignment methods utilizing cluster detected features can be used. Some alignment methods combine both approaches, *e.g.*, by first conducting a retention time mapping between runs and then clustering detected features using corrected retention times.

8.2.3 Normalization and Quantification

Normalization of the metabolomics data is required to correct for possible methodological variations, such as due to variations during sample preparation, analysis and detection of the metabolites. Often, a set of internal standards is added to each sample before the sample preparation and preferably also in additional steps during the analysis, and their response is used for the correction of the variation. The internal standards should be selected so that they represent the metabolites covered in the analysis in terms of their physicochemical features. In quantitative analysis, labelled versions of each individual metabolite are used, whereas in non-targeted analysis this is not possible although functional group-specific standards are commonly used.

Another normalization strategy is to use specific scaling factors for each sample based on the complete data set. The median or average of intensities of selected metabolites could be an example of the scaling factor. However, this statistical approach suffers from the lack of an absolute concentration reference for different metabolites. In addition, constraining the data to a specific norm based on a total signal affects its covariance structure, therefore requiring special caution when pursuing multivariate analysis of such data.[16]

In 2007, a normalization approach using optimal selection of multiple internal standards (NOMIS) was proposed.[17] The idea of the NOMIS approach is that monitoring of multiple standard compounds across multiple sample runs may help determine how the standards are correlated, which variation is specific to a particular standard and which patterns of variation are shared between the measured metabolites and the standards so that they can be removed. Based on this premise, a statistical model was developed which models the systematic variation of metabolites as a function of variation of standard compounds.

Among the available software packages, SuperHirn,[18] MetaboAnalyst[19] and MZmine 2[8] provide methods for the normalization of the data. SuperHirn uses a modified version of the central tendency normalization method, MZmine 2 allows normalization using internal standards and linear normalization, and MetaboAnalyst can normalize row-wise to allow comparisons among samples and column-wise to allow the comparison of the magnitudes of features, using different scaling factors.

8.3 Quality Control

As in other 'omics' approaches, quality control (QC) in metabolomics is essential. As part of QC, the quality of data processing must be taken into account together with the methodological QC. The methodological variation can be assessed by analysing suitable set of standards, blank samples and technical replicates of pooled samples. However, evaluation of the quality of data processing, such as the quality of automated peak detection, identification, peak integration and alignment, is more complex. Erroneous data processing can cause several problems with the data, such as missing peaks, false-positive peaks, and errors in (semi)quantitative values of the metabolites.

Comparison of data processing algorithms has shown that different methods can give different results. For example, a comparison of peak detection algorithms of LC-MS data using centWave,[20] matched filter implemented in XCMS[7] and MZmine[21] showed that there was only a partial overlap of the results obtained with these methods and a number of peaks were only detected by one software (not overlapped). In a more detailed comparison of six freely available software packages, including XCMS, MZmine, msInspect and OpenMS, it was shown that no single software perfectly aligned the datasets.[22] In a similar way, three commercially or freely available software packages (ChromaTOF, AnalyzerPro and AMDIS) have been evaluated for the data processing of GC-TOFMS data.[12] This study also clearly showed distinct differences in the results obtained with each software package. One particular feature of the present ChromaTOF and AMDIS deconvolution software packages is that they tend to generate artefacts and to produce duplicate or multiple peak assignments that clearly correspond to a single chromatographic peak and chemical entity. On the other hand, the results from AnalyzerPro showed the detection of a greater number of false negatives.

Even when using a single software package, different parameter settings may produce different results. It is therefore essential that for each analytical approach, the data processing method and related parameter settings are carefully tested and optimized prior to routine use.

8.4 Processing of Large-scale Metabolomics Data – Handling of Batch Effects

In metabolomics, large-scale studies are often necessary and, in principle, current analytical methodologies can be applied for large-scale studies involving many samples. However, in the analyses of hundreds or even thousands of samples, proper control of the possible instrumental and methodological variation is essential.

The large-scale analyses are often performed in batches (each including a few hundred samples). This can cause the so-called batch effects, *i.e.*, variation of the results due to variation in laboratory conditions, reagent lots or intra-batch variability due to instrumental or methodological reasons. For example, injection order effects are frequently observed in LC-TOFMS data due to a build-up of contaminants that were not being removed during the analysis of a batch of samples.[23] In order to avoid incorrect findings due to batch effects or injection order effects, it is important to have consistent measurements with the least amount of error from instrumental drift or offset and sophisticated methods for correction of these batch effects.

In typical targeted analytical methods, internal standards are used for the correction of systematic biases in sample preparation and/or analysis. By selecting a suitable internal standard (ISTD), with physicochemical properties identical with or similar to those of the target metabolites, systematic batch-to-batch differences can be removed after calibrating with the ISTD. It should be noted, however, that even with the ISTD calibration, overlapping chromatographic peaks and ion suppression effects may introduce biases, especially when the sample matrices are very complex.[24] Additionally, in non-targeted profiling, it is not possible to have ISTDs for all of the metabolites. The most typical centring methods used for correction are generally not well suited to non-targeted profiling because when the abundance variance of each peak from different batches is unequal, application of the centring method will not be sufficient.

Several methods are available for scaling of the data, including autoscaling, pareto scaling and range scaling. All these methods use a specific scaling factor to adjust each peak abundance for systematic variance. Autoscaling is the most often applied scaling method and uses the standard deviation of each peak of all samples as the scaling factor. Pareto scaling and range scaling, on the other hand, use the square root of the standard deviation of each peak of all samples and the abundance range of each peak of all samples as the scaling factors. In a GC-MS-based study, different scaling methods were evaluated and autoscaling and range scaling were concluded to perform better than the other methods.[25] However, the standard deviation and abundance range of each peak must be calculated from all samples in order to obtain scaling factors. Thus, scaling methods have the general disadvantage that a recalibration must be performed between batches when obtaining a new data set.

Another way to correct for batch effects is to use linear normalization, where each value is divided by a value which can be directly calculated from each sample, hence no recalibration is required after acquiring new data. For example, linear normalization methods were applied to calibrate non-biological systematic biases by dividing by the average of overall peak abundance, the median of peak abundance or the abundance of a peak.[8,26] The unit-norm method[27] calibrates each sample by setting the sum of the *n*th power of the abundances of all peaks of each sample equal to one. Quantile normalization[28] is frequently used in the analysis of transcriptomics data. In this method, the expression of an entire data set is calibrated with a reference expression distribution. Draisma *et al.*[29] developed an improved quantile normalization for LC-TOFMS metabolomics data, which corrects each peak for the linear and the non-linear differences in the distribution among each batch for metabolomics studies. Another common normalization approach is to calibrate the LC-TOFMS data using ISTDs. For example, a single-point regression method with pooled QC samples and multiple ISTDs was used to calibrate batch and injection order effects.[30]

8.5 Data Visualization and Analysis

Visualization techniques are useful in QC and in the identification and interpretation of metabolomics data. In QC, chromatograms and spectra are visualized as they are, *e.g.*, individual chromatogram curves can be superimposed and features such as consistency of peak positions, peak overlap, symmetries, smoothness and intensities can be checked just by eye. Also, fragmentation patterns, noise levels, contaminants and other prominent features are often effectively found by just looking at the spectra and chromatograms. In addition, the visualization of intensities of the standards and blank samples gives a good overall view of whether intensity estimates are dependent on the order in which they have been analysed. All these QC checks help in selecting proper preprocessing methods, such as alignment, deconvolution and noise removal algorithms by which an automated or semi-automated preprocessing and often also normalization can be carried out as described earlier.

After preprocessing, the data are usually in the form in which they are to be statistically analysed in order to find answers to the actual study-specific questions. At this stage, it is common to use projection methods by which the high-dimensional metabolomics data are visualized in two dimensions to gain an overall view of the sample behaviour in the data. For example, principal component analysis (PCA) is a widely used projection method,[31] with which it is possible to detect outlying samples and check whether the samples form clusters or show dependences on any external or internal variables. If the study does not contain too many samples or too many compounds, it is also possible to visualize the whole dataset by a heatmap, which colour codes the concentration estimates of each compound in each sample. In ideal cases, study-specific patterns can be observed from such data, which may facilitate the selection of statistical methods for further analysis.

Many metabolites share similar pathways and thus are being controlled by the same enzymes, leading to a high degree of co-regulation among the metabolites.[32] One therefore cannot assume that all measured metabolites are independent. For this reason, clustering of the data such as by using the Bayesian model-based clustering[33] is a useful tool also for surveying the overall changes of metabolome. Such an approach decomposes the metabolome into specific clusters of co-varying metabolites. This approach was applied recently, for example, in metabolomics studies of schizophrenia[34] and Alzheimer's disease.[35] Needless to say, such an approach may miss individual metabolites which vary strongly between the biological groups of interest but are not well represented by any of the average cluster profiles. Therefore, multivariate analysis such as clustering should always be complemented by univariate analyses that focus on metabolites as single variables. In such analyses, common principles and considerations of analysis of high-dimensional data apply.[36]

8.6 Conclusion

High quality of data processing is essential for any successful metabolomics analysis. Each step of the data processing pipeline has its challenges. The modular design of some open-source software packages such as MZmine allows for the introduction and comparison of novel methods for specific stages of data processing and for the optimization of their parameter settings for specific analytical methods. It is important to understand, however, that no software application for metabolomics data processing can be considered as a 'black box.' For each analytical method, comprehensive testing with different processing parameters is needed before they are set for routine analyses.

Owing to the inherent challenges of MS-based analyses of large sample series and the increased need for such analysis in biomedical applications, there is a great need for improved data processing methods that account for systemic variations such as batch effects. Also, meta-analyses across different metabolomics datasets often need to be performed, requiring similar statistical solutions.

Despite many advances in metabolomics data processing over the past years, the continuous improvements and innovations in analytical technologies combined with higher and higher expectations from metabolomics in the life sciences and medicine are likely to continue to provide new challenges and opportunities for the field of metabolomics data processing and analysis. One of the keys to the success of this domain will be the availability of flexible software environments available for data exchanges and for introducing novel algorithms.

References

1. G. A. Theodoridis, H. G. Gika, E. J. Want and I. D. Wilson, *Anal. Chim. Acta*, 2012, **711**, 7–16.
2. J. van der Greef, P. Stroobant and R. v. d. Heijden, *Curr. Opin. Chem. Biol.*, 2004, **8**, 559–565.
3. M. Oresic, *Nutr. Metab. Cardiovasc. Dis.*, 2009, **19**, 816–824.
4. M. Katajamaa and M. Oresic, *J. Chromatogr. A*, 2007, **1158**, 318–328.
5. L. W. Sumner, E. Urbanczyk-Wochniak and C. D. Broeckling, *Methods Mol. Biol.*, 2007, **406**, 409–436.
6. S. Castillo, P. Gopalacharyulu, L. Yetukuri and M. Orešič, *Chemometr. Intell. Lab. Syst.*, 2011, **108**, 23–32.
7. A. Nordstrom, G. O'Maille, C. Qin and G. Siuzdak, *Anal. Chem.*, 2006, **78**, 3289–3295.
8. T. Pluskal, S. Castillo, A. Villar-Briones and M. Oresic, *BMC Bioinformatics*, 2010, **11**, 395.
9. A. D. Patterson, H. Li, G. S. Eichler, K. W. Krausz, J. N. Weinstein, A. J. Fornace Jr, F. J. Gonzalez and J. R. Idle, *Anal. Chem.*, 2008, **80**, 665–674.
10. E. Melamud, L. Vastag and J. D. Rabinowitz, *Anal. Chem.*, 2010, **82**, 9818–9826.
11. L. Martens, M. Chambers, M. Sturm, D. Kessner, F. Levander, J. Shofstahl, W. H. Tang, A. Rompp, S. Neumann, A. D. Pizarro, L. Montecchi-Palazzi, N. Tasman, M. Coleman, F. Reisinger, P. Souda, H. Hermjakob, P. A. Binz and E. W. Deutsch, *Mol. Cell. Proteomics*, 2011, **10**, R110 000133.
12. H. Lu, Y. Liang, W. B. Dunn, H. Shen and D. B. Kell, *Trends Anal. Chem.*, 2008, **27**, 215–227.
13. T. Yu, Y. Park, J. M. Johnson and D. P. Jones, *Bioinformatics*, 2009, **25**, 1930–1936.
14. M. Bellew, M. Coram, M. Fitzgibbon, M. Igra, T. Randolph, P. Wang, D. May, J. Eng, R. Fang, C. Lin, J. Chen, D. Goodlett, J. Whiteaker, A. Paulovich and M. McIntosh, *Bioinformatics*, 2006, **22**, 1902–1909.
15. S. Castillo, I. Mattila, J. Miettinen, M. Oresic and T. Hyotylainen, *Anal. Chem.*, 2011, **83**, 3058–3067.
16. J. Aitchison, *The Statistical Analysis of Compositional Data*, Blackburn Press, Caldwell, NJ, 2003.
17. M. Sysi-Aho, M. Katajamaa, L. Yetukuri and M. Oresic, *BMC Bioinformatics*, 2007, **8**, 93.
18. L. N. Mueller, O. Rinner, A. Schmidt, S. Letarte, B. Bodenmiller, M. Y. Brusniak, O. Vitek, R. Aebersold and M. Muller, *Proteomics*, 2007, **7**, 3470–3480.

19. J. Xia, R. Mandal, I. V. Sinelnikov, D. Broadhurst and D. S. Wishart, *Nucleic Acids Res.*, 2012, **40**, W127–W133.
20. R. Tautenhahn, C. Bottcher and S. Neumann, *BMC Bioinformatics*, 2008, **9**, 504.
21. M. Katajamaa, J. Miettinen and M. Oresic, *Bioinformatics*, 2006, **22**, 634–636.
22. E. Lange, R. Tautenhahn, S. Neumann and C. Gropl, *BMC Bioinformatics*, 2008, **9**, 375.
23. L. Lai, F. Michopoulos, H. Gika, G. Theodoridis, R. W. Wilkinson, R. Odedra, J. Wingate, R. Bonner, S. Tate and I. D. Wilson, *Mol. Biosyst.*, 2010, **6**, 108–120.
24. H. Redestig, A. Fukushima, H. Stenlund, T. Moritz, M. Arita, K. Saito and M. Kusano, *Anal. Chem.*, 2009, **81**, 7974–7980.
25. R. A. van den Berg, H. C. Hoefsloot, J. A. Westerhuis, A. K. Smilde and M. J. van der Werf, *BMC Genomics*, 2006, **7**, 142.
26. W. Wang, H. Zhou, H. Lin, S. Roy, T. A. Shaler, L. R. Hill, S. Norton, P. Kumar, M. Anderle and C. H. Becker, *Anal. Chem.*, 2003, **75**, 4818–4826.
27. M. Scholz, S. Gatzek, A. Sterling, O. Fiehn and J. Selbig, *Bioinformatics*, 2004, **20**, 2447–2454.
28. B. M. Bolstad, R. A. Irizarry, M. Astrand and T. P. Speed, *Bioinformatics*, 2003, **19**, 185–193.
29. H. H. Draisma, T. H. Reijmers, F. van der Kloet, I. Bobeldijk-Pastorova, E. Spies-Faber, J. T. Vogels, J. J. Meulman, D. I. Boomsma, J. van der Greef and T. Hankemeier, *Anal. Chem.*, 2010, **82**, 1039–1046.
30. F. M. van der Kloet, I. Bobeldijk, E. R. Verheij and R. H. Jellema, *J. Proteome Res.*, 2009, **8**, 5132–5141.
31. S. Wold, K. Esbensen and P. Geladi, *Chemometr. Intell. Lab. Syst.*, 1987, **2**, 37–52.
32. R. Steuer, J. Kurths, O. Fiehn and W. Weckwerth, *Bioinformatics*, 2003, **19**, 1019–1026.
33. C. Fraley and A. E. Raftery, *J. Stat. Soft.*, 2007, **18**, 1–13.
34. M. Oresic, J. Tang, T. Seppanen-Laakso, I. Mattila, S. E. Saarni, S. I. Saarni, J. Lonnqvist, M. Sysi-Aho, T. Hyotylainen, J. Perala and J. Suvisaari, *Genome Med.*, 2011, **3**, 19.
35. M. Oresic, T. Hyotylainen, S. K. Herukka, M. Sysi-Aho, I. Mattila, T. Seppanan-Laakso, V. Julkunen, P. V. Gopalacharyulu, M. Hallikainen, J. Koikkalainen, M. Kivipelto, S. Helisalmi, J. Lotjonen and H. Soininen, *Transl. Psychiatry*, 2011, **1**, e57.
36. D. Broadhurst and D. Kell, *Metabolomics*, 2006, **2**, 171–196.

CHAPTER 9

Applications

PÄIVI PÖHÖ,[a] MAARIT KIVILOMPOLO,[a]
MONICA CALDERON-SANTIAGO,[b] SIRKKU JÄNTTI,[a]
SUSANNE K. WIEDMER[a] AND
TUULIA HYÖTYLÄINEN*[a]

[a] VTT Technical Research Centre of Finland, Tietotie 2, P.O. Box 1000, 02044 VTT, Espoo, Finland; [b] Department of Analytical Chemistry, University of Córdoba Agroalimentary Excellence Campus, Campus of Rabanales, 14071 Córdoba, Spain
*Email: tuulia.tyotylainen@vtt.fi

9.1 Introduction

This chapter presents selected applications of metabolomics. The metabolites include compounds from the key metabolic pathways, including lipids, acyl-carnitines, acyl-coenzyme A, amino acids, bile acids, carbohydrates, eicosanoids, fatty acids, nucleotides, steroids, sterols and citric acid cycle metabolites. The chapter is divided based on the type of analytes into lipids and polar metabolites, as these groups typically require separate sample preparation methods. Naturally, the type of sample also plays a major role; however, the final analytical methods are typically the same, independent of the type of the sample. The major differences then lie in the sample preparation. Typical illustrative examples are given in the following sections.

The two most common separation techniques used in metabolomics are liquid chromatography (LC) and gas chromatography (GC), while capillary electrophoresis (CE) has been used to a lesser extent. Microfluidic techniques are still under development and have not yet been widely applied for

RSC Chromatography Monographs No. 18
Chromatographic Methods in Metabolomics
Edited by Tuulia Hyötyläinen and Susanne Wiedmer

Published by the Royal Society of Chemistry, www.rsc.org

Table 9.1 Suitability of GC, LC and CE techniques for metabolic applications: –, not suitable; +, can be used; not optimal; ++, suitable; +++, highly suitable.

Analyte	*GC-MS*	*GC×GC-MS*	*LC-MS*	*CE-MS*
Targeted				
Glycerophospholipids	–			
Cholesterol esters	–			
MG/DG/TG[a]	+	+	+++	–
Sphingolipids	–	–		
Fatty acids	+++	+++	+	+
Eicosanoids	+	+	+++	++
Steroids	++	++	+++	++
Sterols	++	++	+++	++
TCA[b] cycle	+++	+++	++	++
Glycolysis cycle	++	++	++	++
Nucleotides	–	–	++	++
Sugars	+++	+++	+	+
Amino acids	++	+++	+++	++
Acylcarnitines	–	–	++	++
Vitamins	+	+	+++	++
Global profiling				
Lipids	–	–	+++	–
Polar metabolites	+	+++	++	+

[a]Mono-, di- and triacylglycerols.
[b]Tricarboxylic acid.

metabolomics, but is a promising approach for diagnostic applications. Both LC and GC have been used to separate the different metabolites present in biological samples (Table 9.1). LC can be used either in reversed-phase (RP) or hydrophilic interaction liquid chromatography (HILIC) mode, thus making LC useful for both polar and non-polar metabolites. On the other hand, GC can only be used for the analysis of volatile metabolites or compounds that can be converted into volatile compounds after derivatization. Despite this fact, GC is widely used as there are many compounds that are difficult to analyse by LC. CE coupled with mass spectrometry (MS) using electrospray ionization (ESI), on the other hand, is particularly suited for the analysis of highly or medium polar and charged compounds, which often show no retention in common RPLC systems.

9.2 Lipids

Lipids are a vital class of essential metabolites that have many key biological functions. They are structural components of cell membranes, energy storage sources and intermediates in signalling pathways.[1,2] Structurally, lipids originate entirely or in part from two distinct types of building blocks: ketoacyl and isoprene groups. Lipids are both functionally and structurally a very diverse group of compounds, in part due to the many possible variations of the lipid building blocks and the different types of non-covalent linkage.

The number of molecular lipid species found in biological systems has been estimated to be of the order of hundreds of thousands.[3] The most widely used classification divides lipids into eight main categories, namely fatty acyls (FAcs), glycerolipids, glycerophospholipids, sphingolipids, sterol lipids, prenol lipids, saccharolipids and polyketides. Each category contains distinct classes and subclasses of lipids, as presented in Table 9.2.

Fatty acyls are the simplest lipid class, which includes various types of fatty acids, eicosanoids, fatty alcohols, fatty aldehydes, fatty esters, fatty amides, fatty nitriles, fatty ethers and hydrocarbons. Many lipids in this group, especially the eicosanoids derived from $n-6$ and $n-3$ polyunsaturated fatty acids (PUFAs), have distinct biological activities. FAcs are also the main lipid building blocks of more complex lipids, such as glycerolipids (GL) which in turn includes monoacylglycerides (MGs), diacylglycerides (DGs) and triacylglyceride (TGs). These lipids are neutral and they have a glycerol backbone with fatty acid chains attached to the glycerol group. Glycerophospholipids (GPs), also referred to as phospholipids (PLs), are key components of the lipid bilayer of cells and are also involved in metabolism and signalling. The most common GPs found in biological membranes are phosphatidylcholines (PCs), phosphatidylethanolamines (PEs) and phosphatidylserines (PSs). Sphingolipids (SLs) are a complex family of lipids. all of which have a sphingoid base backbone. Sterols are a class of lipids that contain a common steroid nucleus of a fused four-ring structure with a hydrocarbon side chain and an alcohol group. The main sterol lipid in animal fat is cholesterol, which also has an important role in the lipid membrane. Other lipids include compounds such as polyketides and saccharolipids.

Lipid analyses include both global profiling of the main lipid classes and targeted methods for specific lipid classes which are usually not covered by the global methods due to either their low concentrations (*e.g.*, eicosanoids, steroids), instability or their other physicochemical features.

9.2.1 Global Profiling of Lipids

Global lipid profiling is challenged by the large number of different types and classes of lipids, present in a wide range of concentrations, in biological and plants samples. Global lipid profiling methods typically allow the analysis of the main classes of lipids, namely cholesteryl esters (ChoEs), phosphatidylcholines (PCs), phosphatidylethanolamines (PEs), ceramides (Cers), mono-, di- and triacylglycerols (MGs, DGs and TGs, respectively), sphingomyelins (SMs) and lysophosphatidylcholines (lysoPCs).

The analytical methods used for the determination of global profiling of lipids are listed in Table 9.3. Extraction is usually based on simple liquid extraction, as described in Chapter 2, using modified Folch or Blight and Dyer extraction.[4,5] For more acidic lipids, such as phosphatidylserines and phosphatidic acids, adjustment of the pH in the aqueous phase is required. The analysis is most typically performed with LC-MS in RPLC mode, with ultrahigh-performance LC (UHPLC) methods gradually replacing the

Table 9.2 Classification of lipids.

Category	*Abbreviation*	*Subcategory*	*Example*
Fatty acyls	FAc	Fatty acids and conjugates Octadecanoids Eicosanoids Docosanoids Fatty alcohols Fatty aldehydes Fatty esters	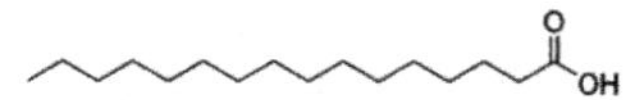 Fatty acid: palmitic acid
Glycerolipids	GL	Monoacylglycerols Diacylglycerols Triacylglycerols	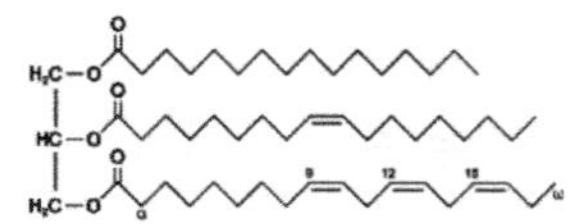 Triglyceride: TG(18:0/18:1/18:3)
Glycerophospholipids	GP	Phosphatidic acids Phosphatidylcholines Phosphatidylserines Phosphatidylglycerols Phosphatidylethanolamines Phosphatidylinositols Phosphatidylinositides Cardiolipins	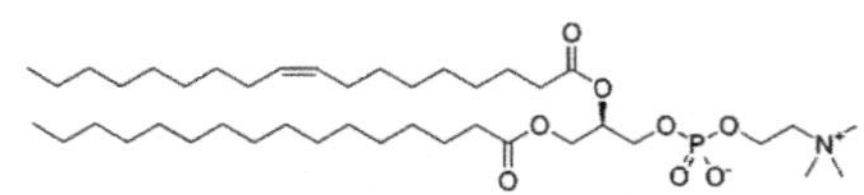 1-Palmitoyl-2-oleoylphosphatidylcholine
Sphingolipids	SP	Sphingoid bases Ceramides Phosphosphingolipids Phosphonosphingolipids Neutral glycosphingolipids Acidic glycosphingolipids	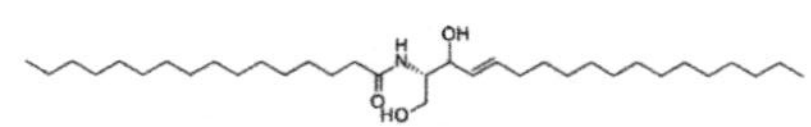 C_{16} ceramide

Sterol lipids	ST	Sterols Steroids Secosteroids Bile acids and derivatives	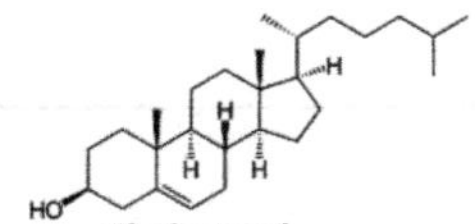 **Cholesterol**
Prenol lipids	PR	Isoprenoids Quinones and hydroquinones Polyphenols	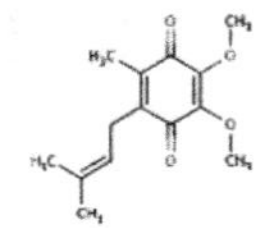 **Ubiquinone Q1**
Saccharolipids	SL	Acylaminosugars Acylaminosugar glycans Acyltrehaloses Acyltrehalose glycans	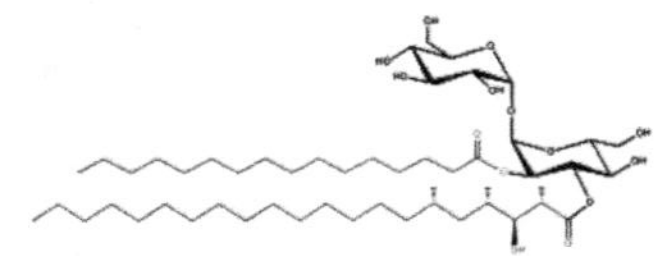 **2,3-Di-*O*-acyltrehalose**
Polyketides	PK	Macrolide polyketides Aromatic polyketides Non-ribosomal peptide/polyketide hybrids	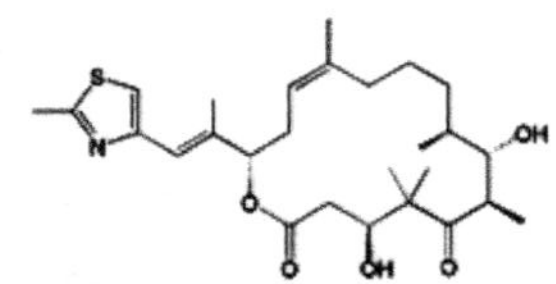

Table 9.3 Typical applications of global profiling of metabolites by chromatographic or electromigration techniques.

Application	*Sample type*	*Sample preparation*	*Analysis*	*Detection*	*LOD/LOQ*	*Analysis time (min)*	*Ref.*
Global lipidomics	Serum	Modified Blight and Dyer extraction	Micro HPLC (C_{18})	MS; ESI, hybrid QTOF	–/–	58	6
Global lipidomics	Rat liver mitochondria	Modified Blight and Dyer extraction, sonication	HPLC (C_{18})	MS; ESI, Orbitrap	–/–	30	7
Global lipidomics	Plasma	Modified Blight and Dyer extraction	UHPLC (C_{18})	MS; ESI, hybrid QTOF	–/–	15	8
Global lipidomics	Rat plasma	Extraction/protein precipitation with ACN	UHPLC (C_8)	MS; ESI, hybrid QTOF	–/–	12	9
Global lipidomics	Plasma, cells, tissues	Modified Folch extraction, sonication, homogenization	UPHLC (C_{18})	MS; ESI, hybrid QTOF	–/–	18	10
Global lipidomics	Lipid droplets	Extraction to MeOH/MTBE, fractionation with HILIC-HPLC	HPLC (C_{18})	MS; ESI, hybrid linear ion trap Fourier transform ion cyclotron resonance	–/–	35	11
Global lipidomics	Rat peritoneal surface layer	Modified Folch extraction, sonication, homogenization	2D HPLC (1D, silica; 2D, C_8)	MS; ESI, hybrid QTOF	–/–	150	12
Global lipidomics	Egg, soya and porcine	Folch extraction, filtration	2D HPLC (1D, HILIC/silica; 2D, C_{18})	MS; ESI, ion trap	–/–	260	13
Global lipidomics	Plasma	Extraction with $CHCl_3$/MeOH	HPLC (C_{18})	MS; ESI, hybrid QTOF		95	14
Global lipidomics/ phospholipids (especially PA and PS)	Cells, rat liver	Modified Blight and Dyer extraction, fractionation with DEAE-cellulose	HPLC (C_{18})	MS; ESI, hybrid linear ion trap Orbitrap	2 fmol/µl (PA and PS)	45	15

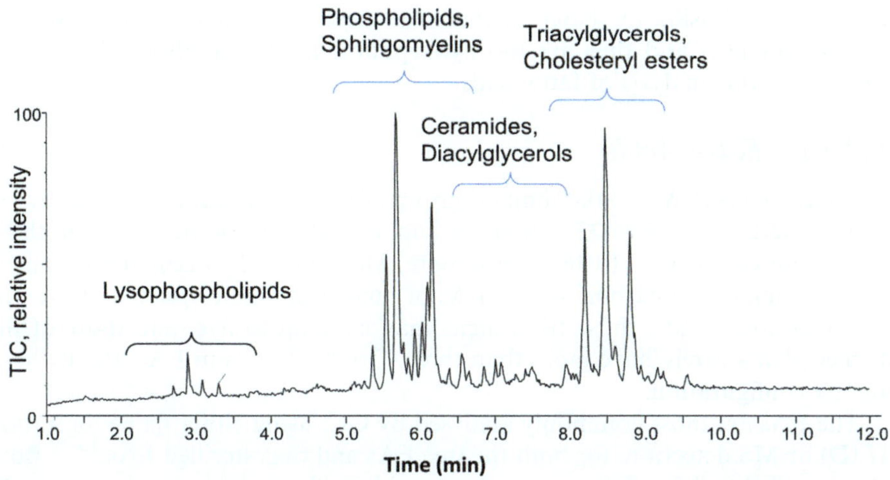

Figure 9.1 Typical UHPLC–QTOFMS trace of a human plasma sample. For more details, see ref. 10.

conventional HPLC methods. High-resolution MS (HRMS) systems, such as time-of-flight MS (TOFMS), quadrupole TOFMS (QTOFMS) and Orbitrap, are needed in the global profiling. UHPLC-HRMS analysis typically allows the detection of several hundred lipids in a single run, with analysis times varying from 10 to 20 min. In lipidomics analyses, the solvent strength is an important parameter and the organic solvent should have a relatively strong solvent strength in order to obtain good peak shapes for late-eluting TGs and also to avoid any significant carryover due to complex biological samples. Often, acetonitrile–2-propanol mixtures are used as the organic solvent and also elevated temperatures can be utilized for the enhancement of the elution of late-eluting lipids.

A typical UHPLC–QTOFMS trace of a human plasma sample obtained in the positive ionization mode is shown in Figure 9.1. In the RP-UHPLC separation, using a C_{18} phase, phospholipids are separated based on their fatty acid composition and degree of desaturation. Therefore, lysophospholipids elute first, with lysoPCs containing one fatty acid attached to the glycerol backbone eluting before diacylphospholipids which have two fatty acids. Also, PLs and SMs have similar elution times, whereas ceramides which do not contain a PC head group as their sphingomyelin counterparts elute after PCs and SMs. Also, DGs elute close to PCs and SMs. The neutral lipids, such as cholesterol esters and TGs, have a strong retention in the C_{18} phase and therefore elute late.

9.2.2 Targeted Lipid Analyses

In targeted lipid analysis, specific lipid classes that are poorly covered by the global profiling methods are usually analysed. These lipids include steroids, sterols, bile acids, fatty acids, signalling lipids such as eicosanoids and ceramides, and also polar lipids and inositol lipids. Most targeted lipid analyses are performed using LC-based methods, as these lipids have too low volatility for GC analysis and they are too hydrophobic for CE analysis. GC is used mainly for the analysis of fatty acids.

9.2.2.1 Fatty Acids

The fatty acids (FA) are the simplest group of lipids. The common FAs vary in chain-length from 14 to 22 in humans and animals, but on occasion the chain length can vary from 2 to 36 or even more. Most naturally occurring FAs have an even number of carbon atoms. FAs of human/animal origin may have one to six double bonds, those from algae may have up to five, and those of the higher plants rarely have more than three. Double bonds in FAs usually have the *cis* configuration.

The FAs are most commonly analysed by GC, using either flame ionization (FID) or MS detection, for both the free FAs and the esterified FAs.[16–19] Both the esterified and free FAs can be recovered by refluxing the sample with a 1 M solution of potassium hydroxide in 95% ethanol for 1 h, although the same reaction can also be carried out at room temperature overnight. After addition of water to the cooled mixture, the FAs can be extracted with hexane–diethyl ether (1:1 v/v). The solvent extract is washed with water and the esterified FAs are recovered in the organic phase. The aqueous phase is acidified with hydrochloric acid and the free FAs are extracted with diethyl ether–hexane (1 : 1 v/v). The FAs are analysed by GC-FID or GC-MS as the fatty acid methyl esters (FAMEs) or after, *e.g.*, silylation or acetylation. Fast GC analysis, combined with automated sample preparation, increases the sample throughput of the FA analysis substantially in comparison with traditional methods.[20,21] Specific GC columns are available for the separation of FAs. The stationary phases for FAs include bonded poly(ethylene glycol), acid-modified poly(ethylene glycol) and poly(alkylene glycol) types. In addition, specific stationary phases have been developed for the separation of geometric positional (*cis*/*trans*) isomers of FAMEs, *e.g.*, a highly polar biscyanopropyl phase.

Methods that allow the determination of the FA composition of plasma or whole blood lipids provide valuable information and are more cost-effective than the analysis of FAs in individual lipid fractions;[22] however, the FA composition of individual lipid fractions gives more detailed information. However, in modern metabolomics, determination of molecular lipids by UHPC-MS is typically a better approach, in terms of both sample throughput and amount of information, than determination of the FA composition of the lipids. Determination of free FAs by GC-MS is still important and it is also clearly faster than the determination of esterified FAs, as simple extraction and derivatization are sufficient. In addition, short-chain FAs can also be analysed by headspace techniques without derivatization, when a more polar GC stationary phase is selected.

9.2.2.2 Steroids and Bile Acids

Steroid hormones have an influence in many physiological functions in the body, for example, in development, maintenance of salt–water balance, fertility and maintenance of pregnancy and stress responses. Hormone steroids can be classified based on their biological activities into estrogens, androgens, progestogens, glucocorticoids and mineralocorticoids. Hormone steroids exist in the body tissues and fluids in both intact and conjugated forms, for example, as glucuronides, sulfates and fatty acid and amino acid conjugates or they can be bound to proteins.

Other structurally similar metabolites, such as bile acids (BAs), cholesterols and D vitamins, differ from hormone steroids with respect to their biological functions. The main function of BAs and their taurine and glycine conjugates are emulsification of nutritional lipids in the digestive tract. Primary BAs include cholic acid (CA) and chenodeoxycholic acid (CDCA) in humans and CA and α-muranocholic acid (αMCA) in mice and they are synthesized from cholesterol in the hepatocytes. Secondary BAs, deoxycholic acid (DCA), lithocholic acid (LCA) and 7-oxo-LCA, are formed by gut microbiota and they are further metabolized to tertiary BAs after enterohepatic circulation (>95% are reabsorbed). The metabolism of BAs includes hydrolysis, oxidation, reduction, isomerization, dehydroxylation and hydroxylation reactions. BAs are excreted mainly as glycine and taurine conjugates, but they also undergo phase II metabolism by sulfation, glucuronidation and glycosylation.[23] Glycine conjugates are predominant in humans and taurine conjugates in mice.

The analytical methods used for the determination of steroids and sterols are listed in Table 9.4. The extraction of steroids, and also sterols, from tissues is typically done by using an organic solvent with or without acid. The analysis of steroids and sterols in biological fluids such as in plasma, serum, urine and cerebrospinal fluid (CSF) demands at least removal of proteins or salts prior to the analysis, for example, using solid-phase extraction (SPE), liquid-liquid extraction (LLE) or on-line column switching techniques. Purification of tissue extracts is almost always required, because biological extracts contain large numbers of interfering compounds causing matrix effects in the analysis. The purification is most commonly performed using LLE or SPE. Many of the current methods in the analysis of steroids in tissue samples include several sequential purification steps by SPE (combination of ion exchange and C_{18} cartridge) or LC fractionation steps. Currently, RPLC-MS is the most common method for the analysis of steroids and BAs, although GC and CE methods are also available. However, derivatization of steroids is necessary before GC-MS analysis and enzymatic or chemical hydrolysis of the conjugates is needed.

9.2.2.3 Signalling Lipids

Lipids, such as phosphoinositides (PIPs), eicosanoids and SLs, are important signalling intermediates involved in cellular processes such as membrane trafficking, secretion, adhesion, migration, cell survival and replication. Most targeted lipid analyses are performed with LC-based methods, as these lipids have too low volatility for GC analysis and they are too hydrophobic for CE analysis. Selected examples are listed in Table 9.5. The analysis of many of these signalling lipids is challenging.

Table 9.4 Typical examples of the analysis of steroids and sterols by chromatographic or electromigration techniques.

Metabolite	*Sample matrix*	*Sample preparation*	*Separation*	*Detection*	*LOD/LOQ*	*Analysis time (min)*	*Ref.*
Steroids							
Steroids	Human urine/serum	Urine: hydrolysis; SPE (HLB); dilution with MeOH/serum: protein precipitation	CE	MS; ESI, triple quadrupole (QQQ)	UV: 10–100 ng mL^{-1}; MS: 0.5–1 μg L^{-1}/ MS: 100 ng mL^{-1}–10 mg mL^{-1}	12	24
Steroids	Human urine	Urine diluted 10-fold in water	CE	MS; TOF	0.4 μM	12	25
46 steroids: androgens, corticoids, pregnanes	Brain, rat	Homogenization (PBS), sonication in EtOH + 3% acetic acid, lipid removal by extraction with isooctane (saturated with EtOH in K_2HPO_4 + acetic acid), SPE (Oasis HLB, MCX), derivatization [MO-TMSI, heptafluorobutyric acid (HFB)]	GC (ZB1 30 m×0.25 mm i.d., d_f 0.25 μm)	qMS	LOD 0.004–1.0 ng g^{-1} for steroidsLOD 0.04–1.5 ng g^{-1} for steroid sulfates	32	26
4 steroids: androgens, pregnanes	Brain, rat	Homogenization (MeOH–H_2O), SPE fractionation (C_{18}) Steroids: LC fraction collection prior to analysis, derivatization (HFB). Sulfates: solvolysis, SPE (C_{18})	GC (BPX35, 30 m×0.25 mm i.d., d_f 0.25 μm)	qMS	LOD 1–5 pg per injection	-	27
5 steroids: androgens, pregnanes	Brain, rat	Homogenization (75% MeOH), SPE (C_{18}), derivatization (PFB)	GC (SPB-1, 15 m×0.32 mm i.d., d_f 0.25 μm)	qMS	LOD 0.25 ng (in 100 mg tissue)	12	28

18 steroids: androgens, pregnanes, cortisol	CSF, serum	Extraction with diethyl ether, derivatization (methoxylamine hydrochloride, TMS-MOX)	GC (Rxi, 15 m×0.25 mm i.d., d_f 0.1 μm	qMS	LOD 0.04–11 pmol L^{-1} (0.6–62 fg)	20	29
65 steroids: all classes	Plasma, human	SPE, ethyl acetate–*n*-hexane (2:3 v/v), derivatization (TMS)	GC (Ultra-1, 25 m×0.2 mm i.d., d_f 0.33 μm)	qMS	LOQ 0.2–2 ng mL^{-1}	50	30
5 steroids: androgens, pregnanes	Plasma, rat	SPE (C_{18}), derivatization (PFB)	GC (SPB-1, 15 m×0.32 mm i.d., d_f 0.25 μm)	qMS	LOQ 0.1 ng in 300 μL of albumin sol	12	28
10 steroids: androgens, estrogens, pregnanes	Brain, rat	Homogenization (MeOH–acetic acid), SPE (C_{18})	LC (C_{18}, 150×4.6 mm i.d, d_f 5 μm)	APCI-MS/MS	LOQ 0.02–0.25 pg per sample	50	31
3 pregnanes, 2 androgens	Brain, rat	Homogenization (MeOH–acetic acid), SPE (Strada X), derivatization (HMP)	LC (C_8, 150×2.0 mm i.d., d_f 5 μm)	ESI-MS/MS	Pregnanes: LOQ 0.25 ng g^{-1} Androgens: LOQ 0.05–0.1 ng g^{-1}	10	32,33
2 pregnanes	Brain, rat	Homogenization (1% AcOH in MeOH), SPE (Oasis HLB, Bond Elut Si), derivatization (NFPH)	LC (ODS H-80, 150×4.6 mm i.d., d_f 4 μm)	ECAPCI-MS/MS	LOD 1–6 pg per sample	10	32
7 steroids: androgens, pregnanes	Brain, rat	Homogenization (EtOH), SPE (C_{18}, cation exchange, anion exchange). Steroids: derivatization (oximes), SPE (C_{18}, cation exchange). Sulfates: SPE (C_{18})	CapLC (C_8, 150×2.0 mm i.d., d_f 5 μm)	ESI-MS/MS	LOD 0.1–3 pg per injection	–	33

Table 9.4 (*Continued*)

Metabolite	*Sample matrix*	*Sample preparation*	*Separation*	*Detection*	*LOD/LOQ*	*Analysis time (min)*	*Ref.*
6 steroids: androgens, estrogens	Serum, mouse, horse, baboon, sheep	Homogenization (EDTA in PBS), hexane–ethyl acetate (3:2), on-line SPE	LC (C_8)	APPI-MS/MS	LOQ 0.5–40 pg per injection (corresponding to 0.01–0.6 ng mL^{-1})	–	34
2 estrogens	Serum, human	Extraction with 1-chlorobutane, SPE (Strata X)	LC (75×4.6 mm i.d. column of Luna-Phenyl-Hexyl, d_f 3 µm, C_{18} 100×4.6 mm i.d., d_f 3.5 µm)	ESI-MS/MS	LOD 0.005 ng mL^{-1}	16	35
31 BAs	Serum, liver	Serum: protein precipitation Liver	UPLC (BEH C_{18})	MS/MS; QQQ	LLOQ 2.5–20 nM (1–8 ppb)	21	36
25 BAs	Serum	Protein precipitation, MeOH	UPLC (C_{18}, 100×2.1 mm i.d., d_f 1.7 µm)	MS: Q-TOF	LLOQ 10 ng mL^{-1} (ppb)	25	37
BAs	Liver, bile, plasma	Freeze-drying, homogenization, extraction with EtOH, SPE (ethanol)	UPLC (C_{18}, 100×2.0 mm i.d., d_f 1.7 µm)	MS: QQQ	na	30	38
18 BAs	Serum, plasma	Acidification, protein precipitation (MeOH)	UPLC (C_{18}, 50×2.1 mm i.d., d_f 1.8 µm	MS: Q-trap	LOD 0.5–10 nM (0.2–4 ppb)	6.5	39
BAs	Liver, bile, plasma (100 µl), urine	Liver: homogenization with ACN (5% NH_4OH in ACN), plasma: protein precipitation, CAN. Bile: SPE (Oasis HLB	UPLC (C_{18}, 100×2.1 mm i.d., d_f 1.7 µm)	MS: QQQ	LOD (S/N = 3) 2–5 ng mL^{-1} (ppb)	25	40

Table 9.5 Typical examples of the analysis of selected lipids by chromatographic or electromigration techniques.

Metabolite	*Sample matrix*	*Sample preparation*	*Separation*	*Detection*	*LOD/LOQ*	*Analysis time (min)*	*Ref.*
Eicosanoids and other bioactive lipids	Chronic lymphotic leukaemia cells	SPE with C_{18} cartridges	UPHLC (C_{18}, 150×2 mm, 3 μm)	MS; ESI hybrid linear ion trap Orbitrap, hybrid quadrupole linear ion trap	–/–	15	41
Eicosanoids	Serum, bronchiolar alveolar lavage fluid (BALF)	SPE (Oasis HLB)	UPHLC (C_{18} 2.0×150 mm, 5 μm	MS; ESIQtrap (API–4000, MRM)	–/0.06–3 nM	21	42
Eicosanoids	Tibiotarsal tissue	Pulverization, extraction into 50% EtOH, SPE	HPLC (2.1×250 mm	MS; ESIQtrap (API–4000, MRM)	LOD 10.5 pg	25	43
Eicosanoids	Intrauterine tissue	Pulverization, extraction into 0.22% formic acid in ethanol, SPE (Xstrata)	HPLC (C_{18} 22.5×250 mm, 5 μm)	MS; ESItriple quadrupole	LLOQ 0.3 ng g^{-1}	31	44
Sphingolipids	Cells	Multiple-step extraction, sonication	HPLC (amino, silica, C_8, C_{18})	MS; ESI triple quadrupole, quadrupole linear-ion trap	–/–	8–10	45
Phospholipids	Urine	Modified Folch extraction	nanoLC (C_{18})	MS; ESI ion trap	–/–	60–90	46
Phosphatidylinositol triphosphate	Cells, tissues	Modified Folch extraction, homogenization, derivatization with TMS	UPHLC (C_4)	MS; ESI hybrid triple quadrupole/linear ion trap	–/–	20	47
Phosphoinositides	Yeast cells	Acidic extraction into butanol	HPLC (silica column)	MS; ESI quadrupole, ion trap	–/–	35	48
Sphingosine 1-phosphate, sphinganine 1-phosphate, lysophosphatidic acid	Plasma	Acidic extraction into butanol	HPLC (silica HILIC)	MS; ESI hybrid triple quadrupole/linear ion trap	2–6 nmol L^{-1}/–	2.5	49

Eicosanoids, including prostaglandins, thromboxanes, leukotrienes, hydroxyeicosatetranoic acid, hydroxyeicosapentanoic acid, epoxyeicosatrienoic acid and lipoxins are a diverse class of bioactive lipid mediators derived from arachidonic acid. This class of metabolites has been of particular interest in clinical studies because of their contributions to inflammatory diseases. Eicosanoids are present at very low concentrations in biological samples and, moreover, many of them are isomers, which makes their identification and quantitation even more challenging. The analyses are typically carried out by LC-ESI-MS. Because most of the eicosanoids are highly unstable, they require careful analytical procedures, starting from sampling. A mixture of enzyme inhibitors and antioxidants should be added to the samples directly after sampling and the samples should be immediately transferred to a freezer, if not analysed immediately. During sample preparation, the samples should be protected from atmospheric oxygen and UV light and the sample preparation should be performed in cold conditions.

The PIP family consists of phosphorylated derivatives of the membrane lipid phosphatidylinositol and they are established regulators of multiple cellular functions. Seven different types of PIPS have been identified in biological samples, namely phosphatidylinositol monophosphates (phosphatidylinositol 3-, 4- and 5-phosphate), phosphatidylinositol bisphosphates (phosphatidylinositol 3,4-, 3,5- and 4,5-bisphosphate) and phosphatidylinositol trisphosphate. The most common FAs of PIPs are stearic acid in the *sn*1 position and arachidonic acid in the *sn*2 position. From an analytical perspective, PIPs are relatively stable, but they are poorly extracted from biological matrices and possible adsorption on the extraction vessel walls and other surfaces is a serious problem. It is also difficult to optimize the analytical conditions for LC-MS, as PIPs tend to show poor chromatographic behaviour with the buffers compatible with MS detection. Most methods are based on LC-MS and both RPLC and HILIC methods have been tested.

Sphingolipids are also a highly diverse and complex class of lipids that serve as both structural components of cellular membranes and signalling molecules capable of causing apoptosis, differentiation and chemotaxis. This class includes ceramides, sphingomyelins, glycosphingolipids and inositol-containing ceramides. Currently, LC-MS/MS is the analytical tool of choice for the determination of various sphingolipids. It has been used for the analysis of free sphingoid bases, free sphingoid base phosphates, ceramides, monohexosylceramides, lactosylceramides, sphingomyelins and even more complex glycosphingolipids.

9.3 Polar Metabolites

Polar, hydrophilic metabolites include various types of compounds, such as amino acids, acylcarnitines, nucleotides, coenzyme A esters, nucleotides and carbohydrates (*e.g.*, simple sugars, sugar phosphates), and these are summarized in Table 9.6.

Table 9.6 Main classes of polar metabolites.

Metabolite class	*Examples of metabolites*	*Structure, example*
Acylcarnitines	Acetylcarnitine, propionylcarnitine, butyrylcarnitine, valerylcarnitine, hexanoylcarnitine, octanoylcarnitine, decanoylcarnitine, dodecanoylcarnitine, tetradecanoylcarnitine, hexadecanoylcarnitine, octadecanoylcarnitine	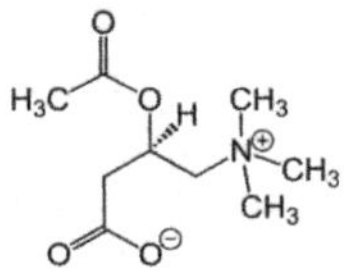
Acyl-CoAs	Acetyl-CoA, malonyl-CoA, acetoacetyl-CoA	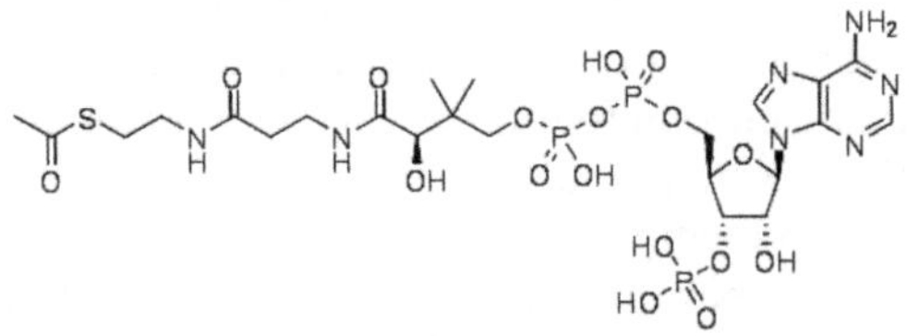
Amino acids	Histidine, alanine, isoleucine, arginine, leucine, asparagine, lysine, aspartic acid, methionine, cysteine, phenylalanine, glutamic acid, threonine, glutamine, tryptophan, glycine, valine, ornithine, proline, serine, tyrosine	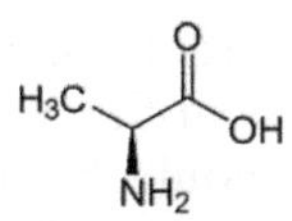
Sugars and sugar derivatives	Glucose, fructose, mannitol, galactose, *myo*-inositol, glucose-6-phosphate, fructose-6-phosphate	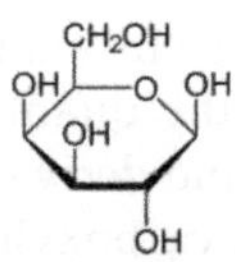
Citric acid cycle metabolites	Citric acid, succinic acid, oxalic acid, isocitric acid, α-ketoglutaric acid	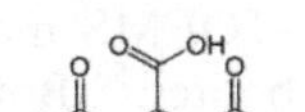
Nucleosides	Cytidine, uridine, adenine, adenosine, guanosine, uracil, thymine, thymidine, inosine	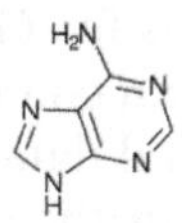
Nucleotides	Adenosine 5′-triphosphate, guanosine 5-triphosphate, cytidine triphosphate, uridine 5-triphosphate	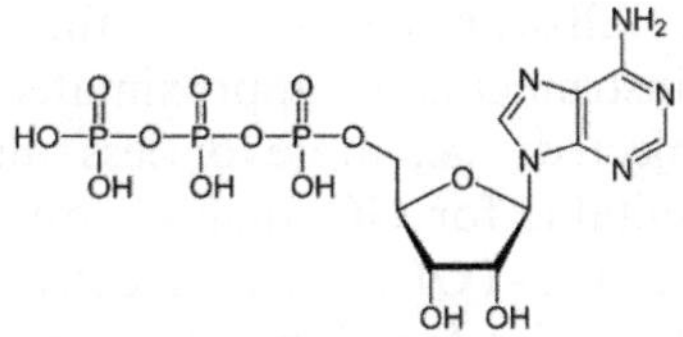

9.3.1 Global Profiling of Polar Metabolites

Global profiling of polar metabolites can be carried out with GC- and LC-based methods, and CE has also been used, but to lesser extent (Table 9.7). As in lipidomics, global profiling requires typically HRMS systems, at least in combination with LC. The sample pretreatment methods for the global profiling methods are often based on liquid extraction (tissues, cells, plant material) or simple protein precipitation (biological fluids such as serum/plasma, CSF) or dilution (urine), as described in detail in Chapter 2.

LC-MS methods have been widely used in non-targeted profiling of polar metabolites. However, very polar compounds are challenging to analyse with RPLC owing to their poor retention and therefore HILIC and ion-pair chromatography have also been applied for global profiling. In a recent study, 22 different stationary phase/mobile phase conditions were screened for their adequate chromatographic selectivity for global profiling of polar metabolites.[50,51] Of the combinations tested, the best results were obtained with a HILIC method employing a zwitterionic ZIC-HILIC column. However, although good separation has been reported with HILIC columns, in practice it is very difficult to optimize the conditions for all types of compounds and a large number of compounds have only moderate or even poor interactions with the chromatographic phase. Ion-pair LC has been proposed for the analysis of polar metabolites and promising results have been obtained. However, ion-pair reagents tend to cause problems in the MS instruments when analysing large numbers of samples, so they should be used with care. Often, two parallel analyses with RPLC and HILIC are used for more comprehensive coverage of the polar metabolites. For example, one LC-MS method based on HILIC chromatography was used in the positive-ionization mode and another method based on reversed-phase ion-pair chromatography was used in the negative-ionization mode. This allowed the quantitation of ~250 water-soluble metabolites, including amino acids and derivatives, sugar phosphates, nucleotides, coenzyme A and derivatives and carboxylic acids.[52]

GC-TOFMS and GC×GC-TOFMS methods have also been used for the global profiling of polar metabolites.[53] Both techniques allow the detection of hundreds of different metabolites in a single analysis and, thanks to the availability of large spectral libraries for GC-EI-MS, combined with retention index data, identification of unknown compounds is possible. Even if the mass spectrum is not present in the spectral database, the fragmentation pattern can be used to obtain information about the compound class of a metabolite. In addition, the full-scan response of the electron ionization (EI) mode for quadrupole instruments is approximately proportional to the amount of compound injected, *i.e.*, more or less independently of the compound, all compounds suitable for GC analysis are detected without discrimination. In particular, GC×GC-TOFMS has an extremely high separation efficiency and it is an excellent choice for global profiling. An example of GC×GC-TOFMS analysis of CSF and serum samples is shown in Figure 9.2. An additional benefit of GC-based methods is that they do not suffer from matrix effects to

Table 9.7 Selected applications of global profiling of polar metabolites by chromatographic or electromigration techniques.

Metabolite	*Sample type*	*Sample preparation*	*Analysis*	*Detection*	*LOD/LOQ*	*Analysis time (min)*	*Ref.*
Polar metabolites	Fermentation broths	Extraction with cold MeOH	HILIC (ZIC-HILIC, 150×4.6 mm id., d_f 5 μm)	QTrap 4000	0.005–0.15 mg L^{-1}	30	51
141 polar metabolites	*Escherichia coli*	Extraction with 80:20 MeOH–water	HILIC (HILIC NH_2, 250×2 mm i.d., d_f 5 μm)	QqQMS, ESI-MS/MS	na	40	56
Central carbon metabolites, amino acid sand nucleotides	*E. coli* cells	Extraction with 80:20 MeOH–water	HILIC (NH_2, 250×2 mm i.d., d_f 5 μm)	QqQMS		50	56
Polar metabolites	*E. coli*, *Bacillus subtilis* and *Saccharomyces cerevisiae* cells	Extraction with $CHCl_3$ (non-polar), extraction with H_2O–MeOH + SPE (IE-SPE)	HILIC (TSK Gel Amide, 100×2 mm i.d., d_f 5 μm), GC-MS (oximation, silylation), ion-pair HPLC (C_{18}, 100×3 mm i.d., d_f 5 μm) with hexylamine	LTQ ion trap MS, qMS	na		57
Polar metabolites	Urine	Dilution with ACN	HILIC (NH_2, 150×2 mm i.d., d_f 5 μm)	LTQ ion trap MS	na	30	58
>1000 polar metabolites	Serum/plasma, CSF, lipoproteins	PP with ACN, MOX/MSTFA derivatization	GC×GC (retention gap 1.5 m×0.53 mm i.d. + Rxi-5MS, 10 m×0.18 mm i.d., d_f 0.18 μm + BPX-50, 1.5 m×0.1 mm i.d., d_f 0.1 μm)	TOFMS	ppb	45	59–61

Table 9.7 (*Continued*)

Metabolite	*Sample type*	*Sample preparation*	*Analysis*	*Detection*	*LOD/LOQ*	*Analysis time (min)*	*Ref.*
polar metabolites	Plasma	PP with water–MeOH, MOX/MSTFA derivatization	GC×GC (Rtx-5MS, 20 m×250 μm i.d., d_f 0.5 μm + Rtx-200 MS, 2 m×180 μm i.d., d_f 0.2 μm)	TOFMS	na	28	62
Low molecular weight compounds	Plasma	PP (methanol), derivatization with MOX/MSTFA	GC (DB5-MS, 10 m×0.18 mm i.d., d_f 0.18 μm	MS	0.1–1 μM	–	63
Polar metabolites	Microorganisms	Chloroform extraction, MOX/MSTFA derivatization	GC (DB5-MS, 30 m×250 μm i.d., d_f 0.25 μm)	qMS	0.1–4.4 mmol g^{-1}	25	64
Central carbon metabolic pathways	*Methylobacterium extorquens* (Gram-negative bacterium)	Extraction with boiling ETOH/acid extraction, for GC×GC-TOFMS derivatization with MOX + MSTFA	HILIC (NH_2, 250×2 mm i.d., d_f 5 μm) RPLC (C_{18}, 100 mm×2 mm i.d., d_f 3 μm) GC×GC (Rtx-5, 20 m×0.25 mm i.d., d_f 0.5 μm + Rtx-200, 2 m×0.18 mm i.d., d_f 0.2 μm	QqQMS, TOFMS	LOD: GC×GC-TOFMS 0.003–0.5 pmol; HILIC-MS 0.5–4 pmol	20 (LC), 38 (GC×GC)	55

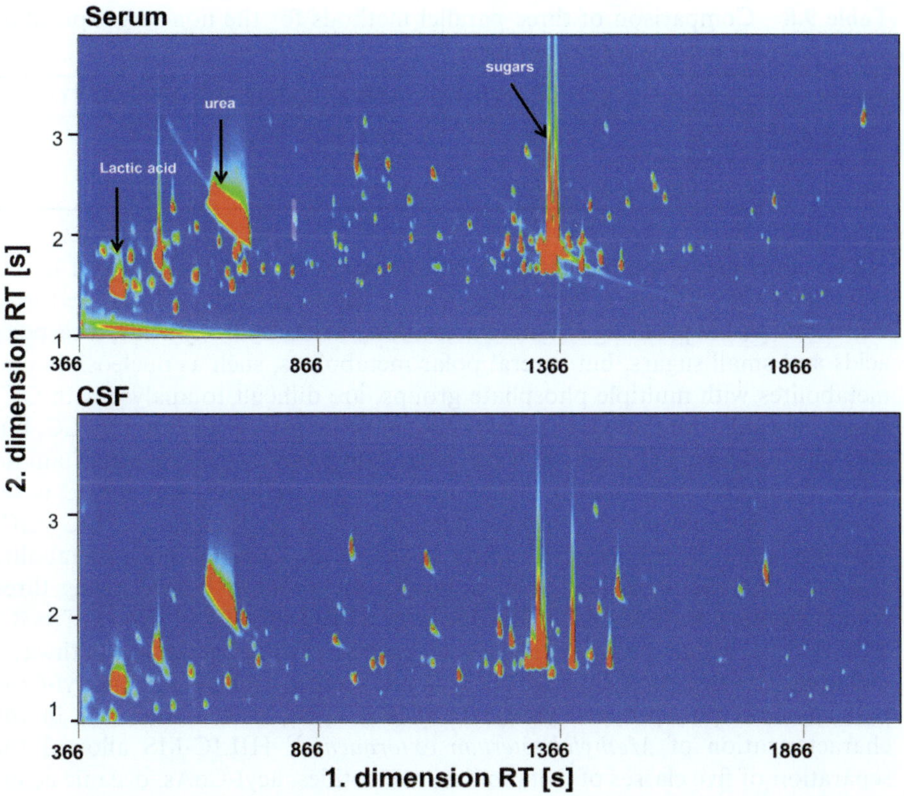

Figure 9.2 Example of GC×GC-TOFMS analysis of CSF and serum samples. For detailed information, see ref. 60.

such an extent as LC-MS techniques. The GC-based methods require derivatization of the polar analytes but, because modern instruments allow automated derivatization, the derivatization step does not really complicate the sample preparation. However, some metabolites may show several peaks due to incomplete derivatization, complicating the interpretation of the results. In addition, several polar metabolites, such as nucleotides, coenzyme A esters and sugar nucleotides, cannot be analysed with GC-based methods.

Both LC- and GC-based methods have their own advantages and limitations in the global non-targeted analysis of polar metabolites. However, only a few reports are available for comparison of LC-MS and GC-MS techniques for the global profiling of polar metabolites. Recently, three platforms, namely GC-TOFMS, GC×GC-TOFMS and UHPLC-QTOFMS, were compared in the profiling of polar metabolites in human serum samples (Table 9.8).[54] With UHPLC-QTOFMS and GC×GC-TOFMS, a large number of peaks could be detected. The GC-TOFMS method allowed the detection of only a relatively small number of peaks. As could be expected, the identification was more efficient with the GC-based methods; however, fewer than 10% of the detected peaks could be identified.

Table 9.8 Comparison of three parallel methods for the non-target profiling of human serum samples.[54]

Parameter	*GC-TOFMS*	*UHPLC-QTOFMS*	*GC×GC-TOFMS*
No. of peaks	278	1601/7973	3024
No. of putative metabolites	200	300/800	1500
No. of identified metabolites	80–90	–	200

The coverage of the metabolome is different with LC and GC, and in LC techniques the coverage is very dependent on the method chosen. GC performs well in the global profiling of amino acids, fatty acids, hydroxy acids, dicarboxy acids and small sugars, but several polar metabolites, such as nucleosides and metabolites with multiple phosphate groups, are difficult to analyse with GC-based methods. All these compounds can, in principle, be analysed by LC, but not with a single LC method. Very polar compounds, such as small amino acids, dicarboxy acids and hydroxyl acids, require HILIC, and RPLC is required for the analysis of semipolar compounds. A combination of GC×GC-TOFMS and LC-MS analysis would be ideal, in terms of both metabolite coverage and the identification of unknown metabolites. Indeed, using three complementary methods, LC–MS/MS (both RPLC and HILIC) and GC×GC-TOF-MS, a comprehensive analysis of central carbon metabolic pathways, including the tricarboxylic acid (TCA) cycle, serine cycle, ethylmalonyl-CoA pathway and poly-β-hydroxybutyrate (PHB) cycle, was carried out in the characterization of *Methylobacterium extorquens*.[55] HILIC-MS allowed the separation of five classes of metabolites: nucleotides, acyl-CoAs, organic acids, small peptides and amino acids. However, significant peak tailing for butyryl-CoA and acetyl-CoA was observed with the HILIC column, therefore RPLC-MS was utilized for the analysis of these metabolites. GC×GC-TOFMS was utilized for the analysis of organic acids and amino acids, using two-step derivatization [methoximation (MOX) + silylation]. Although HILIC-MS allowed the separation of several carboxylic acids, some metabolites were either not reproducibly detected (methylmalonic and mesaconic acid) or not detected at all (glyoxylic acid), probably owing to insufficient sensitivity of the HILIC method. In addition, some isomers, such as methylsuccinic, ethylmalonic, malic and fumaric acid, could not be reliably determined. These compounds were then determined with high sensitivity with GC×GC-TOFMS. Metabolites with low volatility (*e.g.*, nucleotides and acyl-CoAs), were ideal for LC-MS/MS analysis, with relative standard deviation (RSD) <20%, whereas low-volatility metabolites (MW <300 m/z) at low concentrations (<500 nM) were better quantified by GC×GC–TOF-MS.

9.3.2 Targeted Analysis of Polar Metabolites

As in the case of lipids, it is also not possible to cover all polar metabolites with the non-targeted screening methods. Many metabolites are present in very low concentrations or their analysis requires specific analytical procedures, either in sample preparation and/or in their separation. Therefore, several targeted methods have been developed for the quantitative analysis of these metabolites.

9.3.2.1 *Acylcarnitines*

Carnitine has a central role in transportation of the FAs into the mitochondrial membrane, where they are metabolized via β-oxidation to produce energy. The human body can synthesize about 25% of L-carnitine from the amino acids in the liver, kidney and brain; the remainder is obtained from the diet. The highest concentrations of carnitine and its acyl metabolites can be found in skeletal and cardiac muscles. The analytes also exist in the liver and in small amounts in the blood. Abnormal concentrations of carnitine and its acyl metabolites can reveal many genetic and acquired disorders, *e.g.*, propionic aciduria and medium-chain acyl-coenzyme A dehydrogenase deficiency (MCADD).[42–44]

The analytical methods used for the determination of acylcarnitines are listed in Table 9.9. The sample pretreatment typically includes multiple extraction and evaporation steps, derivatization and redissolution of the extract in a solvent suitable for LC-MS analysis. In LC analysis, the polar nature of carnitine makes direct RPLC analysis challenging. The common procedure to improve the retention of the carnitine is to add an ion-pair reagent such as heptafluorobutyric acid (HFB) to the eluents.[45–48] However, ion-pair reagents have a tendency to accumulate on the detector and to contaminate it, hence these types of methods are not well suited for the analysis of large numbers of samples. With UV detection, the acylcarnitines are detected at a wavelength of 210 nm, but the sensitivity is very low. To improve the detectability, the analytes have been derivatized to form pentafluorophenacyl esters.[25,49] In LC-MS/MS analyses, the compounds have been typically analysed as butyl ester derivatives.[50,51] However, preparation of acylcarnitine butyl esters has been observed to partially hydrolyse acylcarnitines, therefore increasing the value of free carnitine. On the other hand, MS detection is possible without derivatization.[58]

Recently, very polar and hydrophilic compounds have increasingly been analysed by HILIC. In HILIC, solvents with a high organic content are used and this usually improves the sensitivity in MS. HILIC has been applied in the analysis of carnitine[51,53,56] and medium-chain[56] and long-chain acylcarnitines.[56–58,67] An example of the fast separation of short- and medium-chain acylcarnitines in human serum by HILIC-MS is shown in Figure 9.3.

9.3.2.2 *Acyl-CoAs*

Acyl-coenzyme A (acyl-CoA) is a group of coenzymes involved in numerous metabolic processes in eukaryotic cells, including β-oxidation within mitochondria and peroxisomes and the biosynthesis/remodelling of lipids (*e.g.*, MGs, DGs and TGs, PLs and SLs). Acyl-CoA is a temporary compound formed when coenzyme A (CoA) attaches to the end of a long-chain FA inside cells. The compound then undergoes β-oxidation, forming one or more molecules of acetyl-CoA. This, in turn, enters the citric acid cycle, eventually forming several

Table 9.9 Typical examples of the analysis of acylcarnitines by chromatographic or electromigration techniques.

Sample matrix	*Sample preparation*	*Separation*	*Detection*	*LOD/LOQ*	*Analysis time (min)*	*Ref.*
Red blood cells	Centrifugation; separated plasma washed, vortexed, centrifuged; ultracentrifugation (3 kDa cutoff); lysate diluted 1:1 with background electrolyte (BGE)	CE	MS; ESI ion trap	–/–	11	65
Urine	Centrifugation; SPE (MCX); evaporation	UPLC (BEH C_{18}, 150×1.0 mm i.d., d_f 1.7 μm	MS; ESI; Triple Quadrupole/Linear Ion trap	–/–	105	66
Skeletal muscle, liver, serum	Homogenization; ACN–MeOH (3 : 1 v/v); centrifugation	HILIC (100×2.1 mm i.d., d_f 1.7 μm)	MS; ESI; triple quadrupole	0.5–5 ng mL^{-1}/–	7	67
Urine	Dilution; acidification; SPE (cation exchange)	HPLC (C_8, ion pair)	MS/MS; ESI; ion trap	–/0.75–5 μmol L^{-1}	24	68
Plasma, skeletal muscle	Extraction with ACN–MeOH, centrifugation; hydrolysis; SPE (Si)	HPLC (C_{18}, 100×4.6 mm i.d., d_f 3 μm)	MS/MS; ESI; ion trap	–/–	75	69
Skeletal muscle	Lyophilization; homogenization with 80% ACN; centrifugation; derivatization (butyl ester)	HPLC (C_{18}, 7.5×4.6 mm i.d., d_f 3.5 μm)	MS/MS; ESI; tandem quadrupole	–/–	5	70

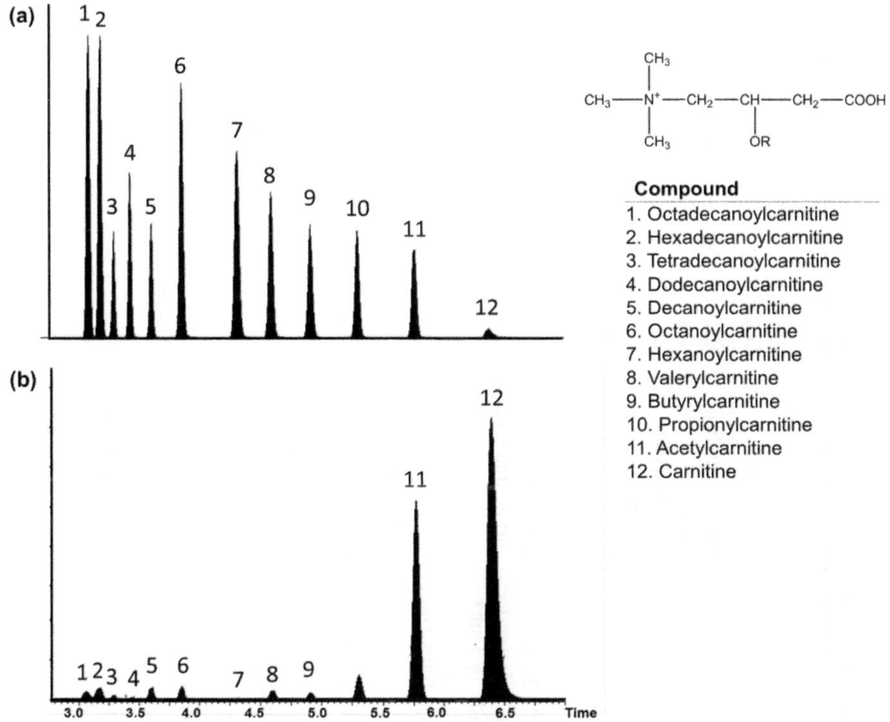

Figure 9.3 Example of the fast separation of short- and medium-chain acylcarnitines in human serum by HILIC-MS. a) Separation of standard compounds and b) analysis of human serum after protein precipitation. For more detailed information, see ref. 67.

molecules of ATP. Structurally, CoA is composed of three chemical moieties, an adenosine triphosphate nucleotide, pantothenate and cysteamine.

Several methods have been developed to measure individual long-chain acyl-CoAs in biological samples. The analytical methods used for the determination of acyl-CoAs are listed in Table 9.10. The sample preparation methods for the Acyl CoAs rely on liquid extraction for tissue and cell samples, typically followed by SPE purification of the extract. The current analytical methods are based on LC-MS, which provides a high level of structural specificity and sensitivity, especially when combined with ESI. Both negative ion mode MS and MS/MS and also positive ion mode MS/MS using a precursor ion scan have been utilized in the detection of acyl-CoAs.

9.3.2.3 Amino Acids

Amino acids are biologically important organic compounds, playing central roles in protein synthesis and regulation of gene expression and protein phosphorylation cascades. Furthermore, they are also a source of energy and serve as precursors for the biosynthesis of neurotransmitters, porphyrins,

Table 9.10 Typical examples of the analysis of acyl-CoAs by chromatographic or electromigration techniques.

Sample matrix	*Sample preparation*	*Separation*	*Detection*	*LOD/LOQ*	*Analysis time (min)*	*Ref.*
Catharanthus roseus cells	LLE; centrifugation (5 kDa); dilution with water	CE	MS; ESI hybrid triple quadrupole linear ion trap	0.25–1.2 µM	14	71
Rat liver	LLE; homogenization; centrifugation; SPE; evaporation; filtration	CE	UV	10–160 pmol	30	72
Biological tissues (skeletal muscle, liver, adipose tissue)	Homogenization with 0.1 M KH_2PO_4–2-propanol (1:1 v/v) and ammonium sulfate–ACN; on-line SPE	HPLC (HLB, 20×2.1 mm i.d. + Zorbax C_{18} Extend, 150×2.1 mm i.d., d_f 3 µm)	MS/MS; triple-stage quadrupole	–	17	73
Biological tissues (liver, heart, skeletal muscle), *Escherichia coli*	Homogenization with 10% TCA; SPE (Oasis HBL)	HPLC (BDS C_{18}, 100×4 mm i.d., d_f 5 µm)	UV and MS; ESI; ion trap	<50 pmol	20	74
	Cell lysis; extraction with ACN–MeOH–H_2O + 0·1% glacial acetic acid (45:45:10)	HPLC (C_{18}, 150×4·6 mm i.d., d_f 3 µm)	MS/MS; ESI; triple quadrupole	–	45	75
Muscle tissue	Homogenization in KH_2PO_4, (pH 4.9) and ACN–2-propanol–MeOH (3:1:1)	UPLC (C_8, 150×2.1 mm i.d., d_f 1.7 µm)	MS/MS; ESI;	–	4.5	76
Plant tissues	Extraction with 10% TCA; homogenization; centrifugation; lyophilization	HPLC (C_{18}, 10 cm×2.1 mm i.d., d_f 3 µm)	UV and MS/MS; ESI; ion trap	< 1.6 nmol/g tissue	30	77
Mouse liver, mouse brown–white adipose tissue, ratliver	Homogenization with KH_2PO_4–2-propanol, extraction with ACN–ammonium sulfate; on-line SPE (HLB)	HPLC (C_{18} Extend, 150×2.1 mm i.d., d_f 3 µm)	Quantum TSQ Ultra AM	na	17	73

polyamines and nitric oxide. Further, glutamate and the non-protein amino acid γ-aminobutyric acid play crucial roles as neurotransmitters and carnitine participates in intracellular lipid transport. Amino acids make up 75% of the human body and are essential to nearly every bodily function. Structurally, amino acids consist of amine (–NH_2) and carboxylic acid (–COOH) functional groups, along with a side-chain specific to each amino acid. About 500 amino acids are known and can be classified either according to the functional groups' locations as alpha- (α-), beta- (β-), gamma- (γ-) or delta- (δ-) amino acids, or by other categories related to polarity, pH level and side-chain group type (aliphatic, acyclic, aromatic, containing hydroxyl or sulfur, *etc.*).

The analytical methods used for the determination of amino acids are listed in Table 9.11. The sample preparation of biological fluids typically involves only protein precipitation and/or dilution; for tissue and plant samples, the sample preparation typically includes liquid extraction with, *e.g.*, acidified ethanol, water or trichloroacetic acid, and sometimes hydrolysis for the determination of total hydrolysable amino acid content. Recently, ionic liquids have also been utilized in the extraction of amino acids.[78] Classical hydrolysis is carried out by heating the sample in 6 M hydrochloric acid under vacuum at 110 °C for 18–24 h. Separation of amino acids can be carried out with GC-, CE- and LC-based methods, with detection using fluorescence detection (FD), UV spectrophotometry or MS. Enantioselective GC can be used for the separation of amino acids enantiomers. A derivatization step is commonly used in the analysis of amino acids, in both GC and LC. In GC, the aim of the derivatization is to modify the polar groups to improve the chromatography, whereas in LC, the goal is both to increase the sensitivity and to increase the retention of very polar amino acids in RPLC separation.

The traditional amino acid analysis has been performed by either cation-exchange LC or RPLC coupled to UV absorbance or fluorescence detection (FD), using either pre- or post-column derivatization. The most common reagents include *o*-phthalaldehyde (OPA), phenyl isothiocyanate (PITC), dimethylaminoazobenzenesulfonyl chloride (DABS-Cl), 9-fluorenylmethyl chloroformate (FMOC-Cl) and 7-fluoro-4-nitrobenzo-2-oxa-1,3-diazole (NBD-F). Ninhydrin is used for post-column derivatization. The advantages of LC-UV or LC-FD are good reproducibility, inexpensive equipment and high sensitivity in the low-picomole range; however, the detection lacks specificity. HILIC-MS and CE-MS permit the direct analysis of amino acids, but the throughput is relatively low and both methods suffer from comparatively poor reliability. Recently, several LC-MS methods have been developed for the determination of amino acids. A commercial method utilizing derivatization with a reactive ester to introduce an isobaric tag, a so-called iTRAQ label, is also available.[79] The method is based on the detection of the cleavable reporter ion of the tag by collision-induced dissociation in the MS/MS mode. Ion-pair LC-MS has been utilized in the analysis of both native and iTRAQ-labelled amino acids. Typically, volatile ion-pair reagents such as perfluorocarboxylic acids are used; however, they can cause contamination of the MS system.

Table 9.11 Typical examples of the analysis of amino acids by chromatographic or electromigration techniques.

Sample matrix	*Sample preparation*	*Separation*	*Detection*	*LOD/LOQ*	*Analysis time (min)*	*Ref.*
Human serum	Serum diluted 10-fold with water, centrifugation (10 kDa cutoff)	CE	MS; triple quadrupole	0.05–0.81 μM/–	14	81
Human urine	Mixed with BGE (1:1), centrifugation (10 kDa cutoff)	CE	MS; TOF	–/–	19	78
Red blood cell	Centrifugation; separated plasma washed, vortexed, centrifuged; ultracentrifugation (3 kDa cutoff); lysate diluted 1:1 with BGE	CE	MS; ESI; ion trap	–/–	13	65
Human urine	Mixed with BGE (1:1), centrifugation	CE	MS; TOF	20–215 nM/ 85–280 nM	13	82
E. coli	Filtration (0.45 μm); LLE; centrifugation (5 kDa cutoff); dilution with water	CE	MS; TOF	–/–	-	83
E. coli	Centrifugation; dilution 20-fold with BGE	CE	MS; ESI; ion trap	1–25 nM/–	12	84
Human serum	Deproteinization (MeOH); automated SPE	HILIC (Luna 3 μm, 100×4.6 mm i.d.)	MS; ESI; triple quadrupole	0.08–6 μM/ 0.2–22 μM	25	85
Human urine, serum and saliva	Deproteinization; centrifugation; derivatization	CE	UV absorbance	0.52–1.7 ppm/1.72–5.72 ppm	25	86
plasma, urine	Deproteinization with MeOH	RPLC (C_{18}, 50×2 mm i.d., d_f 3 μm)	QqQMS, ESI	0.1 –10 μM	20	88
Collagens	Hydrolysis with 6 M HCl; lyophilization	HILIC (TSK-Gel Amide 80, 15 cm×2.0 mm i.d., d_f 5 μm)	QqQMS, ESI	nmol	50	89
Serum	Dilution, ethyl chloroformate derivatization	GC (DB-5MS, 30 mm×250 μm i.d., d_f 0.25 μm)	qMS	LOD 0.2–0.5 ppb	40	90
Biological fluids	Direct derivatization with propyl chloroformate	GC (ZB-AAA, 15 m×0.25 mm i.d., d_f 0.1 μm)	qMS	LOQ 0.3–30 μM	25	91
Filamentous fungi	Filtration, dissolution in sodium hydroxide solution–MeOH–pyridine; derivatization with methyl chloroformate	GC (J&W1701, 30 m×250 μm i.d., d_f 0.10 μm)	qMS	LOD 0.04–4.5 mM	30	92
Biomass	Hydrolysis with HCl; dissolution in water; evaporation; TBDMS derivatization	GC (DB-1701, 30 m×0.25 mm i.d., d_f 0.25 μm)	qMS	na	40	93

Derivatization of amino acids for GC analysis can be carried out with a variety of reagents, the most common ones being silylation agents. Both trimethylsilyl and tributylsilyl derivatives have been used, the latter being more stable. Further, acylation/esterification using various anhydride–alcohol combinations (*e.g.*, pentafluoropropyl anhydride and 2-propanol) has been applied for the derivatization of amino acids, and also indirect alkylation via chloroformates in the presence of an alcohol and pyridine. Comparison of two common derivatization procedures [*N*-methyl-*N*-(*tert*-butyldimethylsilyl)trifluoroacetamide (MTBSTFA) and propyl chloroformate (PCF)] suggested that MTBSTFA is more broadly applicable to metabolomics because it allows the detection of other relevant metabolites in the same analysis.[80] PCF derivatization is targeted at amino acids and the reaction time is shorter than with silylation. However, some amino acids, such as glutamic acid, pyroglutamic acid, serine and histidine, could not be quantified reliably when using PCF owing to coelution or poor peak shapes.

A recent study compared three different methods (LC-MS, GC-TOFMS and GC×GC-TOFMS) for the determination of amino acids in human serum sample (Table 9.12).[80] Most amino acids could be determined using both LC and GC techniques and the results agreed well with each other for most of the amino acids. However, arginine, cystine and ornithine could not be quantified with GC–based methods. The problem with arginine is that the derivatization agents do not stabilize the guanidino group in the side chain and the molecule is thermally unstable, thus converting arginine to ornithine in the GC injector port prior to separation. This makes it impossible to distinguish between the portion of the signal due to arginine and the portion due to ornithine. The results show that LC allows more comprehensive analysis of the amino acids. However, GC-MS may be preferred if unknown amino acids should also be covered or in the enantioselective analysis of amino acids.

9.3.2.4 Nucleosides, Carbon Cycle Metabolites

Nucleosides are a structural subunit of nucleic acids, the heredity-controlling components of all living cells. Structurally they consist of a molecule of sugar linked to a nitrogen-containing organic ring compound. In the most important nucleosides, the sugar is either ribose or deoxyribose and the nitrogen-containing compound is either a pyrimidine (cytosine, thymine or uracil) or a purine (adenine or guanine).[94,95] Analytical methods for the determination of nucleosides are listed in Table 9.13. In the sample preparation, both liquid extraction and SPE have been applied. Solid-phase extraction using phenylboronic acid sorbent is selective for nucleosides owing to the vicinal hydroxyl groups in their structure. However, it is not selective for nucleosides without *cis*-diol groups, and for this type of nucleosides a cation-exchange phase is more suitable. C_{18} cartridges have also been used in the extraction. The main problem in dealing with the analysis of nucleotides is that it is difficult to separate them by LC owing to the presence of a phosphate group. Either ion-pair chromatography[87,95,96] or the presence of a phosphate buffer in the mobile phase[97] is

Table 9.12 Comparison of platforms for the analysis of amino acids in certified serum sample[a].

	LC–MS/MS		*GC-TOF-MS (MTBSTFA)*		*GC×GC-TOF-MS (MTBSTFA)*		*GC-TOF-MS (MTBSTFA)*		*Literature average*	
Amino acid	*Average MF*[b] (μg g^{-1})	*SD*[c]	*Average MF* (μg g^{-1})	*SD*	*Average MF* (μg g^{-1})	*SD*	*Average MF* (μg g^{-1})	*SD*	*Average MF* (μg g^{-1})	*SD (n)*[c]
L-Alanine	24.4	0.3	26.6	1.6	25.6	1.8	28	2.1	29	6.5 (1)
L-Arginine	13.9	0.2	–	–	–	–	–	–	14.3	9.5 (5)
L-Cysteine	5.63	0.64	–	–	11.9	3.7	4.9	1.2	2.6	3.2 (3)
L-Cystine	1.83	0.05	–	–	–	–	–	–	24.9	–
L-Glutamic acid	8.46	0.06	10.9	1.1	11	0.8	–	–	7.6	3.0 (2)
Glycine	18.6	0.4	15.7	0.8	18	1.7	18.6	1.3	17.3	8.1 (4)
L-Histidine	10.8	0.2	9.02	0.49	11.3	0.7	–	–	12.9	3.1 (3)
L-Isoleucine	6.92	0.13	7.13	0.48	7.61	0.4	7.1	0.57	9.1	4.0 (4)
L-Leucine	12.6	0.2	13.6	1.8	13	1.8	12.9	1.6	18.9	6.1 (3)
L-Lysine	19	0.3	20.6	1.7	20.7	1.1	20.5	2.7	27.2	8.0 (3)
L-Methionine	3.07	0.03	3.4	0.32	3.24	0.19	3.4	0.58	4.4	1.5 (4)
L-Ornithine	6.74	0.21							6.7	4.3 (4)
L-Phenylalanine	8.13	0.39	6.31	0.72	8.7	0.66	9.03	0.79	9.2	2.5 (4)
L-Proline	19.7	0.3	19.7	1.4	20.7	1.2	21.9	1.7	21.6	11.8 (3)
L-Serine	9.8	0.57	10.3	0.9	9.6	0.59			12.1	7.0 (7)
L-Threonine	13.9	0.2	14.7	1	14.2	0.8	13.5	1.8	17.1	6.0 (3)
L-Tyrosine	9.99	0.23	10.4	0.7	10.5	0.6	10.1	1	11.4	4.1 (3)
L-Valine	20.2	0.2	21.6	1.2	21.3	1.2	21.1	1.6	26.5	7.5 (3)

[a]Literature averages were taken directly from the Human Metabolome Database (http://www.hmdb.ca/) as a resource overview of many published values. Only matrix-matched values were considered (blood drawn from healthy, adult patients).
[b]MF = mass fraction. SD = standard deviation.
[c](*n*) denotes the number of distinct literature values averaged for the calculation of the standard deviation.
Source: Ref. 80.

Table 9.13 Typical examples of the analysis of other metabolites by chromatographic or electromigration techniques.

Metabolite	*Sample matrix*	*Sample preparation*	*Separation*	*Detection*	*LOD/LOQ*	*Analysis time (min)*	*Ref.*
Neutral sugars, sugar phosphates, sugar alcohols	*Arabidopsis thaliana*	Homogenization with $CHCl_3$/MeOH, extraction with ACN–water (50 : 50 v/v)	HILIC (zwitterionic ZIC1-HILIC stationary phase, 3.5 µm, 150×2.1 mm i.d.; SeQuant)	Ion trap MS	LOD 0.2–1.0 mM	20	101
Sugar alcohols, glucose, fructose, sucrose, raffinose, inulin	Grass species *Lolium perenne*	Extraction with water	HPAEC (CarboPac PA 10, 250×4.6 mm i.d.)	PAD	<0.5 mg L^{-1}	30	102
Sugar alcohols, glucose, fructose, sucrose, raffinose, starch	Needles, twigs, wood	Extraction with water	HPAEC (CarboPac)	PAD	22–317 nmol	30	103
Arabinose, fructose, galactose, glucose, maltose, sucrose, raffinose, xylose	Norway spruce needles	Extraction with MeOH/$CHCl_3$/H_2O (12:5:3 v/v/v)	HILIC (Zorbax-NH_2, 250×4.6 mm i.d., d_f 5 µm)	RI, ITMS	na	25	104
Nucleotides	*E. coli*	Filtration (0.45 µm); extraction with MeOH/$CHCl_3$/H_2O; centrifugation (5 kDa cutoff); dilution with water	CE	MS; ESIQuadrupole	0.5–1.7 µM/–	17	105
	Bacillus subtilis	Filtration (0.45 µm); extraction with MeOH/$CHCl_3$/H_2O; centrifugation (5 kDa cutoff); dilution with water	CE	MS; ESIQuadrupole	0.3–11 µM/–	20	106

Table 9.13 (*Continued*)

Metabolite	*Sample matrix*	*Sample preparation*	*Separation*	*Detection*	*LOD/LOQ*	*Analysis time (min)*	*Ref.*
	C. roseus cells	extraction with $CHCl_3$/MeOH/H_2O; centrifugation (5 kDa cutoff); dilution with water	CE	MS; ESI hybrid triple quadrupole linear ion trap	0.040–8.8 μM/–	16	71
	Rat plasma	SPE or plasma precipitation	CE	MS; ESI ion trap	0.05–5.0 μM/ 2.0–20 μM	14	107
	Human urine	SPE (Strata X)	Ion-pair RP-LC (Polar-RP 80, 250×4.6 mm i.d. with 10 mM TBAHS)	UV absorbance	0.89–4.56 nM/ 2.97–15.22 nM	35	96
	Neuro-2a mice cell line	Homogenization with 6% TCA; neutralization	Ion-pair LC (C_{18}, 150×4.6 mm i.d., d_f 5 μm, with 1 mM TBAHS)	UV absorbance	0.74–0.98 nM/–	30	95
	Cauliflower	Centrifugation with a filter; dilution	Ion-pair LC (HSS T3, 100×2.1 mm i.d., d_f 1.8 μm, with DHAA)	MS; ESI	0.05–1 ng/ 0.1–2.1 μM	50	94
	CSF	SPE (Strata X); lyophilization	LC (18-DB, 250×4.0 mm i.d. with 10 mM TBA)	UV	0.1 pM/–	18	97

	Rat liver	Homogenization and extraction with MeOH	LC (Hypercarb. 150×2.1 mm i.d., d_f 5 μm)	MS; ESI triple quadrupole	0.5–1.5 μM/ 0.5–1.6 μM	20	87
	HepG-2 cells	Extraction with perchloric acid (0.3 M) + Na-EDTA (1 mM); neutralization; centrifugation	LC (Hypercarb, 150×2.0 mm i.d., d_f 5 μm)	MS; ESI Ion Trap	0.06–0.18 μM/ 0.22–0.59 μM	–	98
TCA cycle	Human serum	Deproteinization (MeOH); cleaning with chloroform; preconcentration; derivatization	GC (VF-5MS, 30 m×0.25 mm i.d., d_f 0.25 μm)	MS; EIion trap	11–113 nM/ 38–378 nM	22	108
	Hamster ovary cells	Extraction with PCA; neutralization	IC (IonPac AS11-HC)	Conductivity	0.12–0.48 μM/ 2–20 μM	55	109
	Pig heart tissue	Extraction with 0.6 M PCA; derivatization; automated SPE (C_8)	LC (C_{18}, 100×2.1 mm i.d., d_f 2.6 μm)	MS; ESI TOF	60–1000 nM/–	9	110

often required for satisfactory LC separation. However, phosphate buffers are not suitable with MS detection because of the low volatility of phosphates. CE-MS has also been used for the analysis of nucleosides. The pH of the mobile phase seems to be a key variable. For example, Wang *et al.* demonstrated the influence of pH on the method sensitivity, which increased on increasing the pH from 0 to 7, whereas above pH 7 the increase was less significant.[98]

The citric acid cycle, which is also known as the tricarboxylic acid cycle (TCA cycle) or the Krebs cycle, is a metabolic pathway used by all aerobic organisms to generate energy through the oxidization of acetate derived from carbohydrates, fats and proteins into carbon dioxide. Structurally, the metabolites in the TCA cycle are short hydroxy acids, such as citric acid, isocitric acid, succinic acid, α-ketoglutaric acid and oxalic acid. They have all been analysed using LC-, GC- and CE-based techniques (Table 9.13). Because these compounds are very polar, they have poor retention in RPLC, hence GC is a better option for their analysis. However, the most common analytical platform to determine these compounds is GC–MS, which requires a derivatization step, typically silylation. Several LC-MS/MS-based methods have been developed for the analysis of TCA cycle intermediates, including ion-pair LC, HILIC and RPLC with derivatization. The use of ion-pair agents can be problematic owing to ionization suppression and memory effects. In HILIC, coelution of highly polar analytes with endogenous salts may cause severe ion suppression. In LC-MS, the detection sensitivity of small carboxylic acids in negative-ion electrospray ionization (ESI) is often problematic and the sensitivity is typically only in the micromolar range.

Carbohydrates are present in the cells of all living organisms and they make up a significant proportion of the metabolome. They can be divided into classes of monosaccharides, oligosaccharides and polysaccharides. All of them are polar molecules, which encompass a wide diversity of molecular size, glycosidic linkage and functionality. Sample preparation is typically carried out by extraction based on LLE or SPE. If needed, hydrolysis of oligosaccharides and polysaccharides can be carried out either chemically (*i.e.*, with acids) or enzymatically. Extraction of carbohydrates from aqueous media is challenging owing to their low solubility in organic solvents. Xylose and other sugars have been extracted from complex matrices using boronic acid extractants, with ion pairs with boronate anion and quaternary ammonium cation.[99] For other carbohydrates, such as water-soluble non-structural carbohydrates, water, alcohols, (ethanol, methanol), chloroform and mixtures of these have been utilized for extraction. SPE using C_{18} cartridges has commonly been used for the purification of carbohydrates prior to their analysis and other sorbents such as styrene–divinylbenzene, aminopropylsilica and ion-exchange phases are also used.[99] Various methods have been used for the analysis of carbohydrates, including LC-, CE- and GC-based methods. Owing to the relatively low volatility of carbohydrates, GC analysis is limited to derivatized sugars of low molecular weight, mainly mono-, di- and trisaccharides. In GC, carbohydrates have to be derivatized before their analysis; methyl ethers, acetates, trifluoroacetates and trimethylsilyl ethers are the most common derivatives used for carbohydrate determination.[100] In LC, carbohydrates have very poor retention in RPLC and anion-exchange chromatography coupled with pulsed amperometric detection (PAD) has been used for the separation of these compounds. However, the high salt content required for the anion-exchange separation hinders MS detection. The use of post-column membrane suppressors that can remove some salt from the mobile phase have been reported, but they are not well suited for rugged high-throughput analyses. Porous graphitic carbon (PGC) stationary phases have also been used for the separation of carbohydrates; however, the elution of highly polar compounds often requires extensive method development. Recently, HILIC has been gaining popularity in the analysis of carbohydrates.

9.4 Conclusion

As this chapter clearly shows, a large number of methods are available for the metabolomics. Selection of the method is not always straightforward and multiple methods are required to cover the wide range of metabolites in the different types of samples. Naturally, the selection is also dependent on practical issues, such as the instrumentation available. Currently, most methods are based on GC-MS and LC-MS, while CE is not used as extensively. However, CE-based methods have potential as simple diagnostic systems that could be applied on-site for validated biomarkers. It should also be noted that the implementation of published methods is not always easy and the methods may not perform well in routine-based analysis as very few publications show data on the analysis of large numbers of samples.

References

1. M. Orešič, V. A. Hänninen and A. Vidal-Puig, *Trends Biotechnol.*, 2008, **26**, 647.
2. E. Fahy, S. Subramaniam, H. A. Brown, C. K. Glass, A. H. Merrill, R. C. Murphy, C. R. H. Raetz, D. W. Russell, Y. Seyama, W. Shaw, T. Shimizu, F. Spener, G. van Meer, M. S. VanNieuwenhze, S. H. White, J. L. Witztum and E. A. Dennis, *J. Lipid Res.*, 2005, **46**, 839.
3. L. Yetukuri, K. Ekroos, A. Vidal-Puig and M. Oresic, *Mol. Biosyst.*, 2008, **4**, 121.
4. J. Folch, M. Lees and G. H. S. Stanley, *J. Biol. Chem.*, 1957, **226**, 497.
5. E. G. Bligh and W. J. Dyer, *Can. J. Biochem. Physiol.*, 1959, **37**, 911.
6. P. Del Boccio, D. Pieragostino, M. Di Ioia, F. Petrucci, A. Lugaresi, G. De Luca, D. Gambi, M. Onofrj, C. Di Ilio, P. Sacchetta and A. Urbani, *J. Proteomics*, 2011, **74**, 2826.
7. S. S. Bird, V. R. Marur, M. J. Sniatynski, H. K. Greenberg and B. S. Kristal, *Anal. Chem.*, 2010, **83**, 940.
8. J. M. Castro-Perez, J. Kamphorst, J. DeGroot, F. Lafeber, J. Goshawk, K. Yu, J. P. Shockcor, R. J. Vreeken and T. Hankemeier, *J. Protein Res.*, 2010, **9**, 2377.

9. P. D. Rainville, C. L. Stumpf, J. P. Shockcor, R. S. Plumb and J. K. Nicholson, *J. Protein Res.*, 2006, **6**, 552.
10. H. Nygren, T. Seppänen-Laakso, S. Castillo, T. Hyötyläinen and M. Orešič, in *Metabolic Profiling: Methods and Protocaols*, ed. T. O. Metz, *Methods in Molecular Biology*, vol. 708, Humana Press, Totowa, NJ, 2011, pp. 247.
11. A. Fauland, H. Köfeler, M. Trötzmüller, A. Knopf, J. Hartler, A. Eberl, C. Chitraju, E. Lankmayr and F. Spener, *J. Lipid Res.*, 2011, **52**, 2314.
12. H. Nie, R. Liu, Y. Yang, Y. Bai, Y. Guan, D. Qian, T. Wang and H. Liu, *J. Lipid Res.*, 2010, **51**, 2833.
13. M. Lísa, E. Cífková and M. Holčapek, *J. Chromatogr. A*, 2011, **1218**, 5146.
14. K. Sandra, S. Pereira Ados, G. Vanhoenacker, F. David and P. Sandra, *J. Chromatogr. A*, 2010, **1217**, 4087.
15. H. Ogiso, T. Suzuki and R. Taguchi, *Anal. Biochem.*, 2008, **375**, 124.
16. S. Abdulkadir and M. Tsuchiya, *J. Exp. Mar. Biol. Ecol.*, 2008, **354**, 1.
17. L. Akoto, R. J. J. Vreuls, H. Irth, R. Pel and F. Stellaard, *J. Chromatogr. A*, 2008, **1186**, 365.
18. C. de La Cruz Garcia, J. Lopez Hernandez and J. Simal Lozano, *J. Chromatogr. A*, 2000, **891**, 367.
19. C. Glaser, H. Demmelmair and B. Koletzko, *J. Lipid Res.*, 2010, **51**, 216.
20. M. A. Masood and N. Salem Jr, *Lipids*, 2008, **43**, 171.
21. G. B. Phillips and J. T. Dodge, *J. Lipid Res.*, 1967, **8**, 676.
22. A. Baylin, M. K. Kim, A. Donovan-Palmer, X. Siles, L. Dougherty, P. Tocco and H. Campos, *Am. J. Epidemiol.*, 2005, **162**, 373.
23. L. S. Costanzo, *Physiology*, 4th edn, Saunders/Elsevier, Philadelphia, 2010.
24. H. Sirén, T. Seppänen-Laakso and M. Orešič, *J. Chromatogr. B*, 2008, **871**, 375.
25. N. L. Kuehnbaum and P. Britz-McKibbin, *Anal. Chem.*, 2011, **83**, 8063.
26. M. J. Ebner, D. I. Corol, H. Havlíková, J. W. Honour and J. P. Fry, *Endocrinology*, 2006, **147**, 179.
27. P. Liere, A. Pianos, B. Eychenne, A. Cambourg, S. Liu, W. Griffiths, M. Schumacher, J. Sjövall and E.-E. Baulieu, *J. Lipid Res.*, 2004, **45**, 2287.
28. M. Vallée, J. D. Rivera, G. F. Koob, R. H. Purdy and R. L. Fitzgerald, *Anal. Biochem.*, 2000, **287**, 153.
29. R. Kancheva, M. Hill, Z. Novák, J. Chrastina, M. Velíková, L. Kancheva, I. Říha and L. Stárka, *J. Steroid Biochem. Mol. Biol.*, 2010, **119**, 35.
30. Y. W. Ha, J.-Y. Moon, H.-J. Jung, B. C. Chung and M. H. Choi, *J. Chromatogr. B*, 2009, **877**, 4125.
31. D. Caruso, S. Scurati, O. Maschi, L. De Angelis, I. Roglio, S. Giatti, L. M. Garcia-Segura and R. C. Melcangi, *Neurochem. Int.*, 2008, **52**, 560.
32. T. Higashi, N. Takido and K. Shimada, *Steroids*, 2005, **70**, 1.
33. T. Higashi, A. Nagahama, N. Otomi and K. Shimada, *J. Chromatogr. B*, 2007, **848**, 188.

34. K. M. McNamara, D. T. Harwood, U. Simanainen, K. A. Walters, M. Jimenez and D. J. Handelsman, *J. Steroid Biochem. Mol. Biol.*, 2010, **121**, 611.
35. P. Caron, E. Audet-Walsh, J. Lépine, A. Bélanger and C. Guillemette, *Anal. Chem.*, 2009, **81**, 10143.
36. J. C. García-Cañaveras, M. T. Donato, J. V. Castell and A. Lahoz, *J. Lipid Res.*, 2012, **53**, 2231.
37. E. J. Want, M. Coen, P. Masson, H. C. Keun, J. T. M. Pearce, M. D. Reily, D. G. Robertson, C. M. Rohde, E. Holmes, J. C. Lindon, R. S. Plumb and J. K. Nicholson, *Anal. Chem.*, 2010, **82**, 5282.
38. M. Hagio, M. Matsumoto, M. Fukushima, H. Hara and S. Ishizuka, *J. Lipid Res.*, 2009, **50**, 173.
39. M. Scherer, C. Gnewuch, G. Schmitz and G. Liebisch, *J. Chromatogr. B*, 2009, **877**, 3920.
40. Y. Alnouti, I. L. Csanaky and C. D. Klaassen, *J. Chromatogr. B*, 2008, **873**, 209.
41. M. Masoodi, M. Eiden, A. Koulman, D. Spaner and D. A. Volmer, *Anal. Chem.*, 2010, **82**, 8176.
42. J. Yang, K. Schmelzer, K. Georgi and B. D. Hammock, *Anal. Chem.*, 2009, **81**, 8085.
43. V. A. Blaho, M. W. Buczynski, C. R. Brown and E. A. Dennis, *J. Biol. Chem.*, 2009, **284**, 21599.
44. J. W. Newman, T. Watanabe and B. D. Hammock, *J. Lipid Res.*, 2002, **43**, 1563.
45. R. L. Shaner, J. C. Allegood, H. Park, E. Wang, S. Kelly, C. A. Haynes, M. C. Sullards and A. H. Merrill, *J. Lipid Res.*, 2009, **50**, 1692.
46. H. Min, S. Lim, B. Chung and M. Moon, *Anal. Bioanal. Chem.*, 2011, **399**, 823.
47. J. Clark, K. E. Anderson, V. Juvin, T. S. Smith, F. Karpe, M. J. O. Wakelam, L. R. Stephens and P. T. Hawkins, *Nat. Methods*, 2011, **8**, 267.
48. T. R. Pettitt, S. K. Dove, A. Lubben, S. D. J. Calaminus and M. J. O. Wakelam, *J. Lipid Res.*, 2006, **47**, 1588.
49. M. Scherer, G. Schmitz and G. Liebisch, *Clin. Chem.*, 2009, **55**, 1218.
50. S. Cubbon, C. Antonio, J. Wilson and J. Thomas-Oates, *Mass Spectrom. Rev.*, 2010, **29**, 671.
51. S. Schiesel, M. Lämmerhofer and W. Lindner, *Anal. Bioanal. Chem.*, 2010, **396**, 1655.
52. W. Lu, B. D. Bennett and J. D. Rabinowitz, *J. Chromatogr. B*, 2008, **871**, 236.
53. M. Almstetter, P. Oefner and K. Dettmer, *Anal. Bioanal. Chem.*, 2012, **402**, 1993.
54. H. Antti, presented at Metabolomics Discussion Meeting, Umeå University, 2012.
55. S. Yang, M. Sadilek, R. E. Synovec and M. E. Lidstrom, *J. Chromatogr. A*, 2009, **1216**, 3280.

56. S. U. Bajad, W. Lu, E. H. Kimball, J. Yuan, C. Peterson and J. D. Rabinowitz, *J. Chromatogr. A*, 2006, **1125**, 76.
57. M. J. van der Werf, K. M. Overkamp, B. Muilwijk, L. Coulier and T. Hankemeier, *Anal. Biochem.*, 2007, **370**, 17.
58. K. Kim, P. Aronov, S. O. Zakharkin, D. Anderson, B. Perroud, I. M. Thompson and R. H. Weiss, *Mol. Cell. Proteomics*, 2009, **8**, 558.
59. S. Castillo, I. Mattila, J. Miettinen, M. Oreŝiĉ and T. Hyötyläinen, *Anal. Chem.*, 2011, **83**, 3058.
60. M. Hartonen, I. Mattila, A.-L. Ruskeepää, M. Orešič and T. Hyötyläinen, *J. Chromatogr. A.*, 2013, **293**, 142.
61. T. Hyotylainen, I. Mattila, S. K. Wiedmer, A. Koivuniemi, M.-R. Taskinen, H. Yki-Jarvinen and M. Oresic, *Mol. Biosyst.*, 2012, **8**, 2559.
62. A. C. Beckstrom, E. M. Humston, L. R. Snyder, R. E. Synovec and S. E. Juul, *J. Chromatogr. A*, 2011, **1218**, 1899.
63. J. A. J. Trygg, J. Gullberg, A. I. Johansson, P. Jonsson, H. Antti, S. L. Marklund and T. Moritz, *Anal. Chem.*, 2005, **77**, 8086.
64. M. M. Koek, B. Muilwijk, M. J. van der Werf and T. Hankemeier, *Anal. Chem.*, 2006, **78**, 1272.
65. K. R. Chalcraft, R. Lee, C. Mills and P. Britz-McKibbin, *Anal. Chem.*, 2009, **81**, 2506.
66. A. Zuniga and L. Li, *Anal. Chim. Acta*, 2011, **689**, 77.
67. M. Kivilompolo, L. Öhrnberg, M. Orešič and T. Hyötyläinen, *J. Chromatogr. A*, 2013, **1292**, 189.
68. L. Vernez, G. Hopfgartner, M. Wenk and S. Krähenbühl, *J. Chromatogr. A*, 2003, **984**, 203.
69. P. E. Minkler, M. S. K. Stoll, S. T. Ingalls, S. Yang, J. Kerner and C. L. Hoppel, *Clin. Chem.*, 2008, **54**, 1451.
70. F. ter Veld, S. Primassin, L. Hoffmann, E. Mayatepek and U. Spiekerkoetter, *J. Lipid Res.*, 2009, **50**, 1556.
71. K. Harada, Y. Ohyama, T. Tabushi, A. Kobayashi and E. Fukusaki, *J. Biosci. Bioeng.*, 2008, **105**, 249.
72. G. Liu, J. Chen, P. Che and Y. Ma, *Anal. Chem.*, 2002, **75**, 78.
73. C. Magnes, M. Suppan, T. R. Pieber, T. Moustafa, M. Trauner, G. Haemmerle and F. M. Sinner, *Anal. Chem.*, 2008, **80**, 5736.
74. P. E. Minkler, J. Kerner, T. Kasumov, W. Parland and C. L. Hoppel, *Anal. Biochem.*, 2006, **352**, 24.
75. J. W. Armando, B. A. Boghigian and B. A. Pfeifer, *Lett. Appl. Microbiol.*, 2012, **54**, 140.
76. A. U. Blachnio-Zabielska, C. Koutsari and M. D. Jensen, *Rapid Commun. Mass Spectrom.*, 2011, **25**, 2223.
77. M. A. D. N. Perera, S.-Y. Choi, E. S. Wurtele and B. J. Nikolau, *J. Chromatogr. B*, 2009, **877**, 482.
78. H.-X Li, Z. Li, J.-M. Yin, C.-P. Li, Y. S. Chi, Q.-S. Liu, X.-L. Zhang and W.-B. Urs, *Green Chem.*, 2012, **14**, 1721.
79. P. L. Ross, Y. N. Huang, J. N. Marchese, B. Williamson, K. Parker, S. Hattan, N. Khainovski, S. Pillai, S. Dey, S. Daniels, S. Purkayastha,

P. Juhasz, S. Martin, M. Bartlet-Jones, F. He, A. Jacobson and D. J. Pappin, *Mol. Cell. Proteomics*, 2004, **3**, 1154.
80. E. A. McGaw, K. W. Phinney and M. S. Lowenthal, *J. Chromatogr. A*, 2010, **1217**, 5822.
81. J.-S. Jeong, S.-K. Kim and S.-R. Park, *Electrophoresis*, 2012, **33**, 2112.
82. R. Ramautar, O. A. Mayboroda, R. J. E. Derks, C. van Nieuwkoop, J. T. van Dissel, G. W. Somsen, A. M. Deelder and G. J. de Jong, *Electrophoresis*, 2008, **29**, 2714.
83. Y. Toya, N. Ishii, T. Hirasawa, M. Naba, K. Hirai, K. Sugawara, S. Igarashi, K. Shimizu, M. Tomita and T. Soga, *J. Chromatogr. A*, 2007, **1159**, 134.
84. R. Lee, A. S. Ptolemy, L. Niewczas and P. Britz-McKibbin, *Anal. Chem.*, 2006, **79**, 403.
85. M. Calderón-Santiago, F. Priego-Capote, J. G. Galache-Osuna and M. D. Luque de Castro, *J. Pharm. Biomed. Anal.*, 2012, **70**, 476.
86. L. Song, Z. Guo and Y. Chen, *Anal. Chim. Acta*, 2011, **703**, 257.
87. Y. Jiang, C. Sun, X. Ding, D. Yuan, K. Chen, B. Gao, Y. Chen and A. Sun, *J. Pharm. Biomed. Anal.*, 2012, **66**, 258.
88. M. Piraud, C. Vianey-Saban, C. Bourdin, C. Acquaviva-Bourdain, S. Boyer, C. Elfakir and D. Bouchu, *Rapid Commun. Mass Spectrom.*, 2005, **19**, 3287.
89. T. Langrock, P. Czihal and R. Hoffmann, *Amino Acids*, 2006, **30**, 291.
90. X. Tao, Y. Liu, Y. Wang, Y. Qiu, J. Lin, A. Zhao, M. Su and W. Jia, *Anal. Bioanal. Chem.*, 2008, **391**, 2881.
91. H. Kaspar, K. Dettmer, W. Gronwald and P. J. Oefner, *J. Chromatogr. B*, 2008, **870**, 222.
92. S. G. Villas-Bôas, D. G. Delicado, M. Åkesson and J. Nielsen, *Anal. Biochem.*, 2003, **322**, 134.
93. M. R. Antoniewicz, J. K. Kelleher and G. Stephanopoulos, *Anal. Chem.*, 2007, **79**, 7554.
94. N. Yamaoka, Y. Kudo, K. Inazawa, S. Inagawa, M. Yasuda, K.-i. Mawatari, K. Nakagomi and K. Kaneko, *J. Chromatogr. B*, 2010, **878**, 2054.
95. Y. Chen, D. Xing, W. Wang, Y. Ding and L. Du, *Biomed. Chromatogr.*, 2007, **21**, 628.
96. A. Contreras-Sanz, T. Scott-Ward, H. Gill, J. Jacoby, R. Birch, J. Malone-Lee, K. G. Taylor, C. Peppiatt-Wildman and S. P. Wildman, *Purinergic Signal*, 2012, **8**, 741.
97. J. Czarnecka, M. Cieślak and K. Michał, *J. Chromatogr. B*, 2005, **822**, 85.
98. J. Wang, T. Lin, J. Lai, Z. Cai and M. S. Yang, *J. Chromatogr. B*, 2009, **877**, 2019.
99. M. L. Sanz and I. Martínez-Castro, *J. Chromatogr. A*, 2007, **1153**, 74.
100. D. R. Knapp, *Handbook of Analytical Derivatization Reactions*, Wiley, New York, 1979.
101. C. Antonio, T. Larson, A. Gilday, I. Graham, E. Bergström and J. Thomas-Oates, *Rapid Commun. Mass Spectrom.*, 2008, **22**, 1399.

102. M. Raessler, B. Wissuwa, A. Breul, W. Unger and T. Grimm, *J. Agric. Food Chem.*, 2008, **56**, 7649.
103. M. Raessler, B. Wissuwa, A. Breul, W. Unger and T. Grimm, *Anal. Methods*, 2010, **2**, 532.
104. J. Cabálková, K.-G. Wahlund and J. Chmelík, *J. Chromatogr. A*, 2007, **1148**, 189.
105. T. Soga, T. Ishikawa, S. Igarashi, K. Sugawara, Y. Kakazu and M. Tomita, *J. Chromatogr. A*, 2007, **1159**, 125.
106. T. Soga, Y. Ohashi, Y. Ueno, Naraoka H., M. Tomita and T. Nishioka, *J. Protein Res.*, 2003, **2**, 488.
107. S. K. Bhowmik and B. H. Jung, *Rapid Commun. Mass Spectrom.*, 2012, **26**, 1426.
108. M. Calderón-Santiago, F. Priego-Capote, J. G. Galache-Osuna and M. D. Luque de Castro, *J. Pharm. Biomed. Anal.*, 2013, **74**, 178.
109. S. Lu, X. Sun, C. Shi and Y. Zhang, *J. Chromatogr. A*, 2003, **1012**, 161.
110. D. Kloos, R. J. E. Derks, M. Wijtmans, H. Lingeman, O. A. Mayboroda, A. M. Deelder, W. M. A. Niessen and M. Giera, *J. Chromatogr. A*, 2012, **1232**, 19.

Subject Index

Locators in **bold type** refer to figures/tables

Common abbreviations used in subheadings:

APCI = atmospheric pressure chemical ionization
API = atmospheric pressure ionization
CE = capillary electromigration
CEC = capillary electrochromatography
CZE = capillary zone electrophoresis
EI = electron ionization
ESI = electrospray ionization
FID = flame ionization detection
GC = gas chromatography
HILIC = hydrophilic interaction liquid chromatography
HPLC = high-performance liquid chromatographic
LC = liquid chromatography
LIF = laser-induced fluorescence
MCE = microchip capillary electrophoresis
MFA = metabolic flux analysis
MS = mass spectrometry
QMS = quadrupole mass spectrometry
RPLC = reversed-phase liquid chromatography
SFC = supercritical fluid chromatography